Nuclear Technologies: Reactors, Instrumentation, Measurement and Applications

Nuclear Technologies:
Reactors, Instrumentation, Measurement and Applications

Edited by
Matt Fulcher

WILLFORD PRESS
www.willfordpress.com

Published by Willford Press,
118-35 Queens Blvd., Suite 400,
Forest Hills, NY 11375, USA

ISBN: 978-1-68285-331-3

Cataloging-in-publication Data

Nuclear technologies : reactors, instrumentation, measurement and applications / edited by Matt Fulcher.
p. cm.
Includes bibliographical references and index.
ISBN 978-1-68285-331-3
1. Nuclear engineering. 2. Nuclear energy. 3. Nuclear reactors. 4. Nuclear engineering--Instruments. I. Fulcher, Matt.
TK9145 .N83 2017
621.48--dc23

For information on all Willford Press publications
visit our website at www.willfordpress.com

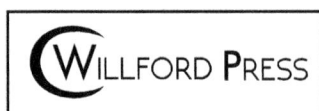

WILLFORD PRESS

Printed in the United States of America.

Contents

Permissions

List of Contributors

Index

Preface

Nuclear technology is a dynamic field and has gained global significance in the past decade. This book elucidates new techniques of nuclear technology and their applications in a multidisciplinary approach. It consists of contributions made by international experts. This text unfolds the innovative aspects of nuclear technology, which will be crucial for the progress of this field in the future. Those in search of information to further their knowledge will be greatly assisted by this book. Scientists and students actively engaged in this field will find this book full of crucial and unexplored concepts. In this book, using case studies and examples, constant effort has been made to make the understanding of the difficult concepts of nuclear technology as easy and informative as possible, for the readers.

The information shared in this book is based on empirical researches made by veterans in this field of study. The elaborative information provided in this book will help the readers further their scope of knowledge leading to advancements in this field.

Finally, I would like to thank my fellow researchers who gave constructive feedback and my family members who supported me at every step of my research.

Editor

1

Possible in-vessel corium progression way in the Unit 1 of Fukushima Dai-ichi nuclear power plant using a phenomenological analysis

Frédéric Payot[1,*] and Jean-Marie Seiler[2]

[1] CEA Cadarache/DTN/SMTA/LPMA, 13108 Saint-Paul-lez-Durance cedex, France
[2] CEA Grenoble/DTN/STCP/LTDA, 17, rue des Martyrs, 38054 Grenoble cedex 9, France

Abstract. In the field of severe accident, the description of corium progression events is mainly carried out by using integral calculation codes. However, these tools are usually based on bounding assumptions because of high complexity of phenomena. The limitations associated with bounding situations ([J.M. Seiler, B. Tourniaire, A phenomenological analysis of melt progression in the lower head of a pressurized water reactor, Nucl. Eng. Des. **268**, 87 (2014)] e.g. steady state situations and instantaneous whole core relocation in the lower head) led CEA to develop an alternative approach in order to improve the phenomenological description of melt progression. The methodology used to describe the corium progression was designed to cover the accidental situations from the core meltdown to the molten core concrete interaction. This phenomenological approach is based on available data (including learnings from TMI2), on physical models and knowledge about the corium behavior. It provides emerging trends and best estimated intermediate situations. As different phenomena are unknown, but strongly coupled, uncertainties at large scale for the reactor application must be taken into account. Furthermore, the analysis is complicated by the fact that these configurations are most probably three dimensional, all the more so because 3D effects are expected to have significant consequences for the corium progression and the resulting vessel failure. Such an analysis of the in-vessel melt progression was carried out for the Unit 1 of the Fukushima Dai-ichi nuclear power plant. The core uncovering kinetics governs the core degradation and impacts the appearance of the first molten corium inside the core. The initial conditions used to carry out this analysis are based on available results derived from codes like MELCOR calculation code [R. Ganntt, D. Kalinich, J. Cardoni, J. Phillips, A. Goldmann, S. Pickering, M. Francis, K. Robb, L. Ott, D. Wang, C. Smith, S. St. Germain, D. Schwieder, S. Phelan, Fukushima Daiichi Accident Study (Status as of April 2012), Sandia Report Sand 2012-6173, Unlimited Release Printed August, 2012]. The core degradation could then follow different ways: axial progression of the debris and the molten fuel through the lower support plate; lateral progression of the molten fuel through the shroud. On the basis of the Bali program results [J.M. Bonnet, An integral model for the calculation of heat flux distribution in a pool with internal heat generation, in *Nureth7 530 Conference Saratoga Springs, NY, USA, September 10–15, 1995* (1995)] and the TMI-2 accident observations [D.W. Ackers, J.R. Wolf, Relocation of Fuel Debris to the Lower Head of the TMI2 Reactor Vessel-A possible scenario, TMI 2 pressure vessel investigation project, in *Proceedings of the Open forum OECD/NEA and USNRCm, Boston, USA, 20–22 October 1993* (1993)], this work is focused on the consequences of a lateral melt progression (not excluding an axial progression through the support plate). Analysis of the events and the associated time sequence will be detailed. Besides, this analysis identifies a number of issues. Random calculations and statistical analysis of the results could be performed with calculation codes such as LEONAR–PROCOR codes [R. Le Tellier, L. Saas, F. Payot, Phenomenological analyses of corium propagation in LWRs: the PROCOR software platform, in *ERMSAR 2015, Marseille, France, 24–26 March, 2015* (2015)].

* e-mail: frederic.payot@cea.fr

1 Introduction

The three accidents (i.e. accidents in Units 1, 2, and 3) led to different degrees of core damage, with Unit 1 being probably the most severely damaged of the three [1]. The first conjectures about the Unit 1 core damage assumed a vessel lower head failure, a core material release into the containment cavity, and core-concrete interactions likely initiated. Units 2 and 3 are likely less damaged [2,3].

The description of corium progression events is mainly carried out by the mechanistic calculation codes. The safety demonstrations (e.g. AP1000 [4]) using these codes are usually based on bounding situations because of high complexity of phenomena. The limitations associated to bounding situations (e.g. problem of the focusing effect during the transient formation and steady state situations) led CEA [5] to develop, together with EDF, an alternative phenomenological approach (so-called "phenomenological approach") in order to supplement the current severe accident calculation codes [6].

The phenomenological approach developed in order to describe the corium progression covers the accidental situations from the core melting propagation down to the Molten Core Concrete Interaction (MCCI) in LWRs. This approach was elaborated from physical models and knowledge concerning the corium behavior, which provides emerging trends and plausible "best estimate" sequences. The analysis is complicated by the fact that phenomena are sometimes unknown and highly coupled at various scales. Moreover, these corium configurations in the lower head are most probably three dimensional, all the more so because local and non-axisymmetric effects are expected to have significant consequences for the vessel failure and corium release conditions into the reactor pit.

These last years, "phenomenological approach" studies were first concentrated on the French 1300 MWe PWR, considering both dry scenarios and the possibility of flooding of the primary circuit and/or the reactor pit. BWR reactors were also studied which provided the piece of information to analyze the in-vessel corium progression scenario in the Unit 1 of Fukushima Dai-ichi nuclear power plant.

Such an analysis of the in-vessel melt progression was carried out for the Unit 1 of the Fukushima Dai-ichi nuclear power plant. In the event timeline, the core uncovering velocity led to the core degradation of the Unit 1. As an assumption, without additional water injection, the first core degradation events are the control rod liquefaction and downward relocation of the B_4C and stainless steel, and fuel debris, in the lower core region. Then, the partial fuel melting could give rise to the appearance of the first corium pool in the centre of the core, as described by the MELCOR calculation code [2]. From that time, the core degradation scenario could follow different ways [7] according to the in-vessel melt progression, i.e.:

– axial progression of the debris and the molten fuel through the lower support plate and/or;
– lateral progression of the molten fuel through the shroud.

It is possible that both previous events did occur sequentially during the accident. In the following, we will develop the scenario based on lateral corium flow through the shroud.

The objective of this paper is to describe the alternative relocation path taking into account local and non-axisymmetric effects. Several issues will be addressed such as the thermal loads on the shroud, the location and time delay to vessel failure and corium configuration in the lower head at vessel failure. Besides, this analysis identifies a number of open issues.

The models which have been derived from this analysis have recently been implemented in the PROCOR Platform [6], which is used for LWR reactor calculations. This calculation tool includes statistical evaluations (probability of occurrence, impact of uncertainties, and identification of most important parameters).

This work was presented in the frame of the OECD/NEA/CSNI Benchmark Study of the Accident at the Fukushima Dai-ichi Nuclear Power Station (BSAF) project [8]. During 2012 and 2014 years, the purpose of this project was both to study, by mean of severe accident codes, the Fukushima accident in the three crippled units, until six days from the reactor shut-down and to give information about in particular the location and composition of core debris.

2 Methodology

The initial conditions used to carry out this analysis are based on partial core melting with an initial corium pool formed in the core. This core degradation state (e.g. amount, location, power, of melt corium pool, etc.) is described by existing codes, as for example, the MAAP and MELCOR calculation codes. The appearance of the first corium pool is strongly dependent on the kinetics of the core uncovery. During the Unit 1 damage sequence, available water levels in the core are not reliable. We will use data provided by the MELCOR calculation code [2].

Then the phenomenological evaluation is conducted step by step following the corium relocation path:

– in the core region:

   • a new situation (compared to computational results) is carried out which corresponds to the kinetics of the melt corium growth, to the relocation in the space between core and shroud and to the ablation of shroud and of the core support plate. These phenomena depend on the presence of water, whose late injection conditions are also not well known;

– in the vessel lower part:

   • evolution of the corium masses released into the vessel lower part, taking into account the occurrence of several corium flows at various time intervals;
   • the formation of debris, the impact of focusing effect, the variations in the thermal loads and heating up of the vessel wall are evaluated in the presence of residual water;
   • the time until vessel failure, thermal loads at this time and failure conditions are also evaluated (location, mass of corium, etc.) for dry ex-vessel situation.

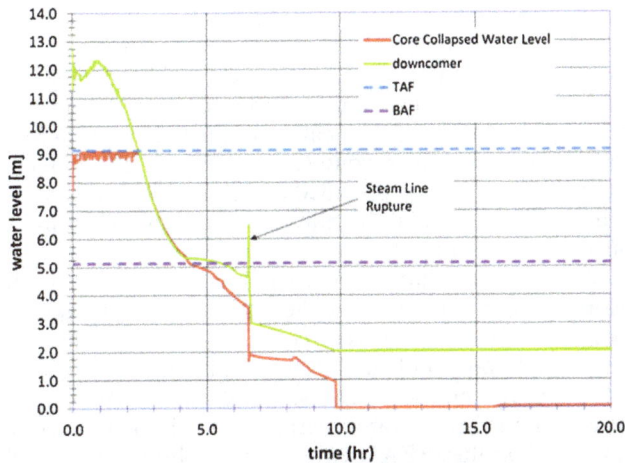

Fig. 1. MELCOR prediction of water level evolution in the reactor core and downcomer regions (Unit 1) [2].

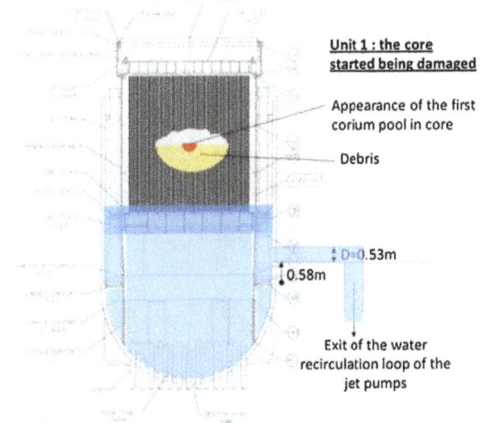

Fig. 2. Illustration of the appearance of the first corium pool in the core: ~4 h after the reactor shutdown.

3 Core degradation and core melt progression

In the Unit 1 accident, as with all of the affected reactors, following the earthquake, the reactor shutdown was accomplished on March 11, 2011 at 14:46.

According to the MELCOR calculations [2], the loss of cooling water leads to core uncover ~2 h 30 after the reactor shutdown, within a short period of time as shown in Figure 1. From there, the core is not sufficiently cooled and the cladding and fuel heatup follows. When the core temperature reaches between 1000 K and 1200 K, cladding failure is possible. Indeed, the temperature of the Zircaloy (Zr) cladding can escalate to its melting temperature, which would cause cladding failure and relocation. The Zr oxidation reaction, once started, leads to fast escalation in fuel temperature. Interaction between fuel, cladding and other structural materials leads to the formation of molten material at temperatures possibly below the individual melting points of the respective materials. This molten material relocates within the core.

In Unit 1, in the absence of adequate core cooling, core degradation leads to a large mass of debris relocating within the lower regions of the core and/or settling on the lower core support plate. Also, molten pools could form within the debris, located in the centre of the core. From the MELCOR results, the appearance of the first liquid corium pool could occur ~4 h after the reactor shutdown as illustrated in Figure 2. At this time, the water level could be located just above the core support plate.

The kinetics of corium pool growth in a debris bed (see [5]) is governed by two contributions:

- debris melting due to the heat flux at the molten pool boundaries (linked to power dissipation in the corium pool (volume power dissipation $q \sim 0.6\,\mathrm{MW/m^3}$);
- heating and melting of the debris under the effect of residual power in the solid debris.

The molten pool tends to propagate radially, due to heat-flux distribution linked to internal natural convection [9]. Indeed, the lateral heat flux of a corium pool is about one order of magnitude higher than downward heat flux, which limits the axial propagation rate of the corium pool (see Appendix A). Axial melt progression rate is, thus, reduced in comparison with lateral melt progression. In this situation, the corium pool could be supported, during the transient melt progression, by the debris bed and solid relocation in the lower part of the core (Fig. 2). Axial propagation of melt is mainly controlled by debris heat-up and corium relocation in the lower part of the core, but the heat flux from the pool does not contribute significantly to axial progression. Besides, it is important to underline that the downward melt progression is also limited by corium freezing and significant formation of debris from the structure degradation in the lower parts of the core (due to the presence of water and low hydraulic diameter).

Typically, the whole core meltdown process (i.e. ~120 t of oxidic corium from fuel and Zircaloy) could take ~5 h under dry conditions (after complete core uncovery).

We consider that the corium pool surface was located at the core center i.e. height equal to ~2 m from the lower support plate (Fig. 2). Due to the tendency of the corium pool to propagate radially, as previously explained, the molten pool could reach the peripheral sub-assemblies before the lower part of the core is molten, as illustrated in Figure 3. When the pool reaches the outer core assemblies, there is no obstacle for the melt to relocate between the core and the shroud (~5 h 40 after the reactor shutdown). A relocated melt pool can thus form in this space, which we will call the Core Annulus Pool or CAP (Fig. 4).

The distance between the external core sub-assemblies and the shroud is azimuthally not uniformly distributed, but is of the order of ~0.1 to ~0.20 m (mean value: ~0.16 m). A significant proportion of core (20–25 t out of ~120 t) could relocate between the core and the shroud. The level of corium in this annular space is supposed to reach the same level as the corium pool level in the core, as illustrated in Figure 4. The duration of this sequence is estimated to be ~40 min.

Meanwhile, we assume that the residual water level reaches the core support plate. In the following section, a

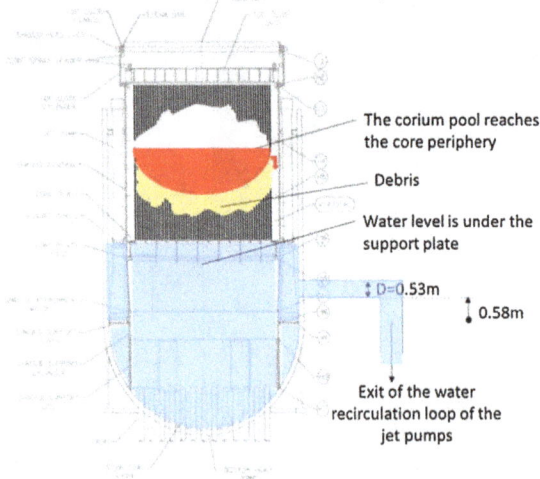

Fig. 3. Illustration of the corium flow in the core annulus pool: ~5 h 40 after the reactor shutdown.

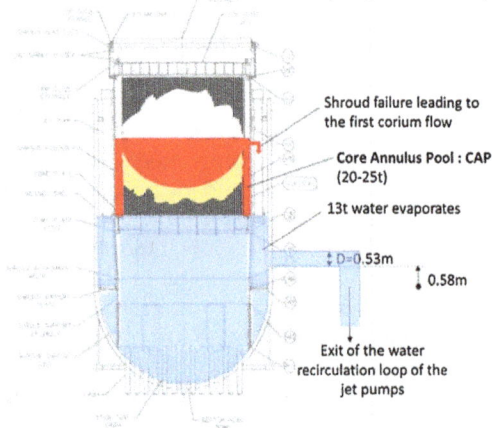

Fig. 4. Illustration of the shroud failure from the core annulus pool: ~6 h 40 after the reactor shutdown.

scenario with the presence of residual water below the core support plate, in the vessel lower head region, is assumed.

4 Core annulus pool formation

The corium accumulation duration in the CAP could take ~40 min. During this period, no water is present on the outside of the shroud. With a small lateral heat flux towards the shroud (order of magnitude[1]: ~0.03 MW/m²;), the shroud thickness ablated over 40 min is evaluated to be ~9 mm out of 38 mm. Besides, the downward heat flux towards the lower support plate is about one order of magnitude lower that the lateral heat flux. During this

[1] Assuming that half of the dissipated power in the core (q) will be used to heat the shroud wall: $\varphi.S = q.V/2$ with: V and S the volume and surface of corium annulus zone.

40 min period, the support plate failure (50 mm thickness) is here excluded from the corium relocated in the CAP. Then, when the corium height in the CAP reaches the corium pool level in the core (~6 h 40 after the reactor shutdown, 40 min after relocation in the CAP), the lateral heat flux towards the shroud (from corium in the core and in the CAP) increases to ~0.3 MW/m². In this situation, we estimate that the shroud failure could take ~20 min, as illustrated in Figure 4.

We estimate that the shroud failure does not occur before the corium height in the CAP reaches the corium pool level in the core. We then consider that the shroud fails locally (hot spot) and that relocation in the outer volume consequently leads to a 3D configuration of corium. After shroud failure, the molten corium flows into the Shroud External Annulus (SEA) (space between shroud and vessel) which is occupied by the walls of the recirculation jet pumps and bounded at its lower part by the recirculation jet pump support plate.

At that time, the corium pools in the CAP and in the core form a single pool. From lateral heat flux distribution considerations in the melt pool, we estimate that the corium mass released from the core region towards the shroud external annulus (SEA) is estimated to ~36 t.

The focusing effect (if any, linked to metal layer stratification above oxidic corium in the core pool) is not expected to have a significant impact on the time required for the transfer of oxidic material to the SEA. Indeed, in the case of a focusing effect, the metal relocates before the oxidic melt in the external volume which does almost not affect the shroud ablation by the oxydic melt.

5 Shroud external annulus pool formation (SEA; jet pump area)

The ~36 t corium mass released from the core region (i.e. core and CAP) towards the shroud external annulus zone (SEA) is expected to occur ~6 h 40 after the reactor shutdown.

We furthermore consider that the duration of corium relocation events is short (a few minutes) in comparison with the whole melting and pool progression time sequence (which takes several hours).

It is worth noticing that an ~13 t water mass could be initially present in the lower part of the SEA area (around the lower part of the jet pumps). We consider that water level is the same level than in the core support plate inside and outside the lower part of the 20 tubes of the jet pumps. Water outside the jet pumps can evaporate from the corium relocation in the SEA which leads to debris formation around the lower part of the jet pumps. We assume that water inside the jet pumps is in connection with the water in the lower head.

The external annulus is bounded at the lower part by the plates supporting the jet pumps. A direct access to the lower head is possible either through the jet pumps (20 mm wall thickness) or through the shroud wall (38 mm thickness). Nevertheless, this presence of water at high pressure (near to 70 bars; heat flux φ_{CHF} ~7.4 MW/m² at

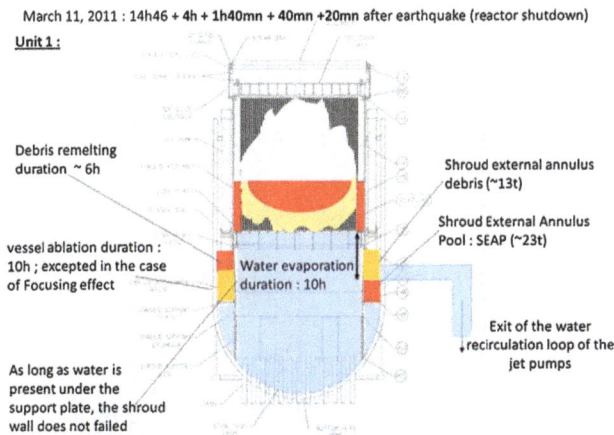

Fig. 5. Illustration of the corium relocation in the SEA: ~6 h 40 after the reactor shutdown.

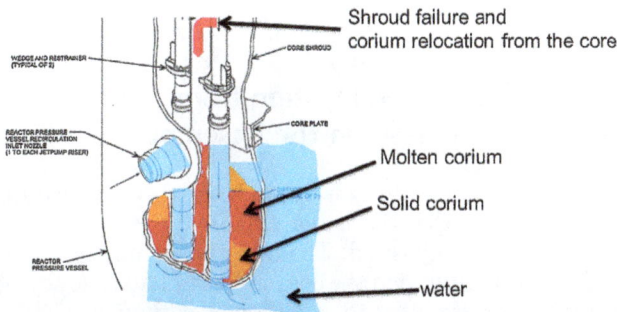

Fig. 6. Zoom of the corium relocation in the SEA: ~6 h 40 after the reactor shutdown.

70 bars[2]) in the lower head and inside the jet pump excludes the wall dry-out.

As illustrated in Figures 5 and 6, corium relocation in residual water in SEA leads to quenching and residual water evaporation. Two situations are possible: complete water evaporation in the SEA or only partial water evaporation.

5.1 Complete water evaporation in the SEA

It would lead to a ~13 t solid corium mass (like debris)[3]. The remaining corium mass would be in liquid/dense form i.e. ~23 t. The dense/liquid corium height could be just

---

[2]The Critical Heat Flux (CHF) varies as a function of vessel pressure P and enthalpy of water evaporation L, as follow: $\varphi_{CHF} \sim \sqrt{\frac{P}{P_0} \cdot \frac{L}{L_0}}$ with P_0 the standard pressure (1.013×10^5 Pa) and L_0 the evaporation enthalpy at 1 bar (2.2×10^6 J/kg). At 70 bars, $L - 1.5 \times 10^6$ J/kg.

If we consider $\varphi_{CHF} \sim 1.3$ MW/m^2 at room pressure, φ_{CHF} ~7.4 MW/m^2 at 70 bars.

[3]The solid corium mass (like debris) is evaluated from the quench potential of residual water on the basis of the evaporation heat. Here, we assume that water is at saturation temperature and that vapor is not superheated.

located below the level of the jet pump recirculation loop, as illustrated in Figure 5. Nevertheless, it cannot be excluded that a small part of liquid corium is released into the jet pump recirculation loop because of the presence of the porosity of the solid corium settled in the SEA and because of 3D effects (local failure of the shroud, local relocation of corium in the SEA space). The re-melting of solid corium (e.g. debris) would then take ~6 h (see [5]). Besides, heat from the corium pool can be transmitted to the structures i.e. shroud, jet pumps and vessel:

– in the presence of water in the lower head, melting of the vertical shroud and jet pump wall can be excluded (heat flux φ_{CHF} ~7.4 MW/m^2 at 70 bars versus φ_{shroud} ~0.04 MW/m^2 provided by the corium pool). Also, regarding the φ_{CHF} and φ_{shroud}, failure by focusing effect can also be excluded. Besides, regarding the heat flux to the shroud, φ_{shroud}, the duration to evaporate water to a level below the support plate in front of the corium pool in the SEA is very long i.e. ~10 h;
– under vessel external dry conditions, melting of the vessel wall would take up to ~10 h. However, if some non-miscible mass of molten metal relocates on top of the oxidic phase, a risk of early local vessel failure exists due to a focusing effect.

5.2 Only partial water evaporation

It is consistent with partial corium quenching and with a limited solid corium mass (like debris) smaller than 13 t. A significant corium pool/dense layer could then accumulate from unquenched molten material (>23 t):

– for this liquid corium mass higher than 23 t, the excess corium could potentially be released into the jet pump recirculation loop, as illustrated in Figure 7;
– the remaining water in this area would evaporate at a rate of ~0.2 t/min. As long as water is present, the debris re-melting can be excluded;
– under vessel external dry conditions, melting of the vessel would also take ~10 h.

6 Second melt relocation from the core

After the first corium relocation from the core region (core and CAP), we estimate that ~40 min are necessary to continue to propagate the pool in the core before next corium flow through the shroud. But 3D effects may also play a role (e.g., local shroud continuous melting and continuous 3D flow).

The follow-on corium mass released from the core region is evaluated to be ~30 t (~7 h 20 after the reactor shutdown namely ~40 min after the first corium flow, as depicted in Figure 7). Given the presence of liquid/solid corium in the SEA (water can be considered to be evaporated from SEA at this time), we point out that the second corium flow from the core is released into the two recirculation loops of the jet pumps. The recirculation loop dimensions are significant

Unit 1 : March 11, 2011 : 14h46 + 4h + 1h40mn + 40mn +20mn +40mn after earthquake

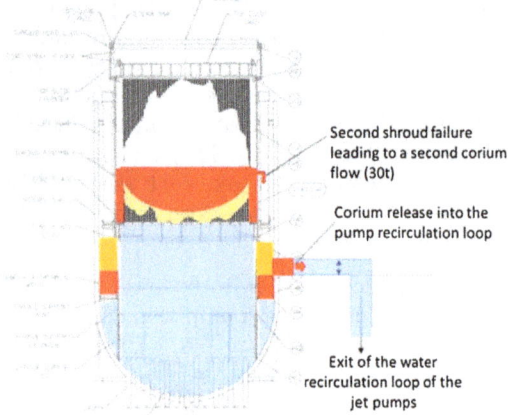

Second shroud failure leading to a second corium flow (30t)

Corium release into the pump recirculation loop

Exit of the water recirculation loop of the jet pumps

Unit 1 : March 11, 2011 : 14h46 + 4h + 1h40mn + 40mn +20mn + 40mn after earthquake

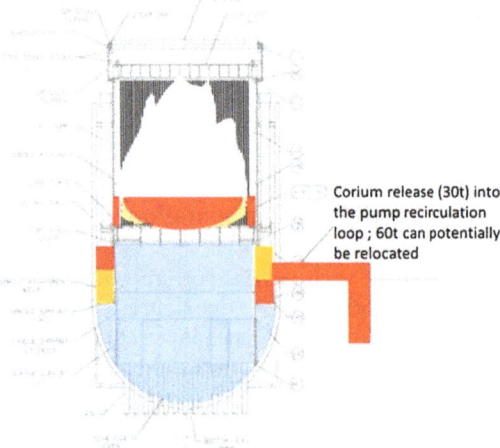

Corium release (30t) into the pump recirculation loop ; 60t can potentially be relocated

Fig. 7. Illustration of the secondary shroud failure from the core annulus pool: $\sim$7 h 20 after the reactor shutdown.

(i.e. 0.53 m diameter and $\sim$16 m length down to the pump) which could accumulate a corium mass up to 60 t. Also, it can be assumed that water was present in these pipes. The water mass in the recirculation loops is estimated to be $\sim$15 t. The delay time corresponding:

– to debris quenching (down to $\sim$800 K);
– to the residual water evaporation;
– to the increase of debris and circuit steel temperatures up to 1700 K (steel melting temperature);

is $\sim$4 h at 70 bars (see Appendix B). So, we cannot exclude a failure of the recirculating pipes $\sim$4 h after corium relocation in these pipes. Thus, a recirculation pipe failure could occur at $\sim$11 h 20 after the reactor shutdown. This is consistent with the decrease of the RPV pressure observed before $\sim$12 h [2].

Then, the corium from the water recirculation pipe of the jet pumps could be released into the dry well on the basemat, outside the pedestal space, as illustrated in Figure 9.

March 11, 2011 : 14h46 : 4h + 1h40mn + 40mn +20mn + 40mn + 90 mn after earthquake

Unit 1 :

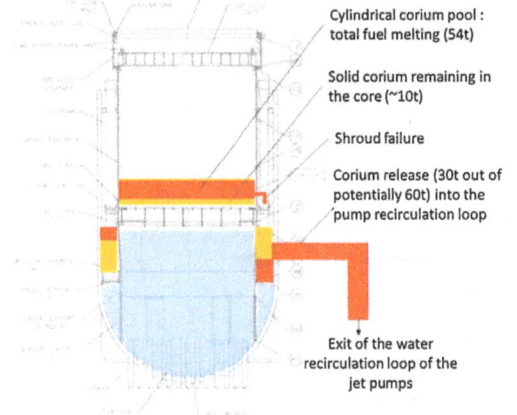

Cylindrical corium pool : total fuel melting (54t)

Solid corium remaining in the core (~10t)

Shroud failure

Corium release (30t out of potentially 60t) into the pump recirculation loop

Exit of the water recirculation loop of the jet pumps

Fig. 8. Illustration of the last corium relocation from the core: $\sim$8 h 50 after the reactor shutdown.

7 Following melt relocation from the core and corium release on the basemat

After these events, $\sim$90 min would be further necessary to melt the rest of the core under adiabatic conditions. Two situations are emphasized. The first one assumes a flow of residual corium through the core support plate. The second situation corresponds to additional corium relocation through the shroud. Our analysis gives the preference to the second situation because the heat flux to the support plate ($\sim$0.01 MW/m^2, Ref. [9]) is estimated to be much less than the lateral heat flux (0.4 MW/m^2).

It is worth noticing that a thermal failure of instrumentation tubes or guide tubes which are in contact with corium debris or a corium pool (during re-melting of corium) inside the core could plausibly precede core support plate failure. As a conclusion, the tube failure through the core support plate cannot be excluded which would lead to relocating a part of corium from the core.

Nevertheless, the lateral relocation (through the shroud) is the conjecture privileged in our scenario analysis. Some part of the remaining corium mass in the core region ($\sim$54 t) could potentially relocate to the SEA, $\sim$9 h after the reactor shutdown.

As regards the corium pool in the core, the lower and upper crust thicknesses are evaluated to be $\sim$8 cm and $\sim$4 cm, respectively. These crust thicknesses correspond to $\sim$10 t of solid corium out of 54 t in the core, as illustrated in Figure 8. The liquid corium part (i.e. 44 t) could flow from the core in the water recirculation loop of the jet pumps (via the SEA). The corium part in crust form is assumed to remain in the core.

Following the failure of the water recirculation loop of the jet pumps ($\sim$11 h 20 after the reactor shutdown), we consider that a part of $\sim$90 t of corium can be relocated on the dry well basemat, as shown in Figure 9. From there, it should be necessary to study the MCCI outside the pedestal space.

March 11, 2011 : 14h46 + 4h + 1h40mn + 40mn +20mn + 40mn +4h ~11h20 after earthquake

Unit 1:

Presence or not of water on the basemat?

Jet pump recirculation failure: ~4h at 70 bars to reach the failure

Recirculation loop pump

Depressurization of the RPV

??

One part of corium is released into the dry well on the basemat

80m²

12,5m

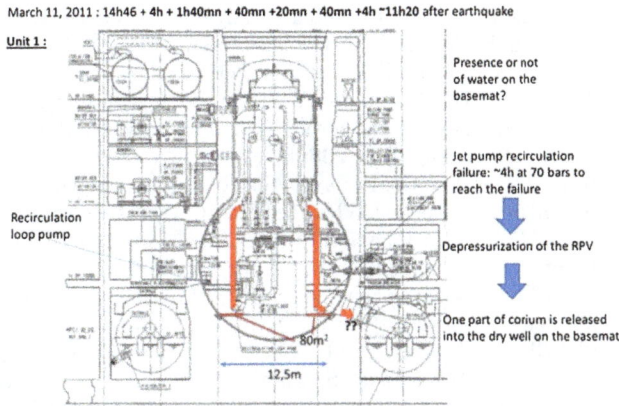

Fig. 9. Illustration of the jet pump recirculation failure from liquid corium settled inside: ~11 h 20 after the reactor shutdown.

8 Conclusion

This study presents an analysis of the in-vessel melt progression in the Unit 1 of the Fukushima Dai-ichi nuclear power plant. Not excluding axial melt progression and core support plate failure, this work focuses on the sequence based on a lateral progression of the molten fuel. The corium could potentially flow through the shroud well before axial draining through the support plate. This scenario leads to the corium accumulation in the core periphery and, from there, in the shroud external annulus, with the jet pumps, before potentially penetrating in the jet pump recirculation loops.

The lateral progression of the molten fuel assumed in this study have been carried out relying on major measurements in the plant during the accident, such as core water level, RPV pressure and PCV pressure. Due to uncertainties caused by the limited information including measurements and physical comprehension, several accident scenarios can reproduce relatively well the measured values. In the frame of the BSAF project, the lateral progression of the molten fuel through the shroud was also predicted by two participants releasing corium in the lateral lower part of the vessel. The other participants predict an axial draining through the support plate.

Appendix A

The analytic expressions for lateral local heat flux distribution (φ) were obtained from BALI results [9] which were qualified for turbulent boundary layer regime and top cooled cavity, in 3D geometry (hemisphere), as written hereafter:

$$\frac{\varphi}{\varphi\mathrm{max}} = \sin(\theta)^{1/3} \bigg/ \left(\frac{1-\cos(\theta)}{k\frac{H}{R}}\right)^{4/3} \quad \text{for } \theta < \arccos(1\text{-}kH/R)$$

$$\frac{\varphi}{\varphi\mathrm{max}} = \sin(\theta)^{1/3} \text{ for arcos } (1\text{-}kH/R) < \theta < \arccos(1\text{-}H/R)$$

with $k = 0.60$ and 0.70 for $H/R = 0.25$ and 1.0, respectively with R, H and θ the radius, the height and the local inclination of the corium pool.

The lateral local heat flux distribution is presented just below.

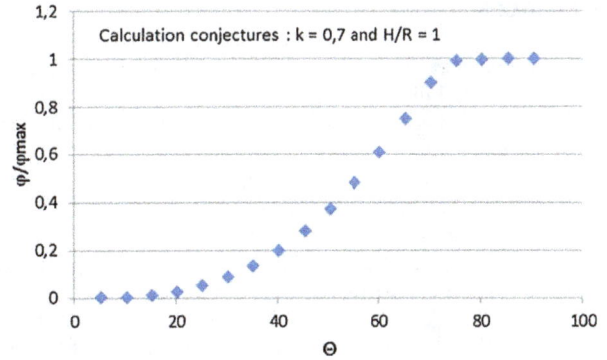

Fig. A.1 The lateral local heat flux distribution for a oxide corium pool from BALI experiments.

Appendix B: Corium relocation in a steel pipe in the presence of water

Objective: evaluation of the failure time of a steel pipe when corium is release inside in the presence of water

Analysis: the corium transfer in a pipe in the presence of water can be described by an energy balance as followed:

$$M_{\mathrm{corium}}\left(L_f + C_p(T_i - T_f)\right) + \frac{M_{\mathrm{corium}} \cdot q_{\mathrm{vol}}}{\rho_{\mathrm{corium}}} \cdot \tau =$$

$$M_{\mathrm{water}} \cdot L_{\mathrm{water}} + M_{\mathrm{steel}} \cdot \left(C_p(T_{\mathrm{melting-steel}} - T_f)\right) +$$

$$M_{corium} \cdot \left(C_p(T_{\mathrm{melting-steel}} - T_f)\right) + \varphi_{\mathrm{lost}} \cdot \tau \cdot S_{\mathrm{pipe}}$$

with

M_{corium}	the corium mass in the pipe
M_{steel}	the steel mass of the pipe
L_f	latent melting heat of corium (2.8×10^5 J/kg)
C_p	specific heat (517 J/kg/K for corium and 600 J/kg/K for steel)
T_f	debris temperature (800 K)
T_i	initial temperature (2800 K)
$T_{\mathrm{melting\_steel}}$	steel melting temperature (1700 K)
q_{vol}	the volumetric core power dissipation (i.e.0.55 MW/m³ for the Unit 1)
τ	the failure characteristic time
M_{water}	the water mass
L_{water}	the water evaporation heat (1.5×10^6 J/kg at 70 bars)
φ_{lost}	the lost heat flux
S_{pipe}	the pipe section

As presented in the main text, it can be assumed that 30 t of corium are released in two steel pipes with a 0.58 m diameter (and a 16 m length) filled with 15 t water. By neglecting the lost energy (i.e. the last term), the steel failure time can be evaluated i.e. ~10 h.

References

1. N. Watanabe et al., Review of five investigation committee's reports on the Fukushima Dai-ichi nuclear power plant severe accident: focusing on accident progression and causes, J. Nucl. Sci. Technol. **52**, 41 (2015)

2. R. Ganntt, D. Kalinich, J. Cardoni, J. Phillips, A. Goldmann, S. Pickering, M. Francis, K. Robb, L. Ott, D. Wang, C. Smith, S. St. Germain, D. Schwieder, S. Phelan, Fukushima Daiichi Accident Study (Status as of April 2012), Sandia Report Sand 2012-6173, Unlimited Release Printed 2012

3. H. Bonneville, A. Luciani, Simulation of the core degradation phase of the Fukushima accidents using the ASTEC code, Nucl. Eng. Des. **272**, 261 (2014)

4. H. Esmaili, M. Khatib-Rahbar, Analysis of In-Vessel Retention and Ex-Vessel Fuel Coolant Interaction for AP1000, NUREG/CR. 6849 ERTNRC O4. 2OI, 2004

5. J.M. Seiler, B. Tourniaire, A phenomenological analysis of melt progression in the lower head of a pressurized water reactor, Nucl. Eng. Des. **268**, 87 (2014)

6. R. Le Tellier, L. Saas, F. Payot, Phenomenological analyses of corium propagation in LWRs: the PROCOR software platform, in *ERMSAR 2015 Marseille, France, 24–26 March 2015* (2015)

7. I. Sato, Experimental program for the understanding of Fukushima-Daïshi phenomena, in *PLINIUS 2 seminar Marseille, 2014* (2014)

8. M. Pellegrini, K. Dolganov, L.E. Herranz Puebla, H. Bonneville, D. Luxat, M. Sonnenkalb, S. Band, F. Nagase, J.H. Song, J.H. Park, T.W. Kim, S.I. Kim, R.O. Gauntt, L. Fernandez Moguel, F. Payot, H. Hoshi, Y. Nishi, Benchmark Study of the Accident at the Fukushima Daiichi Nuclear Power Plant Phase I, Final Report, OECD/NEA BSAF project, 2015

9. J.M. Bonnet, An integral model for the calculation of heat flux distribution in a pool with internal heat generation, in *Nureth7 Conference Saratoga Springs NY. USA, September 10–15 1995* (1995)

2

Cooling the intact loop of primary heat transport system using Shutdown Cooling System in case of LOCA events

Diana Laura Icleanu[*], Ilie Prisecaru, and Iulia Nicoleta Jianu

Polytechnic University of Bucharest, Splaiul Independentei, nr. 313, Bucharest, 060042, Romania

Abstract. The purpose of this paper is to model the operation of the Shutdown Cooling System (SDCS) for CANDU 6 nuclear power plants in case of LOCA accidents, using Flowmaster calculation code, by delimiting models and setting calculation assumptions, input data for hydraulic analysis and input data for calculating thermal performance check for heat exchangers that are part of this system.

1 Introduction

Power and energy industries have their unique challenges but they all need to rely on the efficient running of their piping systems and, therefore, optimum design and continual effective maintenance are essential. The ability to ensure accurate delivery of a product and raw materials, especially over long distances and significant elevation changes, is vital to the overall operation and success of a process plant. For such analysis, Flowmaster is a useful code. This code has been applied for analyzing the systems of CANDU reactors due to the user's possibility of defining the incompressible and compressible fluids and also the solid materials based on thermodynamic and thermophysical properties of these materials [1] stored in the corresponding generic database of the program.

Considering this, the following paper has analyzed the failure operation modes in case of loss of coolant accidents (LOCA), described in the design documentation.

The first chapter of the study provides an overview of the Shutdown Cooling System (SDCS) and an overview of the operating regimes of the system. In this section, general considerations and aspects of nuclear safety related to the LOCA accidents are also presented.

Furthermore, modeling the operation of the SDCS is performed using Flowmaster [2], by delimiting the models and developing supportable computing assumptions of the geometric configuration. It also requires introducing the input data and the calculation assumptions for the hydraulic analysis and for the thermal calculation in order to verify the functioning of the heat exchangers that are part of this system.

Abnormal operating conditions [3] for the SDCS were analyzed using Flowmaster [4] calculation code and a comparison of the results was made with data obtained from a series of models developed in Pipenet.

From the results of the thermal-hydraulic analysis and the comparison with data from the compilings performed with Pipenet, it was found that in all operating conditions of the system, in case of a LOCA type accident, performance requirements specified in the design documentation are confirmed by the analysis. After modeling the SDCS, its functionality was demonstrated by achieving the required performance.

2 Overview of the Shutdown Cooling System and the computer code used for analysis

The SDCS is provided for cooling the Primary Heat Transport System (PHTS) from 177 °C to 54 °C and holding the system at 54 °C for an indefinite period of time.

During normal operation with the reactor at power, the SDCS is kept full with heavy water at 38 °C (100 °F) temperature and a pressure equal to or just above atmospheric pressure. Figure 1 reveals the simplified network of the SDCS coupled with the PHTS in normal operation.

There are two cool down options available. The initial phase of both options is similar and involves the use of the CSDVs (Condenser Steam Discharge Valves) to lower the PHTS temperature from 260 °C, at the rate of 2.8 °C/min. During this phase, the PHTS pumps circulate the coolant through the steam generators. If the SDCS pumps are to be used in the next cool down phase, the PHTS temperature first has to be brought down to 149 °C by means of the CSDVs. Cool down to 54 °C at the rate of 2.8 °C/min is

*e-mail: icleanud@router.citon.ro

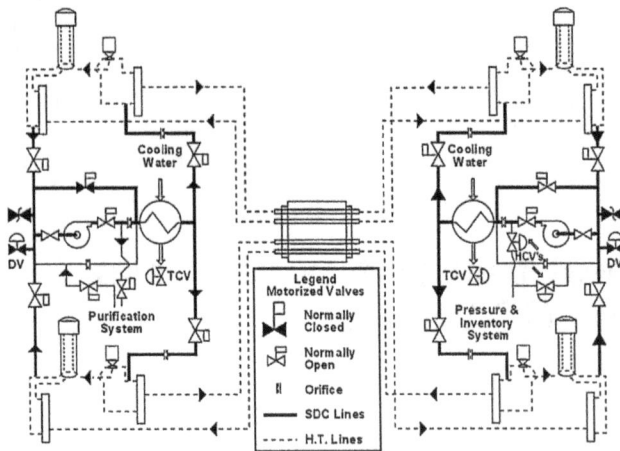

Fig. 1. Simplified network of SDCS coupled with PHTS. (Source: https://canteach.candu.org/Content%20Library/19930204.pdf).

carried out using the SDCS pumps and heat exchangers (HX).

2.1 Operating the SDCS in case of LOCA

Following a large LOCA, with or without Class IV power, the SDCS is required to cool the PHTS intact loop. For the first 900 seconds (15 minutes), upon receipt of the LOCA signal, the Moderator Temperature Control (MTC) program controls the "moderator rapid cool down". Following the first 900 seconds after LOCA, the recirculated cooling water flow rate of 200 L/s is made available to the SDCS by limiting the opening of the large temperature control valves (an MTC program action) to limit the flow towards the heat exchangers.

In this case, in order to cool down the PHTS intact loop, the operator has to bring in the SDCS manually, following a large LOCA, which will act as a backup heat sink for the thermo-siphoning of the intact loop [5].

This paper analyses the case of cooling the intact loop of the PHTS, using the SDCS, 15 or 30 minutes after the initiation of LOCA, using both the pumps and the heat exchangers of the SDCS or only one pump and one heat exchanger (Class IV or Class III available) [6].

LOCA [7] are the most severe challenges for all security systems, requiring that they operate at the best performance levels.

2.2 Fundamentals in Flowmaster

For the development of the thermal-hydraulic analysis of the SDCS, the computing program Flowmaster was used, a one-dimensional thermal-hydraulic calculation code for dimensioning, analyzing and verifying the operation of the pipeline systems.

This code provides a graphical virtual working environment and enables the design, redefinition and test of the whole fluid flow system.

Steady state or transient modules of Flowmaster code for single-phase fluids were designed specifically in order to

model the heat transfer effects in many application areas. The modules enable users to develop transient analysis for such kind of events.

Each component of Flowmaster is a mathematical model for a piece of equipment that is included in a facility.

Selected components are connected via nodes in order to form a network which constitutes a computerized model of the system.

A Flowmaster network contains a number of components (pipes, tubing, pumps, fans, flow and pressure sources, etc.) and the links between them.

The points in which components are linked to one another are called nodes.

When a network is prepared for simulation, each component and node must have a unique label. Filling the entire schematic representation (Flowmaster network) is an essential part of any simulation.

The nodal diagram (Flowmaster network) achieved consisted of a sequence of segments separated by nodes, which represent portions of pipe trails sections, without diameter or branch variations along them. Various equipments or components (except for retaining tabs) are represented by pressure loss coefficients.

For simulating using Flowmaster code, the heavy water flow in the SDCS in order to determine the variation of pressure and flow at various points of the circuit, a nodal scheme – Flowmaster network was done.

The Flowmaster computing code [8] was verified with an exact calculation on the thermal part of the analysis and with Pipenet program on the hydraulic part of it.

3 Application of Flowmaster code in thermal-hydraulic analysis of the SDCS

3.1 Models and computing hypotheses

In order to develop the thermal-hydraulic analysis of the SDCS in case of LOCA, the following calculation models that cover the requirements of the design theme were done.

3.1.1 Model I

Hydraulic calculation model for the SDCS, when operating under a LOCA failure mode, model in which the cooling of the PHTS is started at 177 °C using heat exchangers, HX1 and HX2, to provide the cold source, while the circulation will be maintained by the SDCS pumps, P1 and P2.

3.1.2 Model II

Hydraulic calculation model for the SDCS, when operating under a LOCA failure mode, model in which the cooling of the PHTS is started at 177 °C using one heat exchanger, HX1 or HX2, to provide the cold source, while the circulation will be maintained by one of the SDCS pumps, P1 or P2.

3.1.3 Model III

Thermal calculation model for the heat exchangers, HX1 and HX2, related to the SDCS, when the SDCS is operating in failure mode. In this case the heat exchangers HX1 and HX2 are cooled with water flow coming from the intermediate cooling water system. The inlet temperature considered on the secondary side of the heat exchangers is 30 °C. In model III, the SDCS is working with its associated pumps, P1 and P2.

3.1.4 Model IV

Thermal calculation model for the heat exchangers, HX1 and HX2, related to the SDCS, when the SDCS is operating in failure mode. The heat exchangers HX1 and HX2 are also cooled with water flow coming from the intermediate cooling water system. The difference between model III and model IV is that the inlet temperature on the secondary side of the heat exchangers is 35 °C. In model IV, the SDCS is working with both SDCS pumps.

3.1.5 Model V

Thermal calculation model for the heat exchangers, related to the SDCS, when the SDCS is operating in failure mode (in which case only one pump and one heat exchanger related to the SDCS are used). For this case, the inlet temperature on the secondary side of the heat exchangers is 30 °C.

3.1.6 Model VI

Thermal calculation model for the heat exchangers related to the SDCS, when the SDCS is operating in failure mode (in which case they only use one pump and one heat exchanger related to the SDCS). The inlet temperature on the secondary side of the heat exchangers according to the manual design of the cooling water system is 35 °C.

For the considered analysis, a set of design assumptions were made. For the hydraulic analysis, the hypotheses are as follows:

- system condition at the baseline of cooling is a state of stationary hydraulic regime;
- hydraulic resistances of PHTS lines are determined taking into account the pressure drop values on these lines, for the nominal operating regime;
- hydraulic resistance of SDCS lines is determined taking into account the dimensional characteristics and their composition (the fittings on these lines);

- pumps that do not work are modeled as lines with hydraulic resistance determined from the characteristic curves for the respective pumps;
- for the heat exchangers and steam generators we consider only the primary circuit, that is modeled as a pipeline with the hydraulic resistance;
- pressure in the system is fixed at one of the output collectors of the reactor by boundary condition;
- interfaces with other systems were neglected, connecting pipes to these systems are not functional for the analyzed regimes.

Assumptions considered for the thermal analysis were also set as follows:

- the energy accumulated in metal tubes and shell is neglected;
- the compressibility of the fluids is neglected;
- in the shell and in the heat exchanger's tubes, the flow is single phase;
- the initial thermal condition is that the temperature in the entire system is considered to be the same;
- the paper does not take into account the preparatory steps aimed at achieving either of the necessary cooling configurations, thus neglecting transient hydraulic regimes preceding the making of either of the cooling schemes analyzed.

For the accomplishment of the hydraulic calculation with the help of the calculating code Flowmaster V7, pressure values have been entered, corresponding to hydrostatic pressure determined at the output collector of the reactor by boundary conditions.

Thus, for all hydraulic calculation models the appropriate pressure values for the inlet/outlet components of the nodal scheme were considered according to Table 1.

3.2 Description of collected data and output files

Output files for the thermo-hydraulic calculation with the Flowmaster program are structured according to the type of simulation (hydraulic or thermal) as follows:

- hydraulic calculation results for each component (flow, velocity, Reynolds number, pressure loss, etc.);
- hydraulic calculation results in each node (pressure);
- thermal calculation results suitable for components in which heat transfer occurs (thermal load, overall heat transfer coefficient, temperature difference between input and output);
- thermal calculation results in each node (temperature).

Results can be filtered by type of component or by characteristic parameters calculated, according to the components that make up the nodal scheme.

Table 1. Boundary conditions for the hydraulic analysis.

Operation mode	Point position	Temperature (°C)	Pressure (bar)
Model I	Output collector from the reactor (pressure source: 314)	177	95
Model II	Output collector from the reactor (pressure source: 314)	177	95

3.2.1 Model I

Hydraulic calculation model for the operation in failure mode of the SDCS, model in which the cooling of the PHTS starts at 177 °C using heat exchangers, HX1 and HX2 (water is circulated by SDCS's pumps, P1 and P2). According to the results, the calculated hydraulic parameter values are shown in Table 2.

3.2.2 Model II

Hydraulic calculation model for the operation in failure mode of the SDCS, model in which the cooling of the PHTS starts at 177 °C using one heat exchanger, HX1 or HX2, to provide the cold source, while the circulation will be maintained by one of SDCS's pumps, P1 or P2.

According to the results, the calculated hydraulic parameter values are listed in Table 3.

3.2.3 Model III

Thermal calculation model for the heat exchangers, HX1 and HX2, related to the SDCS, for the operation in failure mode of the SDCS. This model concerns the time evolution of the temperature in PHTS.

According to the results, the parameter values for the heat transfer of the heat exchangers HX1 and HX2, at the moment of achieving the cooling requirement for PHTS (temperature in PHTS must be 54 °C) are shown in Table 4.

Table 2. Hydraulic analysis. Model I.

Component	Flow rate (L/s)
Flow through P1 SDCS	115
Flow through P2 SDCS	114
Flow through HX1 SDCS	104
Flow through HX2 SDCS	103
Flow through inlet collectors HD6, HD2, HD4, HD8	HD6: 104
	HD2: 1.4×10^{-10}
	HD4: 2.27×10^{-9}
	HD8: 103
Flow through outlet collectors HD5, HD1, HD3, HD7	HD5: 104
	HD1: 1.85×10^{-9}
	HD3: 1.73×10^{-9}
	HD7: 103
Flow through P1, P2, P3, P4 PHTS	P1: 4×10^{-10}
	P2: 1.55×10^{-9}
	P3: 54
	P4: 54
Flow through fuel channels R1, R2, R3 and R4	R1: 2.15×10^{-10}
	R2: 50
	R3: 1.79×10^{-9}
	R4: 50

Table 3. Hydraulic analysis. Model II.

Component	Flow rate (L/s)
Flow through P1 SDCS	115
Flow through P2 SDCS	1.7
Flow through HX1 SDCS	104
Flow through HX2 SDCS	1.9
Flow through inlet collectors HD6, HD2, HD4, HD8	HD6: 104
	HD2: 7.75×10^{-12}
	HD4: 7.3×10^{-12}
	HD8: 1.9
Flow through outlet collectors HD5, HD1, HD3, HD7	HD5: 104
	HD1: 4.14×10^{-12}
	HD3: 4.11×10^{-12}
	HD7: 1.9
Flow through P1,P2, P3, P4 PHTS	P1: 4.97×10^{-12}
	P2: 5.2×10^{-12}
	P3: 67
	P4: 34.8
Flow through fuel channels R1, R2, R3 and R4	R1: 6.9×10^{-12}
	R2: 36.7
	R3: 5×10^{-12}
	R4: 36.7

Table 4. Thermal analysis. Model III.

Thermal load of SDCS HX1/HX2	10.99 MW(th)
Outlet temperature for D_2O of PHTS	54.06 °C
Outlet temperature for the cooling water of SDCS HX1 and HX2	30.165 °C

Figure 2 is a plot of temperature decrease of PHTS coolant for the inlet and outlet of the SDCS heat exchangers.

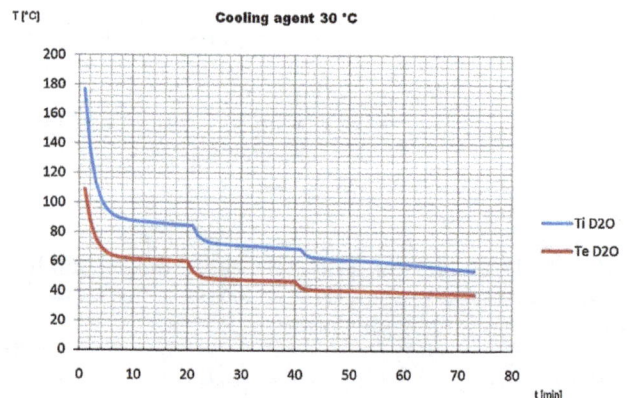

Fig. 2. Examination for 2 HX for cooling agent 30 °C.

Table 5. Thermal analysis. Model IV.

Thermal load of SDCS HX1/HX2	8.95 MW(th)
Outlet temperature for D₂O of PHTS	54.16 °C
Outlet temperature for the cooling water of SDCS HX1 and HX2	41.32 °C

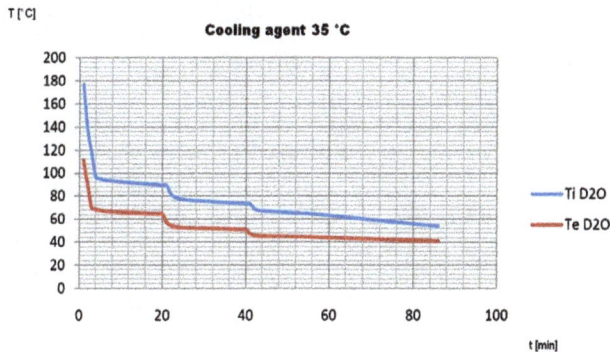

Fig. 3. Examination for 2 HX for cooling agent 35 °C.

Table 6. Thermal analysis. Model V.

Thermal load of SDCS HX1/HX2	5.4 MW(th)
Outlet temperature for D₂O of PHTS	52.43 °C
Outlet temperature for the cooling water of SDCS HX1 and HX2	41.18 °C

3.2.4 Model IV

Thermal calculation model for the heat exchangers, HX1 and HX2, related to the SDCS, for the operation of the SDCS under failure mode (LOCA). The inlet temperature on the secondary side of the heat exchangers is 35 °C. In model IV, the SDCS is working with its own pumps, P1 and P2. The results are shown in Table 5.

Figure 3 is also a plot of temperature decrease of PHTS coolant for the inlet and outlet of the SDCS heat exchangers. The difference between model III and model IV is the inlet temperature of the cooling water that passes through the heat exchangers.

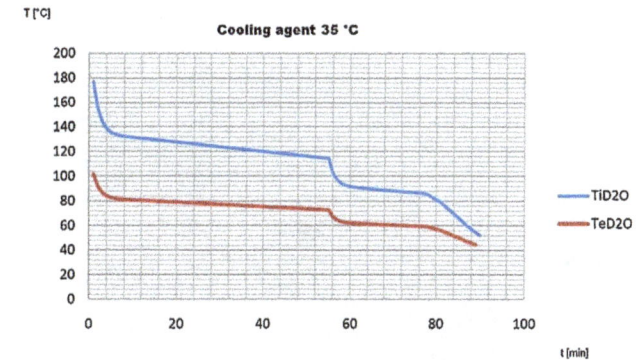

Fig. 4. Examination for 1 HX for cooling agent 30 °C.

Table 7. Thermal analysis. Model VI.

Thermal load of SDCS HX1/HX2	4.17 MW(th)
Outlet temperature for D₂O of PHTS	52.2 °C
Outlet temperature for the cooling water of SDCS HX1 and HX2	43.5 °C

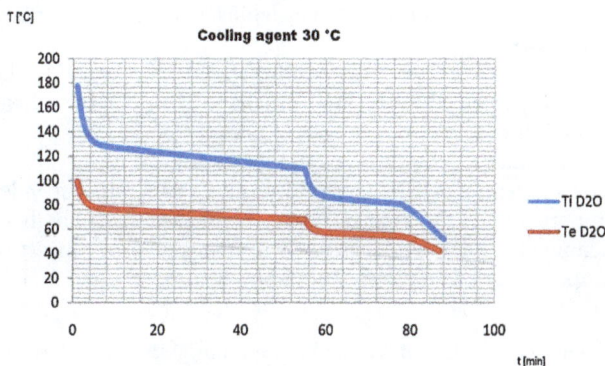

Fig. 5. Examination for 1 HX for cooling agent 35 °C.

3.2.5 Model V

Thermal calculation model for the heat exchangers, HX1 or HX2, related to the SDCS, for the operation in failure mode of the SDCS (LOCA).

Table 8. Comparative results from the hydraulic analyses. Model I.

	Data obtained by using Pipenet (L/s)	Data obtained by using Flowmaster (L/s)
Flow through P1	118.4	115
Flow through P2	118.4	114
Flow through HX1	106.6	104
Flow through HX2	106.6	103
Flow through inlet collectors	HD6: 106.6	HD6: 104
	HD2: 0	HD2: 1.4×10^{-10}
	HD4: 0	HD4: 2.27×10^{-9}
	HD8: 106.6	HD8: 103
Flow through outlet collectors	HD5: 106.6	HD5: 104
	HD1: 0	HD1: 1.85×10^{-9}
	HD3: 0	HD3: 1.73×10^{-9}
	HD7: 106.6	HD7: 103
Flow through PHTS pumps	P1: 0	P1: 4×10^{-10}
	P2: 0	P2: 1.55×10^{-9}
	P3: 58.6	P3: 54
	P4: 58.6	P4: 54
Flow through fuel channels	R1: 0	R1: 2.15×10^{-10}
	R2: 53	R2: 50
	R3: 0	R3: 1.79×10^{-9}
	R4: 53	R4: 50

Table 9. Comparative results from the hydraulic analyses. Model II.

	Data obtained by using Pipenet (L/s)	Data obtained by using Flowmaster (L/s)
Flow through P1	118.4	115
Flow through P2	0	1.7
Flow through HX1	106.6	104
Flow through HX2	0.3	1.9
Flow through inlet collectors	HD6: 106.6	HD6: 104
	HD2: 0	HD2: 7.75×10^{-12}
	HD4: 0	HD4: 7.3×10^{-12}
	HD8: 0.3	HD8: 1.9
Flow through outlet collectors	HD5: 106.6	HD5: 104
	HD1: 0	HD1: 4.14×10^{-12}
	HD3: 0	HD3: 4.11×10^{-11}
	HD7: 0.3	HD7: 1.9
Flow through PHTS pumps	P1: 0	P1: 4.97×10^{-12}
	P2: 0	P2: 5.2×10^{-12}
	P3: 71	P3: 67
	P4: 35.4	P4: 34.8
Flow through fuel channels	R1: 0	R1: 6.9×10^{-12}
	R2: 38	R2: 36.7
	R3: 0	R3: 5×10^{-12}
	R4: 37.9	R4: 36.7

In model V, the SDCS is working with one heat exchanger and one of the SDCS pumps, P1 or P2. Table 6 provides the main results for this case of operation mode.

Figure 4 is also a plot of temperature decrease of PHTS coolant for inlet and outlet of the SDCS heat exchangers.

3.2.6 Model VI

Thermal calculation model for the heat exchangers, HX1 and HX2, related to the SDCS, for the operation under failure mode of the SDCS (LOCA). In this model, the SDCS is working with one heat exchanger and one of the SDCS pumps, P1 or P2.

The inlet temperature on the secondary side of the heat exchangers is 35 °C. Table 7 shows the results of this analysis.

Figure 5 is also a diagram of the temperature decrease for PHTS coolant.

For the hydraulic analyses, two Pipenet models were considered in order to verify the results obtained in Flowmaster. The results are presented in Tables 8 and 9.

4 Conclusions

This paper presents a thermal-hydraulic analysis of the simultaneous operation of the SDCS and of the primary heat transport system, associated to a CANDU 6 NPP (nuclear power plant), operating in LOCA accident regime, using Flowmaster calculation code.

The modelling of heavy water flow through the SDCS and primary heat transport system was performed to determine the distribution of flow rates and pressure in various areas of the hydraulic circuit and the pressure loss corresponding to the components, but also in order to calculate the heat of the heat exchangers related to the system.

The configurations corresponding to the SDCS coupled to the primary heat transport system are in accordance with the thermo-mechanical schemes of the systems similar to those at Cernavoda NPP.

Within this work, complex hydraulic/thermo-hydraulic analyses were performed for the SDCS coupled with the primary heat transport system. Hydraulic analyzes developed using Flowmaster program aimed at the verification of the hydraulic models as well as the determination of flow and pressure loss in baseline cooling processes in degraded mode.

The results of the thermo-hydraulic analysis show that in all cases analyzed, for the LOCA accident regime, the performance requirements are satisfied according to the analysis.

The heat exchangers of the SDCS have the ability to perform the cooling of the primary heat transport system from 177 °C to 54 °C in approximately 79 minutes, if the inlet temperature of reactor cooling water (RCW) is 30 °C.

After 79 minutes, the residual heat necessary to be extracted from the primary circuit by means of both heat exchangers of the SDCS is approximately 10 MW and the thermal load of the heat exchanger is 10.99 MW.

If the inlet temperature of the RCW heat exchangers is 35 °C, then the cooling from 177 °C to 54 °C of the primary heat transport system will be made in approximately 86 minutes.

After 86 minutes, the residual heat necessary to be extracted from the primary circuit by means of a heat exchanger of the SDCS is approximately 4.85 MW and the thermal load of the heat exchanger is 8.95 MW.

For the model in which the PHTS cooling is provided only by one of the heat exchangers of the SDCS, if the inlet temperature of the RCW heat exchangers is 30 °C, the residual heat necessary to be extracted has a value of 3.71 MW. By means of the heat exchanger having the heat load of 5.4 MW, cooling from 177 °C to 54 °C is achieved in approximately 88 minutes.

If the inlet temperature of the operating heat exchanger is 35 °C, then cooling from 177 °C to 54 °C of the primary heat transport system will be achieved in approximately 90 minutes.

After 90 minutes, the residual heat necessary to be extracted from the primary circuit by means of a heat exchanger of the SDCS is approximately 2.87 MW and the thermal load of the heat exchanger is 4.17 MW.

As a result of this thermal analysis wherein the inlet temperature of the intermediate cooling water at the heat exchangers is 35 °C, a series of differences were observed compared to the data sheets of the heat exchangers, HX1 and HX2, namely:

– thermal load taken by the heat exchangers is smaller, but above the necessary;

– primary heat transport system may be cooled to a temperature of 54 °C, but it would take longer;
– intermediate cooling water temperature at the outlet of the heat exchanger has a higher value.

Another observation is that both in the case when two heat exchangers are operation, as well as the case with one heat exchanger, in LOCA accident regime, PHTS cooling can be achieved by using the SDCS. The only difference noticed between the two models considered is that for the operation with a heat exchanger instead of two, cooling is done in a longer but covering time.

Regarding the temperatures, a normal evolution in the PHTS cooling process is found, but it cannot be measured accurately and precisely because of the lack of information on the conditions under which the analyses were developed, as the basis for the evolution curves of the residual heat present in the reactor after 15 minutes, and 30 minutes, respectively, from the start of the accident of the coolant loss type.

By analyzing the parameters of the cooling system for all cooling processes considered, it was found that the values obtained for thermal-hydraulic parameters, as well as the duration up to reaching specified limits fall within the design values of the system. Cooling speeds are situated below the value of 2.8 °C/min at the reactor outlet for all cooling regimes in case of LOCA accidents.

The work has been funded by the Sectoral Operational Programme Human Resources Development 2007-2013 of the Ministry of European Funds through the Financial Agreement POSDRU/159/1.5/s/134398.

References

1. A. Leca, I. Prisecaru, *Thermo-physical and thermodynamic properties of solid, liquid and gas* (Ed. Tehnică, Bucharest, 1994)
2. D.S. Miller, *Internal flow systems*, 2nd edn. (Miller Innovations, 2008)
3. Nuclear Regulatory Commission, *Theoretical possibilities and consequences of major accidents in large nuclear power plants*. (Rep. WASH 740, US Govt Printing Office, Washington, DC, 1957)
4. Flowmaster v7, New user training, version 10
5. Requirements for the safety analysis of CANDU nuclear power plants, AECB Consultative Document C-6, 1980
6. Safety analysis: event classification, www.iaea.org
7. International Atomic Energy Agency, Incorporation of advanced accident analysis methodology into safety analysis reports, IAEA-TECDOC-1351, 2003
8. Haestad Methods Water Solutions, *Computer applications in hydraulic engineering*, 7th edn. (Bentley Institute Press, 2006)

Helium behaviour in implanted boron carbide

Vianney Motte[1,4*], Dominique Gosset[1], Sandrine Miro[2], Sylvie Doriot[1], Suzy Surblé[3], and Nathalie Moncoffre[4]

[1] CEA Saclay, DEN-DANS-DMN-SRMA-LA2M, 91191 Gif-sur-Yvette cedex, France
[2] CEA Saclay, DEN-DANS-DMN-SRMP-JANNuS, 91191 Gif-sur-Yvette cedex, France
[3] CEA Saclay, DSM-IRAMIS-LEEL, 91191 Gif-sur-Yvette cedex, France
[4] CNRS-IN2P3, IPNL, Université Lyon 1, 69622 Villeurbanne cedex, France

Abstract. When boron carbide is used as a neutron absorber in nuclear power plants, large quantities of helium are produced. To simulate the gas behaviour, helium implantations were carried out in boron carbide. The samples were then annealed up to 1500 °C in order to observe the influence of temperature and duration of annealing. The determination of the helium diffusion coefficient was carried out using the ^{3}He(d,p)^{4}He nuclear reaction (NRA method). From the evolution of the width of implanted ^{3}He helium profiles (fluence 1×10^{15}/cm^{2}, 3 MeV corresponding to a maximum helium concentration of about 10^{20}/cm^{3}) as a function of annealing temperatures, an Arrhenius diagram was plotted and an apparent diffusion coefficient was deduced ($E_{a} = 0.52 \pm 0.11$ eV/atom). The dynamic of helium clusters was observed by transmission electron microscopy (TEM) of samples implanted with 1.5×10^{16}/cm^{2}, 2.8 to 3 MeV ^{4}He ions, leading to an implanted slab about 1 μm wide with a maximum helium concentration of about 10^{21}/cm^{3}. After annealing at 900 °C and 1100 °C, small (5–20 nm) flat oriented bubbles appeared in the grain, then at the grain boundaries. At 1500 °C, due to long-range diffusion, intra-granular bubbles were no longer observed; helium segregates at the grain boundaries, either as bubbles or inducing grain boundaries opening.

1 Introduction

With a high neutron absorption efficiency, a good availability and a relatively low cost, boron carbide is used in almost all types of nuclear power plants. It is also widely used as grinding tools or armors, thanks to its mechanical properties: boron carbide is a light (2.52 g/cm^{3} for a fully dense material) super-hard (HV ~40 GPa) ceramic [1,2]. It has a high stiffness (Young modulus ~ 450 GPa) and a high strength (~450 MPa) but is brittle (K_{IC} ~ 6 MPa$\sqrt{m}$). It is a semiconductor material with a thermal conductivity varying as 1/T, about 30 W/m.K at room temperature. Those electrical and thermo-mechanical properties come from the interatomic bonding, which is mainly covalent. But its weak thermo-mechanical properties lead to early damage and short life-cycle when used as a neutron absorber.

The crystalline structure of boron carbide, shown in Figure 1, is now known [1–4] as rhombohedral (most often represented in a hexagonal frame). At the carbon-rich limit, the composition is very close to B$_{4}$C. The unit cell is built with a central chain, mainly C-B-C, and 8 icosahedra mainly constituted of B$_{11}$C situated at the corners, giving

the general formula B$_{4}$C, which is one of all the polytypes of the boron carbide phase (from ~B$_{4}$C to B$_{10}$C).

Boron carbide has a high atomic density, leading to a boron content of about 10^{23}/cm^{3}. Boron is naturally composed of ^{10}B and ^{11}B isotopes with a natural concentration of

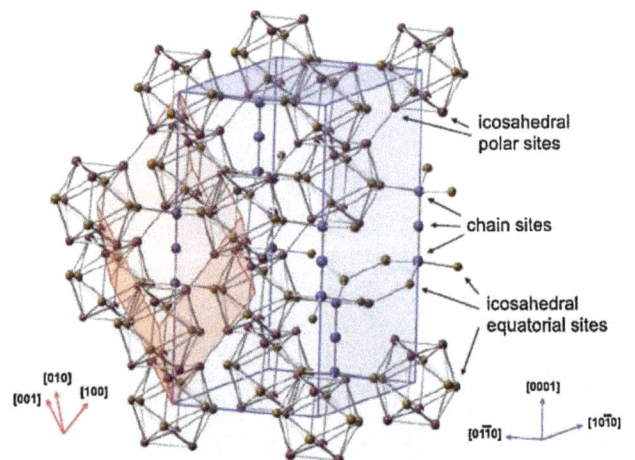

Fig. 1. Cell structure of boron carbide B$_{4}$C (from Ref. [1]).

* e-mail: vianney.motte@cea.fr

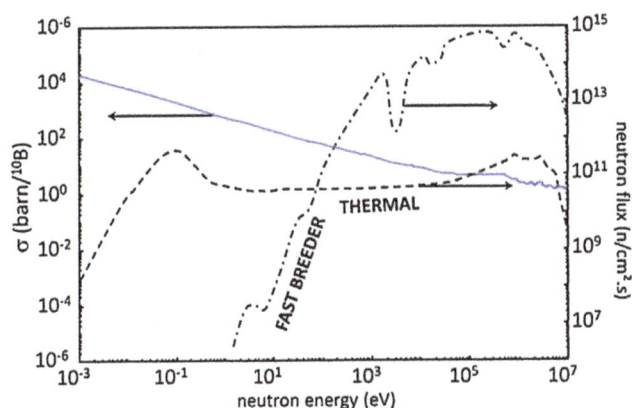

Fig. 2. In blue solid line: neutron absorption cross-section for the [10]B isotope (from Ref. [5]), superimposed to the neutron energy distribution (in black) in a pressurised water (- - thermal) and a fast neutron (- · - fast breeder) reactor (from Ref. [6]).

$\sim$20 at.% [10]B, which can be modified from 1 to 99 at.% depending on the application. The boron-10 isotope is a very efficient neutron absorber because of its high neutron absorption cross-section as shown in Figure 2.

As a material used in nuclear plants, many studies were conducted for a better understanding of the B_4C behaviour under irradiation. Two main phenomena happen in the reactors: atomic displacements leading to high point defects concentration, for which structural consequences are actually not well known (possibly amorphisation, at least at low temperature); and helium production that leads to damage in the micro-structural stability.

Amorphisation in boron carbide under irradiation has been observed with light ions at low [7] or high [8] temperatures. Recent studies [9] have shown amorphisation under slow, heavy ion irradiation, for which most of the damage is in the ballistic regime: at high damage (4×10^{15}/cm^2 Au 4 MeV, about 2 to 4 dpa), amorphisation was partial and heterogeneous in the damaged front zone, with the formation of nanometre-scale amorphous zones, and fully amorphous in the gold implantation zone, as shown in Figure 3.

Helium production arises from neutron capture by the $^{10}B(n,\alpha)^7Li$ reaction, which is highly exothermic (about 2.6 MeV per neutron capture). Helium accumulates in flat, high pressure and parallel bubbles (mainly parallel to the (111) planes [10–13] and also to the (100) and (110) planes [12–15] of the rhombohedral structure). In fast neutron reactors, the combination of heat release and helium production induces strong radial thermal gradients and extensive cracking of the absorber pellets [4,16,17] as shown in Figure 4.

The first steps of the formation of the helium clusters and the diffusion of the gas are not well known. In this context, we have launched a program aiming to study the dynamics of helium in irradiated boron carbide. Here, we present preliminary results about these two topics. This work is part of a systematic study of the behaviour of the gases in boron carbide used as a neutron absorber, aiming at a better description of the evolution of the material under neutron irradiation.

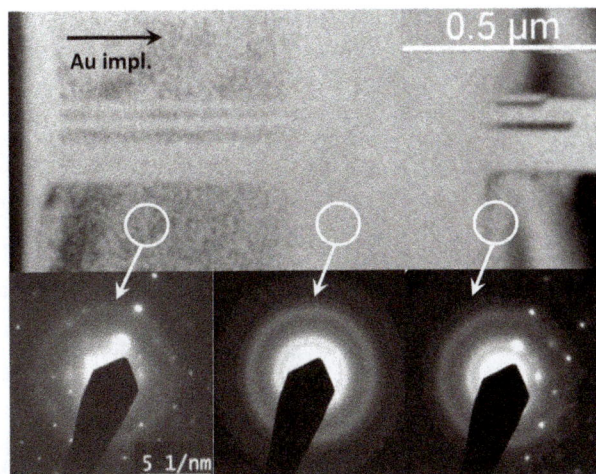

Fig. 3. TEM pictures of B_4C irradiated at 4×10^{15}/cm^2 4 MeV Au ions. An amorphous zone appears in the (centre) implanted zone. Partial amorphisation was observed in the (left) front, damaged zone. Diffraction pictures: (left) at the middle of the speckled front zone; (centre) at the middle of the amorphous zone; (right) at the right amorphous-crystalline boundary (from Ref. [9]).

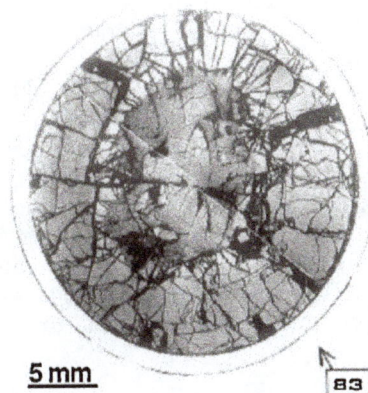

Fig. 4. B_4C pellets (from Ref. [4]) irradiated in the French LMFBR Phenix. 1.2×10^{22} capt./cm^3 (about 12 at.% total boron).

2 Experiments

In order to overcome the difficulties of handling actual materials that have been irradiated in nuclear plants, we simulated the production of helium by implanting it in B_4C pellets (collected from hot pressed boron carbide from CEA records) at different temperatures, energies and fluences. Subsequent thermal annealing treatments allowed us to determine the influence of the temperature on the behaviour of the gas in the material. The studies are then carried out using two techniques for investigation:

- determination of the helium diffusion coefficient by Nuclear Reaction Analysis (NRA);
- observation of helium clusters by Transmission Electron Microscopy (TEM).

2.1 Diffusion coefficient determination

The principle of this experiment is to implant ^{3}He as a surrogate of ^{4}He in B_4C pellets at a known depth, then apply different annealing treatments and then analyse the samples with a nuclear microprobe, from which we observe the evolution of the helium profiles using the ^{3}He(d,p)^{4}He reaction as a final step. The helium profiles were obtained from the proton energy profiles measured by the detector [18].

To proceed, helium-3 was implanted at room temperature at an energy of 3 MeV to obtain a profile with a projected ion range Rp $\sim$ 9 μm and ΔRp $\sim$ 120 nm (as given by SRIM [19] calculations). The chosen fluence was 10^{15} at/cm^2 (about 4×10^{19} at/cm^3 at Rp), which was high enough for detecting helium, while expected to remain low enough to avoid the formation of helium clusters. Annealing treatments were carried out between 15 min and 2 h at 900 °C and 1 h between 500 and 1000 °C in 100 °C steps. This temperature range corresponds to the temperatures that the material is exposed to in a fast breeder reactor.

The ^{3}He(d,p)^{4}He NRA measurements were performed using the nuclear microprobe facility of the *Laboratoire d'Étude des Éléments Légers* in CEA Saclay (CEA/DSM/IRAMIS/LEEL). It is a 3.75 MeV single-ended Van de Graaff accelerator, which can supply proton, deuteron, helium-3 and helium-4 ion beams in the energy range from 400 keV to 3.75 MeV (further descriptions of the facility can be found in Ref. [20]).

Based on SRIM calculations (Fig. 5), a 1300 keV energy for the deuterons was chosen with a 5 nA flux and a 50×50 μm^2 beam spot, which is large enough to mask channelling effects (average grain size of 5 μm). The energy of the deuterons was chosen in order to have the best yield for the (d,p) reaction cross-section. An absorber foil (123 μm thick Mylar foil) was placed in front of the annular detector, in order to stop the backscattered deuterons and slow down the 19 MeV protons, in order for them to stop in the Si

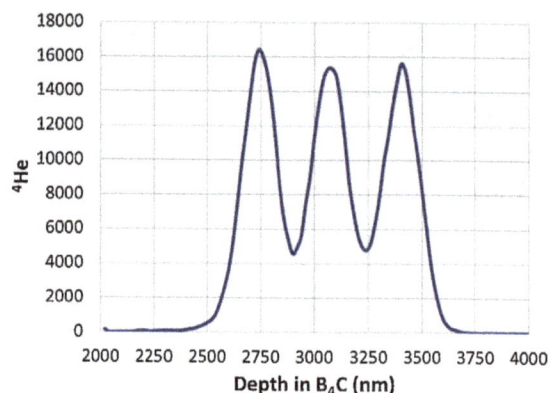

Fig. 6. Helium implantation in B_4C given by SRIM [19]: ^{4}He, 2.8–2.9–3.0 MeV, 1.5×10^{16} at/cm^2 with a 6 μm thick aluminium foil placed in front of the sample.

detector. The obtained proton energy profiles were then converted to helium depth profiles, this allowed an analysis of their evolution and thus enabled us to deduce the apparent helium diffusion coefficient in boron carbide.

2.2 Helium clusters observations

The purpose of this experiment is to observe directly the behaviour of helium (formation of clusters, migration . . .) using Transmission Electron Microscopy (TEM). Helium was implanted in B_4C pellets along a known profile, and annealing treatments were then performed.

To proceed, we implanted helium-4 at 500 °C at three different energies (2.8–2.9–3.0 MeV) to get a wider helium distribution. To move the implantation distribution peak closer to the surface, which is required for the preparation of the samples by the focused ion beam (FIB) method, a 6 μm aluminium foil was set in front of the sample. This setup led to a helium distribution between 2.65 and 3.55 μm from the surface of the pellet (from SRIM calculations, as shown in Fig. 6). We used a fluence of 1.5×10^{16} at/cm^2, leading to a maximum helium concentration of about 10^{21}/cm^3, high enough to allow the formation of bubbles. Subsequent annealing treatments were performed in the temperature range of 900–1500 °C.

The thin-foil specimens were prepared by FIB: classical electrolytic methods cannot be used here, and due to B_4C brittleness, small samples are required. The samples are then about 8 μm large, 6 μm deep and 200 nm thick. TEM observations were performed at the *Service de Recherches Métallurgiques Appliquées* in CEA Saclay (DMN/SRMA/LA2M) on a Jeol 2010F with a Field Emission Gun (FEG) and on a Jeol 2100, both operating at a 200 kV voltage.

3 Results

3.1 Helium diffusion coefficient determination

The helium profiles obtained by NRA were assumed to be Gaussian for simplicity. In that case, the theory of the

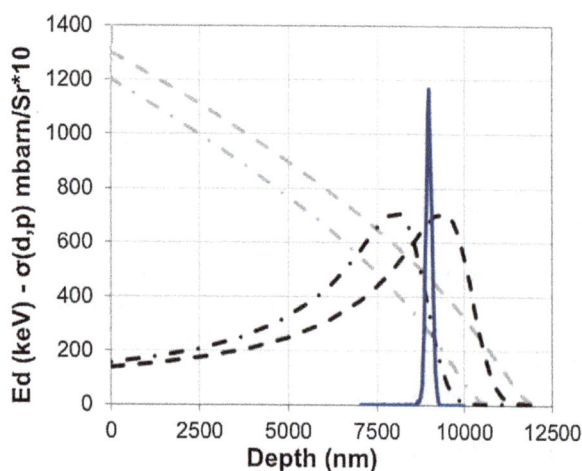

Fig. 5. Calculations for the choice of the energy of the deuterons (between 1200 keV: - · - and 1300 keV: - -) for the (d,p) reaction. Grey: energy of the deuterons versus depth into the material, from SRIM [19]. Black: cross-sections curves according to the initial deuterons energy and along the depth in the material. Blue solid line: implantation profile of helium-3 at 3 MeV in B_4C.

diffusion in the grain (pure diffusion, single mechanism, without any formation of clusters) gives:

$$\sigma_T^2 = \sigma_0^2 + 2 \cdot D_T \cdot t, \tag{1}$$

where σ_T is the standard deviation obtained after an annealing treatment of duration t at the absolute temperature T, σ_0 the standard deviation before annealing and D_T, the diffusion at temperature T defined by:

$$D_T = D_0 \cdot \exp\left(-\frac{E_a}{kT}\right), \tag{2}$$

with D_0, the pre-exponential factor, E_a, the activation energy and k, the Boltzmann constant (8.617×10^{-5} eV/K).

To reach the D_0 and E_a values, we have to measure the standard deviation of the Gaussian profiles, then use equation (1) to find D_T. If the D_T values are aligned in an Arrhenius diagram (log (D_T) vs. $1/T$), then the D_0 and E_a values can be deduced.

The experimental NRA spectra were given in channels as a function of a number of counts. To convert channels into depth, we evaluated the depth at which helium had been implanted by using the SRIM profiles, from which we deduced a linear channel-depth conversion. This approximate conversion can then be used to perform preliminary evaluations of the diffusion coefficients. More accurate calculations taking into account the full setup design [18] are in progress.

We proceeded to carry out two annealing sessions: one at different temperatures over 1 h to draw the Arrhenius diagram, and another at 900 °C from 15 min to 2 h. The latter then allowed a better estimation of the diffusion coefficient at 900 °C for the Arrhenius diagram.

Because of a low statistic (around 300 events for a profile), complex helium profiles cannot be observed and we assumed Gaussian profiles. Some of the results obtained from the one-hour annealing process are plotted in Figure 7.

As shown in Figure 7, the profiles broadened after annealing. We also observed that the area of the profiles was constant (by integration of the curves). This shows

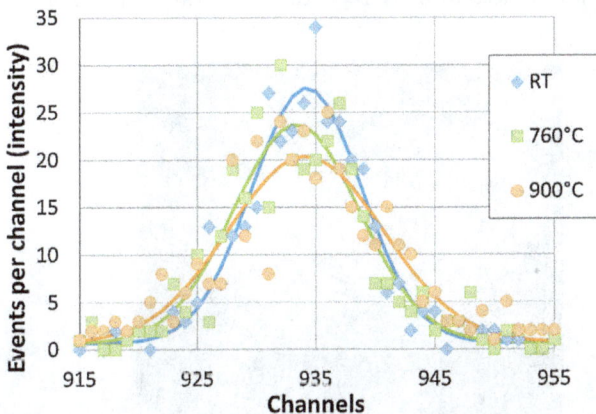

Fig. 8. ^{3}He profiles in B$_4$C analysed by NRA with deuteron energy of 1300 keV. Samples were annealed at 900 °C for different durations (s) before the analysis.

that diffusion occurred in the material without loss of helium: these two points are required in order to calculate a diffusion coefficient.

The 1000 °C curve was not shown in Figure 7 because its width was narrowed and its intensity reduced as compared to the 900 °C curve. It may imply that a part of helium not only diffused on long distances, with concentrations lower than the detection limit of the experiment, but also formed clusters close to the implanted zone. Thus, this data point was not taken into consideration in the Arrhenius diagram.

The profiles obtained from the annealing experiments at 900 °C during different durations (Fig. 8) also broadened after annealing. From 15 min to 1 h, the broadening is quite monotonous so it allowed us to obtain better accuracy for the value of D_T at 900 °C for the Arrhenius diagram. But the sample annealed for 2 h had a profile similar to the one observed after the annealing at 1000 °C: the apparent width and the intensity decreased, so it was not taken into consideration in the Arrhenius diagram.

Afterward, we inserted all the values of the isochronal annealing up to 900 °C in an Arrhenius diagram (Fig. 9) and

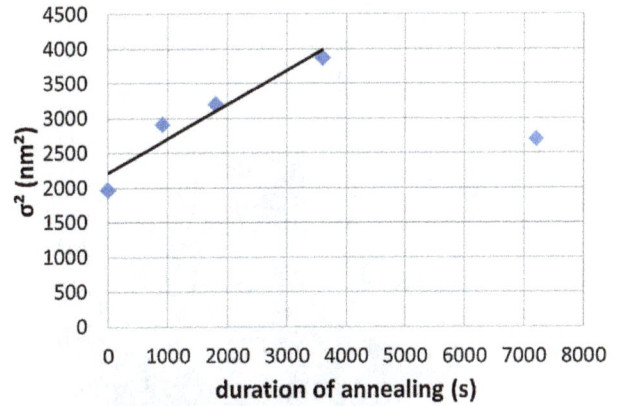

Fig. 7. Gaussian fitting (solid lines) of the ^{3}He profiles in B$_4$C analysed by NRA with deuteron energy of 1300 keV. Samples were annealed over 1 h at different temperatures (°C) before the analysis.

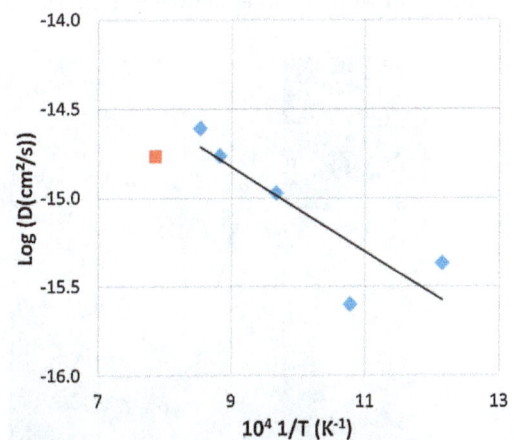

Fig. 9. Arrhenius diagram of the diffusion coefficient of ^{3}He in B$_4$C. The 1000 °C point (in red) was excluded for the linear fitting.

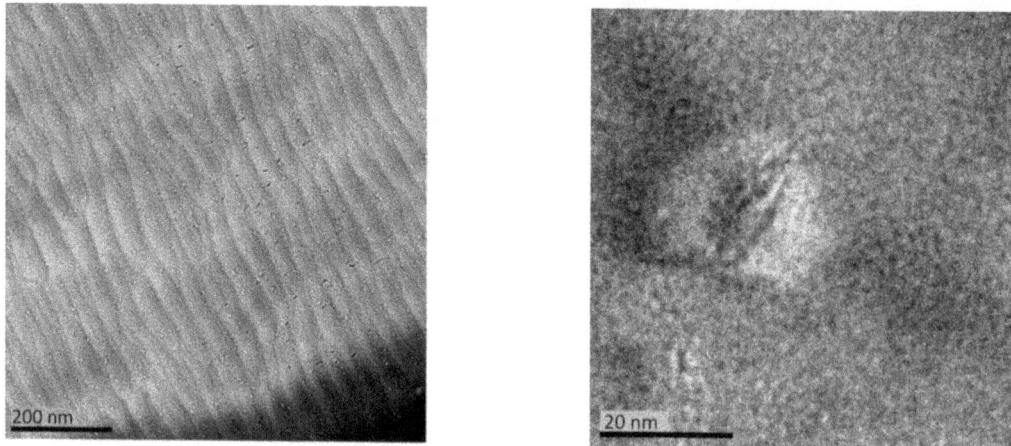

Fig. 10. ^{4}He implanted in B$_4$C then annealed at 900 °C. Left: intra-granular bubbles band (in black, the bubbles; riddles and white dots are artefacts due to FIB thinning). Right: strain field around a bubble.

the data point at 900 °C resulted from the analysis of the isothermal annealing except the 2 h data.

As shown in Figure 9, the helium profiles which have the same intensity (Fig. 7: from RT to 900 °C included) are correctly aligned in the Arrhenius diagram. This shows that parameters fitting to a diffusion law can be estimated. From a linear fitting, we deduced:

− $D_0 = 1.19 \times 10^{-12}$ cm^2/s;
− Ea = 0.523 ± 0.107 eV/atom.

3.2 Helium clusters observations

For the TEM observations, all the samples were implanted at the same fluence (1.5×10^{16} at/cm^2) at 500 °C, and then annealed at different temperatures.

For the as-irradiated sample, no clusters were observed. Helium clusters may have nucleated but these were then too small to be observed (only a few atoms).

For the 900 °C annealed sample (Fig. 10), a bubble band was observed. Surprisingly, the band was only 400 nm wide (instead of a 1 μm wide band, as shown in Fig. 6). Clusters

were very small (between 3 and 20 nm). The smallest clusters were ellipsoidal. The larger bubbles tended to grow in a flat shape and be orientated in parallel. It was difficult to orientate those flat bubbles with respect to the crystal structure, because they need to be on the edge for the observation (which is not exactly the case here), and the sample could not be correctly oriented because the sample was too far from a zone axis. We can notice the presence of strain fields around the clusters as a pattern of "butterfly's wings", which was a consequence of the high pressure of the gas inside the cluster [13].

For the 1100 °C annealed sample (Fig. 11), the same band was observable. However, in this case, all bubbles were plate-like and parallel to each other, showing that strong orientation constraints were acting in the material. As they were on the edge, it became possible to find their habit plane. Two methods can be used: either by recording a diffraction pattern then indexing it, or performing high resolution observations by measuring the distance between atomic planes then deducing their Miller indexes. Both methods led to the same result: the bubbles were oriented along the (111) rhombohedral plane (or (0003) hexagonal plane), as was already reported in literature [10–13].

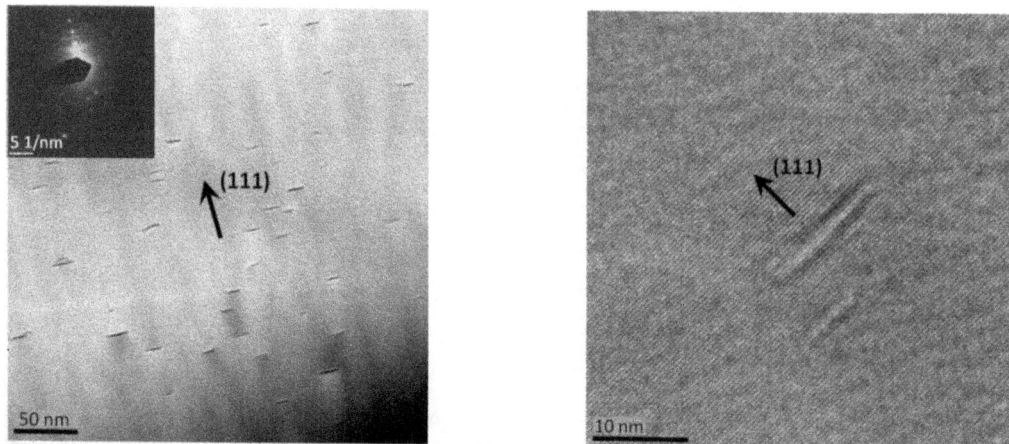

Fig. 11. ^{4}He implanted in B$_4$C then annealed at 1100 °C. Left: parallel plate-like intra-granular bubbles band with strain fields and the corresponding diffraction pattern. Right: two oriented bubbles in high resolution observation.

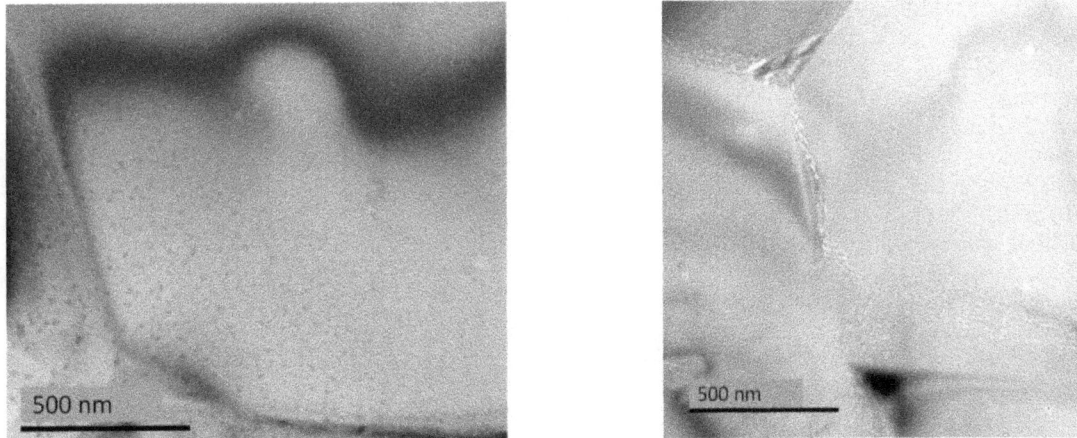

Fig. 12. ^{4}He implanted in B$_4$C and annealed at 1300 °C. Left: intra-granular plate-like bubbles and inter-granular bubbles (bubbles appear in white; the black spots are artefacts due to FIB preparation). Right: inter-granular bubbles (two triple points).

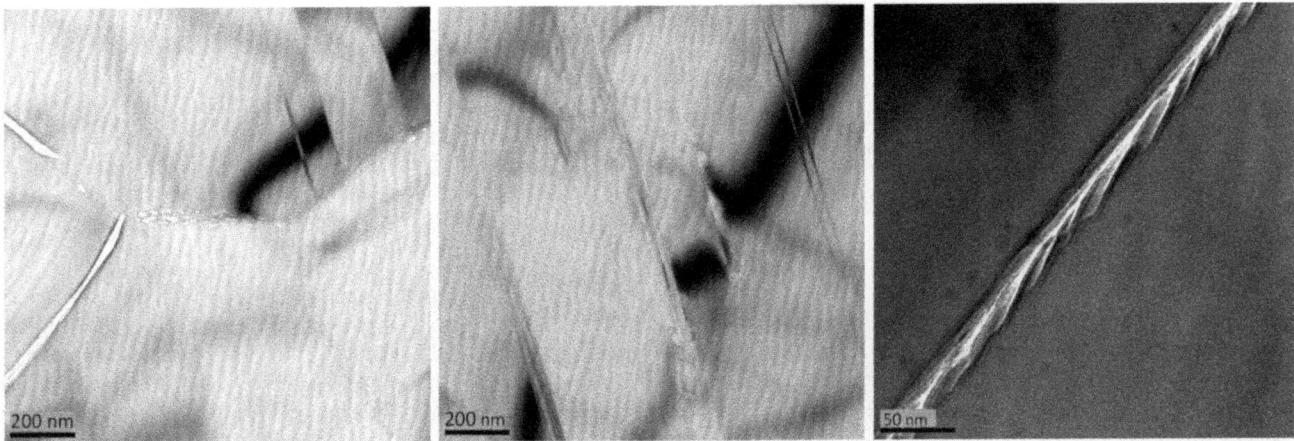

Fig. 13. ^{4}He implanted in B$_4$C and annealed at 1500 °C. Left: two triple points. Helium was trapped in the grain boundaries in the form of bubbles, or the grain boundaries are opened. Middle: helium bubbles located at twin boundaries (riddles due to FIB thinning). Right: opened grain boundaries with a shape of "stair steps".

For the 1300 °C annealed sample (Fig. 12), we can notice a different helium behaviour. The flat parallel bubbles band was still visible, and close to the centre of the maximum calculated by SRIM, but only a few bubbles are present. Large quantities of helium have diffused through long distances in the grain boundaries where bubbles were also formed. Instead of having a 1 μm large profile at around 3 μm from the surface (according to SRIM), bubbles were found between 0.5 and 4.2 μm from the surface, most of them in the grain boundaries. Also, it was noticed that most of the helium bubbles were found between the surface, and the original maximum of the helium distribution, rather than beyond the implantation peak. Different mechanisms were thus activated between 1100 °C and 1300 °C.

For the 1500 °C annealed sample (Fig. 13), the behaviour of helium is different again, and may be due to the specific temperature of 1500 °C, which is close to the brittle-plastic transition of boron carbide [2]. No helium bubbles were observed in the grains, except in some defects such as twin boundaries, which are typical defects in boron carbide. All visible helium was observed as bubbles with various shapes in the grain boundaries, over distances much longer than the peak width calculated by SRIM. No bubbles were observed in the grains. Some grain boundaries were opened in front of the implantation maximum. As was noted for the B$_4$C annealed at 1300 °C, instead of having a 1 μm large helium band at about 3 μm depth, a much larger band, that is at 0.4 to 5.3 μm from the surface was observed. When observed at high resolution, we noticed that the opened grain boundaries have a shape of "stair steps".

4 Discussion

4.1 Diffusion coefficient determination

When annealed for 1 h at a temperature lower than 700 °C, helium did not diffuse significantly, so the microprobe's depth resolution may not be sharp enough to detect profile broadening. That was why the 550 °C and the 655 °C data points were so far apart from the linear fitted line in the Arrhenius diagram. This dispersion allowed us to estimate the error inherent in the value of the activation energy that we have measured.

The helium profile in the B_4C sample annealed at 1000 °C had narrowed, and its intensity had lowered, showing that helium was not diffusing progressively anymore. This may be due to the conjunction of the formation of clusters in the implanted slab, and long-range diffusion up to and in the grain boundaries. In that case, due to the low yield of the method, the foot of the profile would be too low to be detected, thus leading to an underestimation of the integral intensity of the peak. For example, we have observed on the TEM images that bubbles would appear around 900 °C, but at higher fluences. It could then be assumed that at 1000 °C (with smaller fluences), helium clusters can also nucleate and affect the intra-granular diffusion. The same behaviour was observed for the sample annealed at 900 °C for 2 h, for which the helium profile had a similar width as the sample annealed at 1000 °C, showing that there were specific annealing temperature and duration thresholds above which intra-granular diffusion was affected by other mechanisms: influence of grain boundaries and formation of clusters.

In order to check the obtained value of the diffusion coefficient, we compared it to the diffusion coefficient of helium in SiC [21], which is a ceramic with properties similar to boron carbide: they both are built with light elements with tight covalent bonding. Moreover, the two materials were prepared by powder sintering in our investigations. We observed (Fig. 14) similar diffusion coefficients for the two compounds.

One problem of this method is the low number of data points, which prevented an accurate analysis of the profiles that would be necessary to observe asymmetric or long-range diffusion: in fact, only a standard deviation and an integral intensity can be evaluated. Indeed, Gaussian fits were based on only a few points, due to the low dose of helium-3 in the samples, and the deep implantation, the small flux of the accelerator and the few hours of experiment to analyse the samples. To get more accurate results, more data points that will contribute to better statistical power of the analysis are needed.

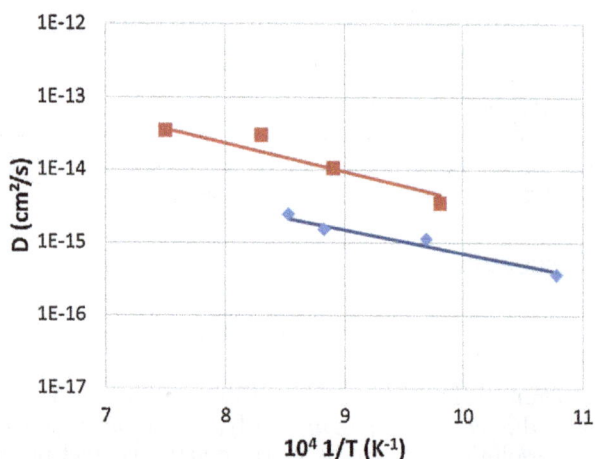

Fig. 14. Comparison of Arrhenius diagrams for the diffusion of helium in B_4C (in blue) and SiC [21] (in red).

4.2 Helium clusters observations

Inside a fast breeder reactor, B_4C is exposed to a temperature range of 500–1200 °C in normal conditions. The TEM observations gave a good scope of the behaviour of helium in B_4C according to the annealing temperature. The gas was found in different forms:

- 500 °C: bubbles were not formed yet;
- 900 °C: bubbles appeared. They were circular and small, and some of them began to orientate parallel to each other;
- 1100 °C: bubbles were all plate-like and parallel in the (111) rhombohedral plane;
- 1300 °C: a few intra-granular plate-like parallel bubbles were observed, and most of helium had diffused in the grain boundaries over large distances. This long-range diffusion appeared asymmetric, since bubbles were mainly observed in front of the implanted zone. This point should be addressed. In particular, the influence of material damage (here produced by helium slowing down) should be analysed;
- 1500 °C: brittle-plastic transition temperature of B_4C. Helium was observed in the grain boundaries over long distances in the form of bubbles with different shapes, or was trapped in structural defects such as twin boundaries or dislocations. The same asymmetric distribution as after annealing at 1300 °C is observed. Some of the grain boundaries were even opened.

Only the sample annealed at 1100 °C had a correct orientation to allow determination of the orientation of the parallel bubbles in the crystal. We found that they all were parallel to the (111) rhombohedral plane. The orientation of the grains in the other samples did not allow us to obtain unambiguous orientations to perform such an analysis.

The results we obtained here were coherent with previous ones [10–15]. However, implanting helium in a thin slab led to different behaviours as compared to those observed in homogeneously implanted or neutron irradiated materials. In particular, there is evidence for long range, to distances much larger than the ones that could be deduced from the diffusion coefficient we obtained. This showed that different mechanisms were competing in the material, such as nucleation of clusters and diffusion, leading to quite different defects (bubbles) distributions. Such complex behaviours should be further analysed. Intra-granular micro-cracks were not observed. This was consistent with previous observations [4], showing that cracking occurs only for a higher helium concentration, around 5×10^{21} at/cm^3.

5 Conclusion

The tests we conducted here gave preliminary results that can be used to guide further studies about the behaviour of helium implanted in boron carbide B_4C. To simulate the production of helium that occurs in a nuclear reactor, we implanted helium in the samples and proceeded to thermal annealing to observe and analyse the diffusion of the gas in the material.

Helium-3 was implanted in small quantities in B_4C. Then the samples were annealed at different temperatures and analysed with a nuclear microprobe by NRA (Nuclear Reaction Analysis). The implantation profiles broadened with an increase in temperature. At low temperatures, the degree of diffusion was lower than the microprobe sensitivity limit. Then up to 900 °C, the gas diffused significantly, at 1000 °C, helium partly escaped to the grain boundaries and possibly began to form clusters. From those results, a diffusion coefficient was deduced ($E_a = 0.52 \pm 0.11$ eV/atom), which was close to the equivalent coefficient of diffusion of helium in SiC.

Helium-4 in larger quantities was implanted at 500 °C in B_4C. Then the samples were annealed until 1500 °C and thinned by FIB to be observed by TEM (Transmission Electron Microscope). At 500 °C, helium clusters were not observable. At 900 °C, helium started to accumulate as pressurised bubbles. Then at 1100 °C, the bubbles were flat, all parallel to each other and oriented along the (111) rhombohedral planes. At 1300 °C, long distance diffusion in gain boundaries occurred and only a few intra-granular bubbles were visible. At 1500 °C, helium was found only in the grain boundaries in the form of bubbles and some parts of the grain boundaries were opened.

We are highly grateful to Benoit Arnal (CEA Saclay, DEN/DMN) who prepared the TEM thin-foil samples.

References

1. V. Domnich et al., J. Am. Ceram. Soc. **94–11**, 3605 (2011)
2. F. Thevenot, J. Eur. Ceram. Soc. **6**, 205 (1990)
3. H. Werheit et al., J. Phys.: Condens. Matter **24**, 305 (2012)
4. D. Gosset, Neutron absorber materials, in *Handbook of nuclear engineering* (Dan Gabriel Cacuci ed., 2010)
5. ENDF database, www-nds.iaea.org/exfor/endf.htm
6. DOE Fundamentals Handbook, *Nuclear physics and reactor theory* (DOE-HDBK-1019/1–93, 1993)
7. K.N. Kushita et al., Microsc. Microanal. Microstruct. **6**, 149 (1995)
8. T. Maruyama et al., Effects of radiation on materials, in *21st International Symposium ASTM STP1447, 2004* (2004), p. 670
9. D. Gosset et al., to be published in NIM-B (2015)
10. G.L. Copeland, J. Nucl. Mater. **43**, 126 (1972)
11. T. Maruyama et al., J. Nucl. Mater. **133&134**, 727 (1985)
12. A. Jostsons et al., J. Nucl. Mater. **44**, 91 (1972)
13. T. Stoto et al., Radiat. Eff. **105**, 17 (1987)
14. V.P. Tarasikov, Atom. Energy **106**, 220 (2009)
15. W.V. Cummings et al., T. Am. Nucl. Soc. **15**, 742 (1972)
16. K. Froment et al., J. Nucl. Mater. **188**, 185 (1992)
17. H. Suzuki et al., J. Nucl. Sci. Technol. **16**, 588 (1979)
18. D. Gosset et al., J. Nucl. Mater. **303**, 115 (2002)
19. J.F. Ziegler, www.srim.org
20. P. Trocellier, Microsc. Microanal. Microstruct. **7**, 235 (1996)
21. Y. Pramono et al., J. Nucl. Sci. Technol. **41**, 751 (2004)

4

Experimental facility for development of high-temperature reactor technology: instrumentation needs and challenges

Piyush Sabharwall[1*], James E. O'Brien[1], SuJong Yoon[1], and Xiaodong Sun[2]

[1] Idaho National Laboratory, PO Box 1625, Idaho Falls, ID 83415-3860, USA
[2] Mechanical and Aerospace Engineering, Ohio State University, Columbus, Ohio, USA

Abstract. A high-temperature, multi-fluid, multi-loop test facility is under development at the Idaho National Laboratory for support of thermal hydraulic materials, and system integration research for high-temperature reactors. The experimental facility includes a high-temperature helium loop, a liquid salt loop, and a hot water/steam loop. The three loops will be thermally coupled through an intermediate heat exchanger (IHX) and a secondary heat exchanger (SHX). Research topics to be addressed include the characterization and performance evaluation of candidate compact heat exchangers such as printed circuit heat exchangers (PCHEs) at prototypical operating conditions. Each loop will also include an interchangeable high-temperature test section that can be customized to address specific research issues associated with each working fluid. This paper also discusses needs and challenges associated with advanced instrumentation for the multi-loop facility, which could be further applied to advanced high-temperature reactors. Based on its relevance to advanced reactor systems, the new facility has been named the Advanced Reactor Technology Integral System Test (ARTIST) facility. A preliminary design configuration of the ARTIST facility will be presented with the required design and operating characteristics of the various components. The initial configuration will include a high-temperature (750 °C), high-pressure (7 MPa) helium loop thermally integrated with a molten fluoride salt (KF-ZrF$_4$) flow loop operating at low pressure (0.2 MPa), at a temperature of ~450 °C. The salt loop will be thermally integrated with the steam/water loop operating at PWR conditions. Experiment design challenges include identifying suitable materials and components that will withstand the required loop operating conditions. The instrumentation needs to be highly accurate (negligible drift) in measuring operational data for extended periods of times, as data collected will be used for code and model verification and validation, one of the key purposes for the loop. The experimental facility will provide a much-needed database for successful development of advanced reactors and provide insight into the needs and challenges in instrumentation for advanced high-temperature reactors.

1 Introduction

Effective and robust high-temperature heat transfer systems are fundamental to successful deployment of Advanced High Temperature Reactor (AHTR) systems for both power generation and non-electric applications. A highly versatile test facility is needed to address research and development (R&D) and component qualification needs. Key activities of this test facility would include (1) qualification and testing of critical components in a high-temperature, high-pressure environment, (2) materials development and qualification, and (3) manufacturer and supplier evaluation and development. A small-scale test loop could provide for early testing of components and design options that require special development tests before finalizing the design of AHTR components and qualifying them for operation in the larger loop or demonstration facility. Since a suitable facility does not exist for testing advanced reactor heat transfer system components (e.g., intermediate heat exchanger [IHX], valves, etc.), reactor internals, or the interface with the heat application plant, a laboratory-directed research and development project was approved to initiate development of such a facility at Idaho National Laboratory. This facility will include three thermally coupled flow loops: a high-temperature He loop, a liquid salt intermediate loop, and a high-pressure water loop. Based on its relevance to advanced reactor systems, the new facility has been named the Advanced Reactor Technology Integral System Test (ARTIST) facility.

AHTR plant designs often include an intermediate heat transfer loop (IHTL) with heat exchangers at either end to

* e-mail: Piyush.Sabharwall@inl.gov

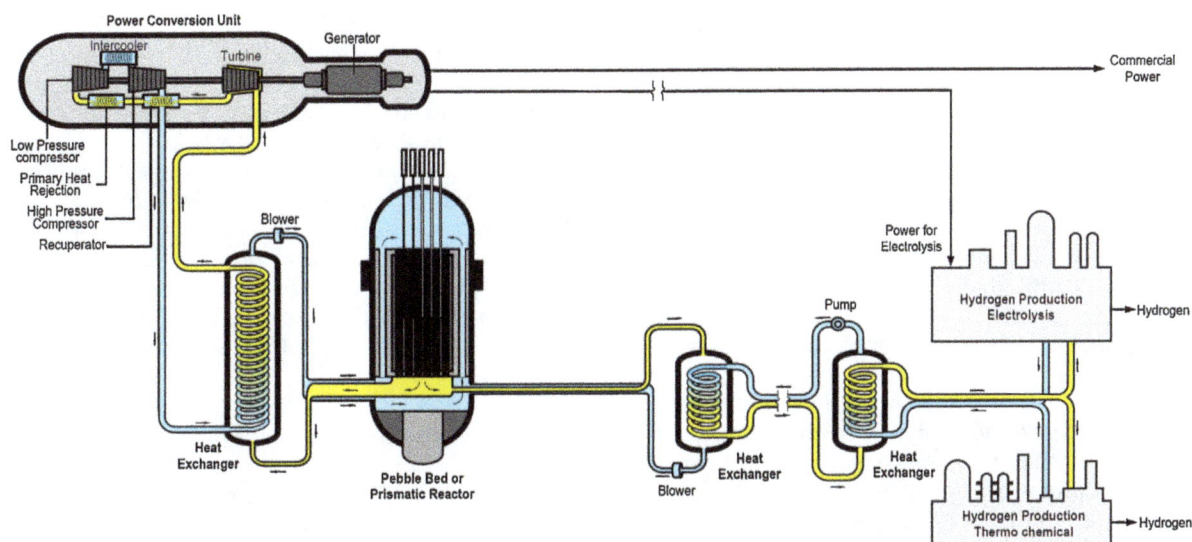

Fig. 1. NGNP power and hydrogen production plant with three IHXs.

deliver thermal energy to the application while providing isolation of the primary reactor system. A conceptual layout for one such plant, the Next Generation Nuclear Plant (NGNP), is shown in Figure 1. This concept indicates the use of a single IHX isolating the secondary power conversion unit working fluid from the primary He reactor coolant. For safety reasons and further isolation of the primary coolant from the process heat application (e.g., hydrogen production), a secondary heat exchanger is included in the process heat loop. In this case, a full intermediate heat transport loop is required, with an appropriate heat transport fluid.

Liquid salts have been identified as excellent candidate heat transport fluids for intermediate loops, supporting several types of advanced high temperature reactors [1–4]. Liquid salts have also been proposed for use as a primary coolant for the Advanced High Temperature Reactor (AHTR) [5] and the Fluoride Salt-cooled High-Temperature Reactor (FHR) [6,7]. Fluoride salt-coolants are eutectic binary or tertiary mixtures of fluoride salts with melting points in the range of 320 to 500 °C. FHRs have reactor outlet temperatures of 600 °C or higher for high-efficiency power generation or process heat applications. Liquid salts exhibit superior heat transfer characteristics compared to He-cooled reactors. FHRs can also take advantage of effective passive natural circulation for decay heat removal.

PCHEs are strong candidate heat exchangers for intermediate heat transport loops due to their very high power density, requiring much less material per unit of heat duty compared to conventional shell and tube heat exchangers. PCHEs are fabricated from individual flat plates into which small flow channels are etched. The plates are stacked into alternating hot/cold layers and are typically diffusion-bonded, yielding a monolithic heat exchanger with strength equal to that of the base material.

With appropriate materials, these heat exchangers can operate at high temperature and high pressure. PCHEs can, however, be susceptible to large thermal stresses during transient thermal hydraulic events [8].

In addition to the heat exchangers, each flow loop in the ARTIST facility will include high-temperature test sections operating at prototypical conditions that can be customized to address specific research issues associated with each working fluid. Possible research topics for the high-temperature helium test section include flow distribution, bypass flow, heat transfer in prototypical prismatic core configurations under forced and natural circulation conditions [9,10], parallel flow laminar instability during pressurized cooldown [11,12], and turbulent heat transfer deterioration [13,14]. Oxidation effects associated with water or air ingress could also be examined [15].

The high-temperature test section in the liquid salt loop can be used for examination of materials issues, thermal stresses, and heat transfer. Metallic materials have been studied extensively in liquid salt environments [1–3], but additional research is needed to evaluate the performance of ceramic and composite materials such as SiC/SiC in liquid salt environments [7]. Fundamental heat transfer issues for liquid salts are related to the fact that these are high-Prandtl-number fluids with high viscosities and specific heats, and relatively low thermal conductivities. Accordingly, prototypical Reynolds numbers are small, in the laminar or transitional flow regimes and heat transfer enhancement strategies (e.g., extended surfaces) may have to be employed in the core and other components. Flow geometries of interest include prototypical prismatic core configurations and pebble beds, as well as heat exchanger flow passages. The high Prandtl number reduces the potential for thermal shock (compared to low-Prandtl-number liquid metal coolants), but the possibility of large thermal stresses still exists [7]. Bypass flow can also be an issue for prismatic reactor core configurations with liquid salt coolants.

The liquid salt loop will include a thermal energy storage (TES) system for support of thermal integration studies. The TES system will be based on freezing and melting of the salt acting as a high-temperature phase change material (PCM). A number of salts have been proposed as high-temperature PCMs for solar energy applications [16,17]. The advantage of using a PCM is that thermal energy can be supplied to the process at a nearly constant temperature, taking advantage of the latent heat of melting.

The high-temperature test section in the steam/water loop will be used primarily for prototypic evaluation of new cladding materials and accident-tolerant fuels. It will be designed to characterize the thermal, chemical, and structural properties of candidate advanced fuel cladding materials and designs under various simulated flow and internal heating conditions to mimic operational reactor conditions prior to in-reactor testing. The capability for out-of-pile mock-up testing of candidate (surrogate) fuel-clad systems is essential for reactor readiness, in particular when innovative fuel cladding will be in direct contact with the test reactor primary coolant system without secondary containment. Careful control of water chemistry will be essential for these studies; a water chemistry control section is included in the design of the loop.

Flow-induced vibration of fuel rod bundles has been identified as an important issue for sodium-cooled reactors [18]. The high-temperature test section of the hot water loop can also potentially be used to study flow-induced vibration of simulated sodium-cooled reactor fuel rod bundles. Hot water at 200 °C and 1.38 MPa matches the density of sodium. This condition is well within the operational range of the proposed loop.

Research conducted in these flow loops will also support verification and validation efforts. Experimental data for validation is required to gain confidence in the existing theoretical and empirical correlations. Development of such an experimental database is needed to advance the technology readiness level of various reactor concepts and high-temperature components (such as heat exchangers). The database will also be used to evaluate the performance of existing models and correlations in predicting thermal hydraulic phenomena. New models and/or correlations will be developed as needed. The facility is designed such that each individual loop can operate independently.

2 Facility description

A process flow diagram for the multi-fluid, multi-loop test facility is shown in Figure 2. The facility includes three thermally interacting flow loops: helium, liquid salt, and steam/water. The helium loop will be initially charged from pressurized gas storage cylinders to the loop operating pressure of 7 MPa. The loop can be evacuated prior to charging for removal of air. This process can be repeated with intermediate gas venting via the deaeration vent to achieve the desired loop He purity level. Helium flow through the loop will be driven by a water-cooled centrifugal gas circulator rated for high-pressure service, with a design flow rate up to

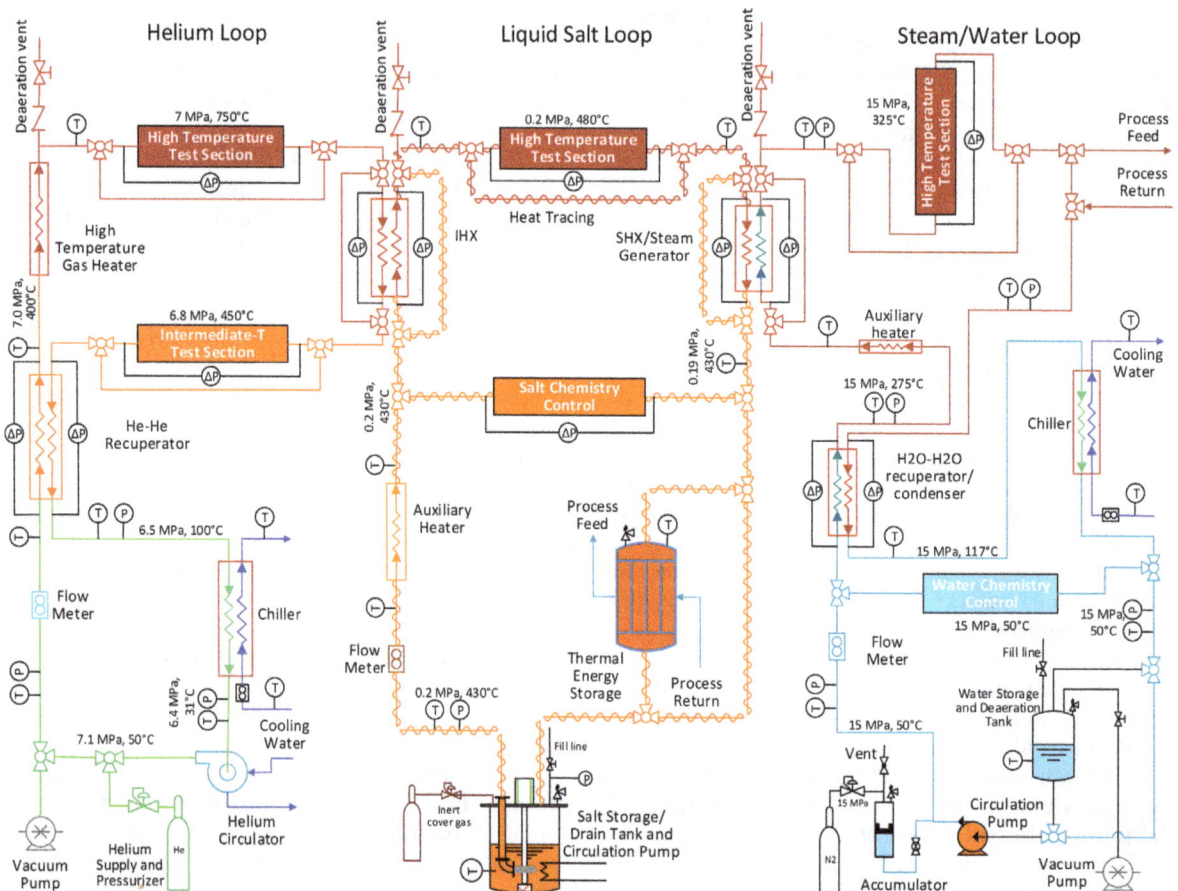

Fig. 2. Schematic of multi-fluid, multi-loop ARTIST thermal hydraulic test facility.

525 LPM at 7 MPa (11,300 SLPM) and a loop pressure drop of 100 kPa. The circulator flow rate will be controlled by means of a variable-frequency drive coupled to the motor. The helium circulator will be designed to operate with a maximum helium temperature of 100 °C. It is therefore located in the low-temperature section of the helium flow loop. The gas is preheated to intermediate temperature by flowing through a helium-to-helium recuperator (60 kW duty) that transfers heat from the intermediate-temperature helium return flow to the low-temperature stream. The high-temperature portion of the flow loop is designed to handle helium temperatures up to 800 °C. This temperature will be achieved using a high-temperature in-line electrical gas heater located downstream of the recuperator. The nominal power requirement for the high-temperature gas heater is 60 kW.

The helium loop will include a high-temperature test section for heat transfer and materials studies. Downstream of the test section, the helium gas flows through a heat exchanger where heat will be transferred to the adjacent liquid salt loop using a scaled version of an IHX. The baseline design for this heat exchanger will be a high-efficiency compact microchannel PCHE with a nominal heat duty of 55 kW. Analysis of a PCHE operating with He as the hot fluid and liquid salt as the cold fluid is provided in reference [8]. Downstream of the IHX, the helium flows through an intermediate-temperature test section and the recuperator to transfer heat back to the inlet stream. Downstream of the recuperator, the helium flows through a water-cooled chiller (10 kW) to cool it down to the gas circulator operating temperature. The baseline design for the He-He recuperator will also be a PCHE. In addition to its heat recuperation role, this heat exchanger simulates an IHX for the case in which He is used as an intermediate heat transfer fluid, albeit at lower operating temperatures. Performance data obtained from this recuperator will provide useful validation data for the reactor system application. The He-He version of the IHX operates with essentially balanced high pressure on both sides, minimizing the possibility of leakage of primary fluid to the secondary side.

The center part of Figure 2 shows the liquid salt portion of the multi-loop facility. The loop will be charged with salt from the salt storage tank. This tank will include a heater designed to heat the frozen salt to a temperature above its melting point. The head space in the salt storage tank will be maintained at slightly elevated pressure with an inert cover gas. The inert gas will prevent in-leakage of air or moisture, minimizing the potential for salt contamination. During startup, liquid salt will drain to the pump inlet by gravity, with assist from the cover gas pressure, as needed. The salt pump will be designed to operate at 450 °C at low pressure (~0.2 MPa). It will provide salt flow rates up to 20 LPM. A standard stainless steel such as SS316 may be suitable for the pump material, but other alloys will also be considered. The entire liquid salt flow loop will be heat-traced to prevent salt from freezing and causing a flow blockage. Downstream of the pump, the salt flow rate will be measured using a high-temperature ultrasonic flow transducer. The salt temperature will be boosted as needed to the desired intermediate temperature using an in-line

electrical auxiliary heater. Careful control of salt chemistry will be critical for successful operation of this loop; a salt chemistry control section will be installed at the intermediate temperature location. The salt temperature will increase to ~480 °C as it flows through the IHX and heat is transferred from the helium loop to the salt loop. Note that the He-salt IHX will have high-pressure helium on one side and low-pressure salt on the other, establishing the potential for high-temperature creep and leakage of primary He into the salt loop, emphasizing the need for demonstrating complete IHX integrity at prototypical conditions.

For independent operation of the liquid salt loop, the IHX will not be required. An IHX bypass will enable salt flow directly to the high-temperature test section without the pressure drop associated with the IHX. The auxiliary heater will be designed to independently heat the salt to the maximum operating temperature of 480 °C even when the IHX is bypassed. Its nominal design heater power will be 75 kW. Downstream of the high-temperature test section, the liquid salt flows through the SHX, transferring heat to the tertiary steam/water loop. A bypass line around the SHX is also provided for cases in which the salt loop will be operated independently of the steam/water loop. The salt can then flow directly back to the pump or it can flow through a thermal energy storage (TES) system for process integration studies.

The right-hand side of Figure 2 shows the steam/water tertiary loop. The SHX can serve as a steam generator or simply a single-phase heat exchanger, depending on conditions to be simulated in the tertiary loop. For most tests, conditions in the tertiary loop will be intended to simulate conditions in the primary loop of a pressurized water reactor (PWR). PWR conditions will be needed for materials/corrosion studies of accident-tolerant fuels, new cladding materials, crud formation, etc. Alternately, at lower operating pressure the tertiary loop can simulate the secondary side of a PWR system, with steam generation for process integration studies. Flow through the hot water loop will be produced by a pump designed to operate at 15 MPa with a nominal water flow rate of 5.7 LPM at 15 MPa and 40 °C.

Downstream of the pump, the water flows through a recuperator designed to recover heat from the high-temperature portion of the loop. The baseline recuperator inlet and outlet water temperatures will be 50 °C and 275 °C, respectively. The water is heated further to 325 °C by heat transfer in the SHX. For cases in which the SHX is not present or is bypassed, an auxiliary heater will be used to achieve the desired 325 °C test section inlet temperature. Note that the SHX will also operate with a large pressure differential between the water side (15 MPa) and the salt side (0.2 MPa), establishing the potential for water leakage into the salt, and emphasizing the need to demonstrate full SHX integrity at these conditions. High-temperature creep should not be a concern at these temperatures. However, at 15 MPa, the maximum water temperature (325 °C) is well below the saturation temperature, so the water remains in the liquid phase throughout the system for the baseline case. The high-temperature test section in the water loop will have a vertical orientation to support boiling (at pressures lower than 15 MPa) and/or natural circulation studies. Simulated PWR core geometries in the test section will support research on new cladding materials, accident-tolerant fuels, etc. Hot water or steam can be supplied to other co-located processes or experiments via the process feed and process return lines. After flowing through the return side of the recuperator, the water temperature is decreased to 117 °C. It is further reduced to the pump operating temperature of 50 °C by means of a water-cooled chiller. Pressure in the hot water loop will be maintained by means of a piston accumulator with regulated nitrogen on the gas side. The water storage and deaeration tank will be plumbed to a vacuum pump to allow for air removal. The water can also be directed to flow through a chemistry control section. This part of the loop will be designed to establish the loop water chemistry. Most often, PWR water chemistry will be established. The chemistry control section will include filtering, a water softener, and a reverse osmosis conditioner for deionization/demineralization. It will also provide the ability to establish the correct pH value to ensure prototypical PWR conditions.

A three-dimensional (3D) computer-aided design (CAD) model of the ARTIST experimental test facility

has been developed using Pro-Engineering software. The CAD model includes all of the major facility components and piping. A rendering of the model is provided in Figure 3, with all of the components labeled. The facility is shown mounted on a large skid, measuring 4.9 m (16 ft) × 9.1 m (30 ft). The highest component is at the 7.0 m (23 ft) elevation. Components in the helium loop are designated with the He- abbreviation, salt loop components with the LS- abbreviation, and water/steam loop components with the WS- abbreviation. Pipe supports and insulation are not shown in these figures. Each component is shown to scale according to the current status of the design. Notable features of the high-pressure helium and water loops include the large flanges on the piping sections. The low-temperature sections of the He loop require class 600 flanges and NPS 2, schedule 160 piping. The high-temperature section of the He loop requires class 2500 flanges. Due to its higher pressure, the water/steam loop utilizes NPS 2, schedule 160 piping and class 2500 valves and flanges throughout. As an alternative to large flanges, the use of Grayloc connectors will also be examined.

The geometry of the He-He recuperator and the IHX shown in Figure 3 is based on a baseline PCHE design, sized

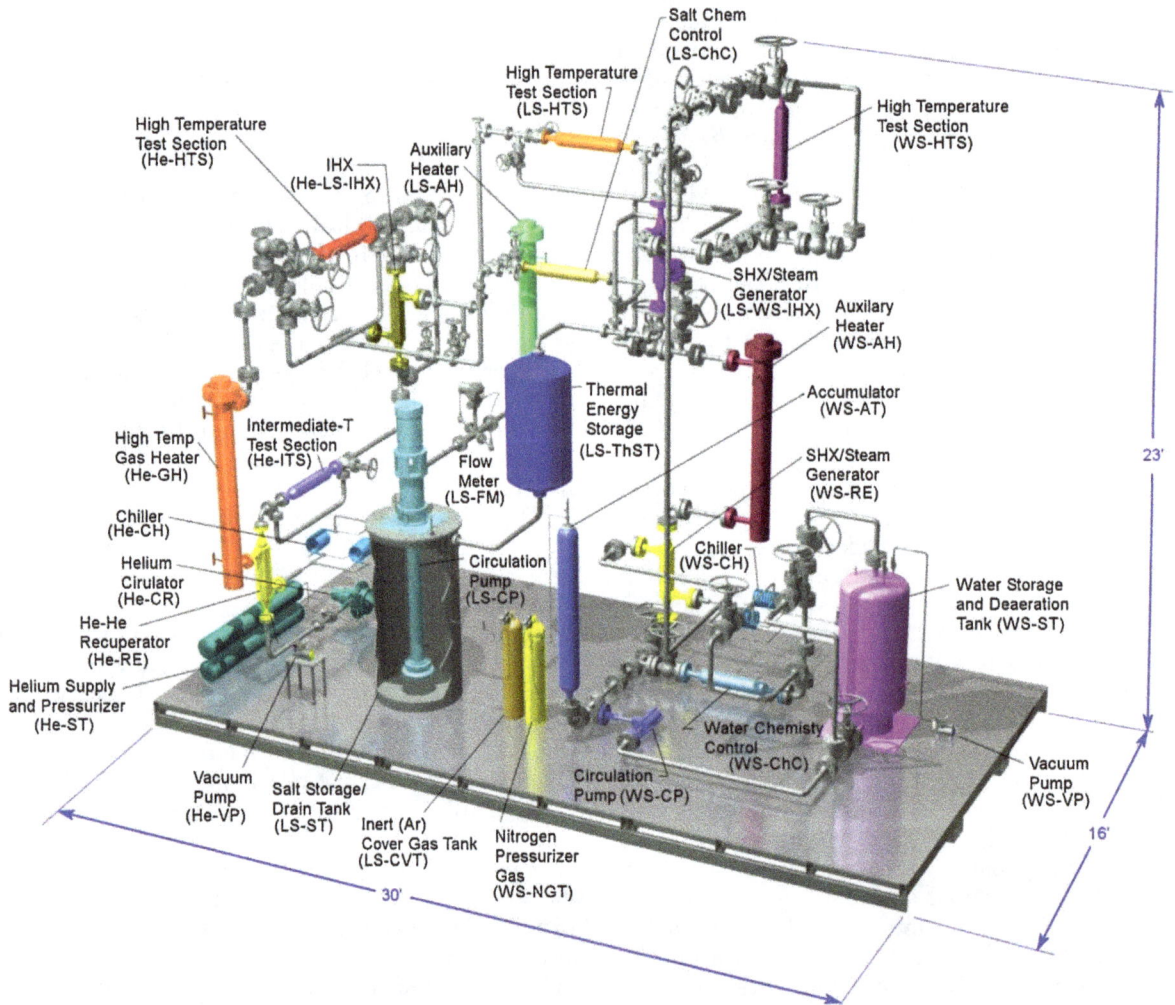

Fig. 3. 3D CAD model of the ARTIST facility.

to support the design helium flow rate and the required heat duty. The exact dimensions of these heat exchangers may change, depending on the final design details and the vendor selected. The high-temperature gas heater is based on a Watlow circulation heater with alloy 800/800H sheaths and ANSI 600 class pressure rating. The valving arrangement shown in Figure 3, downstream of the high-temperature gas heater, allows for bypass of the high-temperature test section. Three valves are shown instead of a single three-way valve because a three-way valve rated for these temperatures and pressures has not yet been identified. The high-temperature and intermediate-temperature helium test sections as shown in Figure 3 are designed to accommodate small tube bundles or other geometries of interest. The instrumentation in the test sections and loops could potentially serve the dual purpose of measuring test parameters and demonstrating new high-resolution advanced instrumentation and control equipment in challenging environments. The test sections will be custom-designed components and may be built in a range of sizes. As shown, the test sections have an outer diameter of 15.2 cm (6 in.) over a length of 0.76 m, with a total length of 1.22 m between flanges.

The He circulator flow rate will be controlled by means of a variable-frequency drive coupled to the motor. The helium circulator will be designed to operate with a maximum helium temperature of 100 °C. The circulator geometry shown in Figure 3 is based on information received from Barber-Nichols for gas circulators with similar requirements. The He chiller is shown as two coiled tube-in-tube counter-flow heat exchangers arranged in parallel. The use of two parallel chillers provides the required flow area with off-the-shelf units. The vacuum pump for removing air from the loop is also shown.

The largest component of the salt loop is the storage/drain tank. The current tank design size is 1 m diameter and 2 m high. This size is large enough to accommodate all of the salt in the loop and in the thermal energy storage system, while maintaining a high enough liquid salt level during operation to avoid any risk of gas entrainment into the pump. The salt storage/drain tank is not a pressure vessel because the salt loop operating pressure will be only slightly elevated above ambient pressure. The salt pump is a vertical cantilever pump, shown with a geometry based on a design for liquid salts available from Nagle. The ultrasonic flow meter (LS-FM) shown in Figure 3 is based on the Panaflow design from GE. The auxiliary heater is based on a Watlow circulation heater design with either alloy 600 or 800H heater element cladding materials.

The thermal energy storage tank in Figure 3 is shown with a diameter of 0.75 m and a height of 1.5 m, providing a volume of 660 L and a thermal capacity of 978 MJ or 272 kW-hr$_{th}$, based on the latent heat of fusion for KF. It will be designed for operation as a high-temperature, phase-change, thermal energy storage system, using the fluoride salt working fluid of the salt loop as the phase-change material.

The water/steam loop is designed to operate at PWR conditions. The water storage and deaeration tank is sized to hold all of the purified water in the loop. It can be isolated from the flow loop and therefore it will not be designed for full loop pressure. Charging and draining of the loop will be performed at atmospheric pressure. As shown in the figure, the water storage tank has a diameter of 0.76 m and a height of 1.5 m, with an internal volume of 695 L, which is about 2.7 times larger than the estimated total loop water volume, therefore providing storage for excess purified water plus a gas space for air removal. The water circulation pump will operate at 15 MPa, providing a loop flow rate of at least 350 kg/h (2.5 gpm) against a loop pressure drop of 20 kPa. Loop pressure will be set and maintained using a bladder accumulator pressurized by nitrogen gas. The commercially available accumulator is 0.24 m in diameter and 2.0 m high. The auxiliary heater will be a custom-engineered circulation heater housed in a class-2500 pressure vessel. The water chiller will consist of two parallel, tube-in-tube, water-cooled, counter-flow heat exchangers rated for 15 MPa service. Depending on the laboratory capabilities, the cooling water will be either once-through tap water or circulated house water cooled by a facility chiller.

3 Instrumentation

A variety of sensors and other diagnostic tools will be employed in the ARTIST facility to continuously monitor system parameters. The instrumentation used in the loop will provide real-time input to the data acquisition system regarding system flow rate, temperatures and pressures at various locations in the test section and composition information for chemistry control. Conditions at several locations in this system will be particularly challenging for instrumentation. A preliminary assessment of instrumentation requirements and suitable hardware (preferably commercially available) has been completed.

Instrumentation utilized for the loop will also provide first-hand information on long-term reliability and drift performance. Most high-temperature loops do not deal with coolants such as fluoride salts, where corrosion can be a significant challenge, especially at higher temperatures. Maintaining a pressurized He system at 750 °C with negligible leakage is also a major long-term challenge.

Instrumentation will be installed at various locations in the loop to monitor temperature, pressure, liquid level, chemistry (where appropriate/needed) and flow conditions for safety of the loop and gather data to support verification and validation efforts. Operating temperatures in the loops will be actively maintained via a closed-loop feedback control system.

3.1 Helium loop

3.1.1 Chemistry/Impurities and moisture

The baseline He loop chemistry is pure helium. However, even ultra-high-purity (UHP) helium contains trace impurities such as H_2O, O_2, hydrocarbons, CO, CO_2, N_2, and H_2. For UHP He, these impurities are in the low parts-per-million (ppm) range and the overall purity is specified as 99.999%. However, additional impurities will enter the gas during loop operation, especially when high-temperature operation is initiated. Furthermore, some gases may have to be intentionally included as additives. For example, low levels of oxygen may be needed to maintain protective oxide scale on metallic components [1]. On the other hand, oxygen can corrode graphite in the vessel core, which can in turn release additional gases such as CO and CO_2. These gases can subsequently form deposits on metallic components. Careful monitoring and control of these gases will be critical for successful long-term operation of the He loop. Monitoring is usually performed by withdrawing a gas sample from the low-temperature part of the loop for analysis using a gas chromatograph (GC) system.

3.1.2 Flow rate

Flow rate in the He loop will be measured in the low-temperature ($\sim$50 °C) leg of the loop just downstream of the circulator. There are several options for measuring helium flow rate including thermal mass flow meter, coriolis meter, venturi meter, or vortex flow meter. For this closed-loop high-flow-rate helium flow system, one of the major costs will be the circulator and its cost will increase with loop pressure drop. Therefore, a low-pressure-drop flow measurement device is desired. The permanent pressure loss for vortex flow meters is quite low, and they provide excellent accuracy with pressure and temperature compensation to yield a true mass flow measurement. Commercial units that meet the pressure and temperature specifications for the helium flow loop are available from several vendors. Thermal mass flow meters with laminar flow elements will also be considered if the pressure drop is low enough.

3.1.3 Pressure: absolute and differential

Pressure instrumentation for the helium loop will include several absolute pressure transducers, as indicated in Figure 2, plus several differential pressure transducers (dP cells). The differential transducers will be used primarily to measure pressure drop across the IHX and the recuperator. In some cases, differential pressure may be needed across the high-temperature test section as well. The differential transducers will be designed to measure relatively small pressure differences while operating at high absolute pressure. Each differential transducer assembly will include a dP cell manifold that allows zeroing of the cell at high absolute pressure.

3.1.4 Temperature

Temperature measurements for the helium loop will mostly be acquired using type K, stainless-steel or inconel-sheathed ungrounded 1/8- or 1/16-in. thermocouples inserted into the flow stream using compression fittings. Inconel sheathing will be used on the high-temperature portion of the loop. Type K thermocouples are rated for service up to 1260 °C. The high-temperature gas heater will be feedback-controlled using the temperature just downstream of the heater as the process variable.

3.2 Salt loop

Instrumentation specification for the salt loop will adapt to lessons learned from researchers at the University of Wisconsin, The Ohio State University, and Oak Ridge National Laboratory, based on their recent experiences in salt loop development.

3.2.1 Chemistry control

Pretreatment of fluoride salts is necessary before introducing them to the flow loop, to remove oxygen, moisture, and other contaminants from the mixture. In addition, continuous monitoring and chemistry control will be necessary to monitor and remove any contaminants such as metal oxides that build up in the mixture as a result of interaction with loop materials or due to air or moisture ingress. Monitoring may include the use of a high temperature electrochemical oxygen sensor based on yttria-stabilized zirconia (YSZ) or yttria-doped thoria (YDT) [19]. Development of a continuous chemistry monitoring system for the salt loop will be an important aspect of the loop design process.

3.2.2 Flow rate

Most standard flow measurements include some kind of probe or sensor in direct contact with the fluid. Many of these are not appropriate for the liquid salt application. The measurement environment is very challenging both in terms of temperature and materials compatibility. Ultrasonic flow meters can be used for this application. Nonintrusive clamp-on ultrasonic flow meters are attached to the outside of a section of pipe with no fluid contact. These are available from several vendors. Wetted ultrasonic flow meters are permanently mounted on a spool piece; they provide higher accuracy [20], but appropriate materials must be selected for the pipe body and the sensor heads.

3.2.3 Pressure and delta-P

A particularly important measurement for this loop will be the pressure drop across the IHX and the SHX. Pressure measurements will be challenging in the salt loop. The minimum requirement is that the transducer can operate at temperatures above the melting point of the salt. Melt pressure transducers operate by hydraulic transmission of pressure through a low-vapor-pressure incompressible liquid from a wetted diaphragm to a measurement diaphragm located away from the high temperature fluid [21]. NaK is commonly used as a hydraulic transmission fluid for high-temperature melt pressure transducers. It has a freezing point that is well below room temperature ($\sim$−12.8 °C) and a boiling point of 785 °C. NaK-filled melt pressure transducers operate over a temperature range up to 538 °C, which is an excellent match for the salt loop operating temperature range. Unfortunately, these transducers are generally only commercially available for high pressure ranges (lowest range is typically 0–10 MPa), whereas the salt loop will operate at low pressure (~0.2 MPa). The molten salt loop at ORNL uses a NaK-buffered pressure transducer that prevents overheating of the transducer electronics [22]. A direct diaphragm displacement pressure measurement probe has also been researched at ORNL and U. of TN for a molten salt loop application [23]. This probe incorporated a nickel diaphragm for a direct capacitance sensor-based measurement. Specification and selection of absolute and differential pressure transducers for the molten salt loop will be a design challenge to be addressed during the detailed design phase.

3.2.4 Temperature

For the liquid salt loop, type K inconel-sheathed ungrounded 1/8- or 1/16-in. thermocouples inserted into the flow stream using compression fittings will be used for most loop temperature measurements. Surface-mounted thermocouples will also be used to provide the process variable measurements required for feedback control of the heat-traced sections of piping and vessel walls in the salt loop.

3.2.5 Liquid level

Liquid level in the salt storage tank will change during system startup and shutdown. Liquid level in the salt storage tank will be obtained using non-contact high-temperature ultrasonic or microwave level transmitters.

3.3 Water/Steam loop

3.3.1 Chemistry control

The water flow loop includes a chemistry control section. Water chemistry control is critical for proper simulation of PWR conditions. Control of water chemistry parameters in an operating PWR is aimed at striking a balance between assuring the integrity of the primary system pressure boundary, the integrity of the fuel cladding, and to minimize out-of-core radiation fields [24]. For example, elevated pH can reduce out-of-core radiation fields, but can also lead to elevated lithium levels that can lead to alloy 600 cracking. Low pH values can lead to increased crud deposits. Operation at pH values of 6.9–7.4 is generally recommended. pH control is achieved by controlling the boric acid (H_3BO_3) concentration (~500 ppm) and the lithium (LiOH) concentration (~2.2 ppm). Corrosion experiments with simulated PWR water chemistry generally follow these guidelines [25]. In addition, minimization of dissolved oxygen to <5 ppb is desired. Small amounts of dissolved hydrogen can be included in PWR water to maintain reducing conditions and to eliminate radiolytically produced oxygen.

Instrumentation for the water loop will include in-line pH and dissolved oxygen sensors. Both of these sensors will be installed in the low-temperature part of the water loop.

These sensors are available from a number of vendors with ranges that are suitable for this application.

3.3.2 Flow rate

Flow rates in the water loop will be measured at full loop pressure (up to 15 MPa) and at low temperature (~50 °C). The biggest challenge for this flow meter is the pressure, which is higher than the standard pressure rating on most off-the-shelf thermal or coriolis mass flow meters and controllers. GE does offer a high-pressure coriolis flow meter (RHM015) that is suitable for pressures up to 70 MPa, with a wide range of flow rates. Alternately, turbine flow meters are available from several manufacturers with standard pressure ratings of 35 MPa or higher over a wide range of flow rates.

3.3.3 Pressure and delta-P

Pressure instrumentation for the water loop will include several absolute pressure transducers, as indicated in Figure 2, plus several differential pressure transducers. The differential transducers will be used primarily to measure pressure drop across the SHX and the recuperator. In some cases, differential pressure may be needed across the high temperature test section as well. The differential transducers will be designed to measure relatively small pressure differences while operating at high absolute pressure. Each differential transducer assembly will include a dP cell manifold that allows zeroing of the cell at high absolute pressure.

3.3.4 Temperature

For the water loop, type K inconel-sheathed ungrounded 1/8- or 1/16-in. thermocouples inserted into the flow stream using compression fittings will be used for most loop temperature measurements.

4 Conclusions

A conceptual design for a new high-temperature, multi-fluid, multi-loop test facility has been presented in this study. This facility will support thermal hydraulic materials, and thermal energy storage research for nuclear and nuclear-hybrid applications. Three flow loops will be included: a high-temperature helium loop, a liquid salt loop, and a hot water/steam loop. The three loops will be thermally coupled through an intermediate heat exchanger (IHX) and a secondary heat exchanger (SHX). The salt loop is representative of an advanced reactor system intermediate heat transfer loop. Advanced reactor systems often include an intermediate heat transfer loop (IHTL) with heat exchangers at either end to deliver thermal energy to the application while providing isolation of the primary reactor system. Liquid salts have been identified as excellent candidate heat transport fluids for intermediate loops, supporting several types of advanced high temperature reactors. Liquid salts have also been proposed for use as a primary coolant for the Advanced High Temperature Reactor (AHTR) and the Fluoride Salt-cooled High-Temperature Reactor (FHR).

Engineering research topics to be addressed with this facility include the characterization and performance evaluation of candidate compact heat exchangers such as printed circuit heat exchangers (PCHEs) at prototypical operating conditions, flow and heat transfer issues related to core thermal hydraulics in advanced helium-cooled and salt-cooled reactors, and evaluation of corrosion behavior of new cladding materials and accident-tolerant fuels for LWRs at prototypical conditions. Research performed in this test facility will advance the state of the art and technology readiness level of high temperature intermediate heat exchangers (IHXs) for nuclear applications while establishing INL as a center of excellence for the development and certification of this technology. The thermal energy storage capability will support research and demonstration activities related to process heat delivery for a variety of hybrid energy systems and grid stabilization strategies.

Fundamental research topics will also be addressed with this facility. Each loop will include a high-temperature test section for this purpose. Research topics that may be studied in the high temperature helium test section include flow distribution, bypass flow, and heat transfer in prototypical prismatic core configurations under forced and natural circulation conditions, parallel flow laminar instability during pressurized cooldown, and turbulent heat transfer deterioration. Oxidation effects associated with water or air ingress could also be examined. The high temperature test section in the liquid salt loop can be used for examination of materials issues, thermal stresses, and high-Prandtl-number heat transfer issues. The high temperature test section in the steam/water loop will be used primarily for prototypic evaluation of new cladding materials and accident-tolerant fuels. It will be designed to characterize the thermal, chemical, and structural properties of candidate advanced fuel cladding materials and designs under a variety of simulated flow and internal heating conditions to mimic operational reactor conditions prior to in-reactor testing. The liquid salt loop will also include thermal energy storage (TES) system for support of thermal integration studies. The TES system will be based on freezing and melting of the salt acting as a high-temperature phase change material (PCM).

Conceptual design will be completed for all three loops as proposed, but the construction of each individual loop will depend on testing needs and future funding.

References

1. D.F. Williams, K.T. Clarno, L.M. Toth, Assessment of candidate liquid-salt coolants for the Advanced High-Temperature Reactor (AHTR) ORNL/TM-2006/12, Oak Ridge National Laboratory, Tennessee, 2006

2. D.F. Williams, K.T. Clarno, Evaluation of salt coolants for reactor applications, Nucl. Technol. **163**, 330 (2008)

3. M.S. Sohal, M.A. Ebner, P. Sabharwall, P. Sharpe, Engineering database of liquid salt thermophysical and thermochemical properties, INL/EXT-10-18297, Idaho, 2010

4. O. Benes, C. Cabet, S. Delpech, P. Hosnedl, V. Ignatiev, R. Konings, D. Lecarpentier, O. Matal, E. Merle-Lucotte, C. Renault, J. Uhlir, Assessment of liquid salts for innovative applications, ALISIA Deliverable (D-50), European Commission, Euratom Research and Training Programme on Nuclear Energy, 2009

5. C.W. Forsberg, The advanced high-temperature reactor: high-temperature fuel, liquid salt coolant, liquid-metal-reactor plant, Prog. Nucl. Energy **47**, 32 (2005)

6. R.O. Scarlat, P.F. Peterson, The current status of fluoride salt-cooled high-temperature reactor (FHR) technology and its overlap with HIF target chamber concepts, Nucl. Inst. Methods Phys. Res. A. **733**, 57 (2013)

7. N. Zweibaum, G. Cao, A.T. Cisneros, B. Kelleher, M.R. Laufer, R.O. Scarlat, J.E. Seifried, M.H. Anderson, C.W. Forsberg, E. Greenspan, L.W. Hu, P.F. Peterson, K. Sridharan, Phenomenology, methods, and experimental program for fluoride-salt-cooled high temperature reactors, Prog. Nucl. Energy **77**, 390 (2014)

8. E. Urquiza, K. Lee, P.F. Peterson, R. Grief, Multiscale transient thermal, hydraulic, and mechanical analysis methodology of a printed circuit heat exchanger using an effective porous media approach, J. Therm. Sci. Eng. Appl. **5**, 041011-1 (2013)

9. R.S. Schultz et al., Next generation nuclear plant methods technical program plan, INL/EXT-06-11804, Idaho, 2007

10. Y. Tung, R.W. Johnson, Y. Ferng, C. Chieng, Bypass flow computations on the LOFA transient in a VHTR, Appl. Therm. Eng. **62**, 415 (2014)

11. E. Reshotko, Analysis of laminar instability problem in gas-cooled nuclear reactor passages, AIAA J. **5**, 1606 (1967)

12. G. Melese, R. Katz, *Thermal and flow design of helium-cooled reactors* (ANS, Illinois, 1984)

13. D.M. McEligot, J.D. Jackson, Deterioration criteria for convective heat transfer in gas flow through non-circular ducts, Nucl. Eng. Design **232**, 327 (2004)

14. J.I. Lee, P. Hehzlar, P. Saha, M.S. Kazimi, Studies of the deteriorated turbulent heat transfer regime for the gas-cooled fast reactor decay heat removal system, Nucl. Eng. Design **237**, 1033 (2007)

15. D. Chapin, S. Kiffer, J. Nestell, *The very high temperature reactor: a technical summary* (MPR Associates, 2004)

16. A. Hoshi, D.R. Mills, A. Bittar, T.S. Saitoh, Screening of high melting point Phase Change Materials (PCM) in solar thermal concentration technology, Solar Energy **79**, 332 (2005)

17. J.C. Gomez, High-temperature Phase Change Materials (PCM) candidates for thermal energy storage applications, NREL Report, NREL/TP-5500-51446, 2011

18. E. Bojarsky, H. Deckers, H. Lehning, P.H. Reiser, L. Schmidt, THIBO experiments – thermohydraulically induced fuel pin oscillations in Na-cooled reactors, Nucl. Eng. Design **130**, 21 (1991)

19. L. Meyer, *Challenges related to the use of liquid metal and molten salt coolants in advanced reactors, TECDOC-1696* (IAEA, Austria, 2013)

20. GE Measurement and Control Brochure, Panaflow HT panametrics ultrasonic SIL flow meter for liquids, 2014

21. Gefran Brochure, Melt pressure transducers and transmitters, 2014

22. G.L. Yoder, A. Aaron, B. Cunningham, D. Fugate, D. Holcomb, R. Kisner, F. Peretz, K. Robb, J. Wilgen, D. Wilson, An

experimental test facility to support development of the fluoride-salt-cooled high-temperature reactor, Ann. Nucl. Energy **64**, 511 (2014)

23. J.A. Ritchie, Pressure measurement instrumentation in a high temperature molten salt test loop, MS Thesis, U. of Tennessee, 2010

24. PWR Primary Water Chemistry Guidelines, EPRI Technical Report, TR-105714-V1R4, 1999

25. T. Terachi, T. Yamada, T. Miyamotot, K. Arioka, K. Fuku, Corrosion behavior of stainless steels in simulated PWR primary water – Effect of chromium content in alloys and dissolved hydrogen, J. Nucl. Sci. Technol. **45**, 975 (2008)

5

Thermodynamic exergy analysis for small modular reactor in nuclear hybrid energy system

Lauren Boldon[1*], Piyush Sabharwall[1,2], Cristian Rabiti[2], Shannon M. Bragg-Sitton[2], and Li Liu[1]

[1] Rensselaer Polytechnic Institute, 110 8th Street, JEC 5046, Troy, NY 12180, USA
[2] Idaho National Laboratory, PO Box 1625, Idaho Falls, ID 8341, USA

Abstract. Small modular reactors (SMRs) provide a unique opportunity for future nuclear development with reduced financial risks, allowing the United States to meet growing energy demands through safe, reliable, clean air electricity generation while reducing greenhouse gas emissions and the reliance on unstable fossil fuel prices. A nuclear power plant is comprised of several complex subsystems which utilize materials from other subsystems and their surroundings. The economic utility of resources, or thermoeconomics, is extremely difficult to analyze, particularly when trying to optimize resources and costs among individual subsystems and determine prices for products. Economics and thermodynamics cannot provide this information individually. Thermoeconomics, however, provides a method of coupling the quality of energy available based on exergy and the value of this available energy – "exergetic costs". For an SMR exergy analysis, both the physical and economic environments must be considered. The physical environment incorporates the energy, raw materials, and reference environment, where the reference environment refers to natural resources available without limit and without cost, such as air input to a boiler. The economic environment includes market influences and prices in addition to installation, operation, and maintenance costs required for production to occur. The exergetic cost or the required exergy for production may be determined by analyzing the physical environment alone. However, to optimize the system economics, this environment must be coupled with the economic environment. A balance exists between enhancing systems to improve efficiency and optimizing costs. Prior research into SMR thermodynamics has not detailed methods on improving exergetic costs for an SMR coupled with storage technologies and renewable energy such as wind or solar in a hybrid energy system. This process requires balancing technological efficiencies and economics to demonstrate financially competitive systems. This paper aims to explore the use of exergy analysis methods to estimate and optimize SMR resources and costs for individual subsystems, based on thermodynamic principles – resource utilization and efficiency. The paper will present background information on exergy theory; identify the core subsystems in an SMR plant coupled with storage systems in support of renewable energy and hydrogen production; perform a thermodynamic exergy analysis; determine the cost allocation among these subsystems; and calculate unit exergetic costs, unit exergoeconomic costs, and first and second law efficiencies. Exergetic and exergoeconomic costs ultimately determine how individual subsystems contribute to overall profitability and how efficiencies and consumption may be optimized to improve profitability, making SMRs more competitive with other generation technologies.

1 Introduction

To assess the inherent value of energy in a thermal system, it is necessary to understand both the quantity and quality of energy available or the exergy. Exergy represents the quality of energy by incorporating the actual energy available to perform work through reversible processes up to the point at which thermodynamic equilibrium with the surroundings is reached. It is the useful work potential in a system [1–4]. Not all energy is created equal, which is why assessing its quality through an exergy analysis is more revealing than simply analyzing energy.

Exergy analysis applications have expanded and often incorporate costs, providing thermoeconomic information on each component within a system. This requires converting economic costs to exergetic costs, allowing for a comparison which was previously not possible. Ultimately, this may be used to determine how individual components contribute to overall system efficiencies and

* e-mail: boldol@rpi.edu

costs and may even help optimize the system. This type of study cannot be performed looking at energy efficiencies alone, as these do not appropriately allocate the costs between different components, subsystems, or streams. Exergy, however, provides information on the true value of the inputs and outputs of a component. In general, a thermodynamic analysis determines a performance criteria or metric for a particular system or uses energy balances to determine where losses are occurring [1]. The former approach is not always useful, as there may be no telling metric; the latter approach neglects differences in energy quality – or considers distinct types of energy as equivalent. The results tend to not include internal losses [1].

Valero and Torres detailed the process of performing a thermoeconomic exergy analysis as a potential method of diagnosing and optimizing thermal energy systems [2]. It may also be used to help determine appropriate prices for any products made by the plant based on thermodynamic properties; to optimize a particular portion of the system in an effort to reduce operating or production costs; to identify inefficiencies and their resulting effects on resource consumption or system economics; and to compare different design features or options [2,3].

In general, exergy analyses are useful in studying potential energy savings, recognizing that the theoretical savings will always be higher than the actual savings enacted as a result of constraints imposed by operational, etc. decisions or limits [5,6].

The thermoeconomics of power plants producing heat and/or electricity may be readily studied with an exergy analysis. This report provides relevant background information on exergy concepts, details the methodology behind an exergy analysis, and provides theoretical first and second law results for a nuclear hybrid energy system with thermal storage.

2 Exergy

The first and second laws of thermodynamics are at the foundation of exergy and provide a mechanism of comparing exergy and energy analysis results. The first law states that energy is never created or destroyed. It is simply transformed to a different form, such that the change in energy ΔE may be determined by the difference in heat added Q and work W, $Q - W = \Delta E$ [4,7]. The second law describes the creation of entropy in an energy transfer due to dissipative energy losses [7]. Understanding these fundamental principles is significant in identifying thermodynamic losses and attempting to minimize these losses.

2.1 Reversible vs. irreversible processes

A reversible process is a process in which properties such as temperature may be altered without generating entropy. In such a process, there is no change to the system or its surroundings. On the other hand, when an irreversible process occurs, entropy increases and exergy

is destroyed or reduced. In a realistic cycle, there will always be some irreversibilities or losses, and equation (1) may be used to determine the destroyed exergy $X_{destroyed}$ per the Gouy-Stodola theorem, where S_{gen} is the generated entropy and T_0 is the temperature of the surroundings [3]. In general, if the destroyed exergy or the generated entropy are equal to zero, the process may be considered internally reversible. If either is greater than zero, the process is irreversible.

$$X_{destroyed} = T_0 S_{gen}. \tag{1}$$

2.2 Exergy within a system

There are distinct types of exergy, including heat, chemical, and work exergy. The exergy of heat energy is dependent upon the component efficiency and the temperatures of the heat source and heat sink or surroundings. The exergy of work energy is equal, as 100% of the work can be directly utilized. Chemical reactions may produce energy, some of which may be turned into heat or other forms. Only a fraction of the chemical energy is exergy. This article focuses on heat and work exergy in nuclear power plant flows.

For a system of heat and work flows, the change in system exergy ΔX_{system} may be calculated from equation (2), where ΔX_{heat}, X_{work}, and $X_{destroyed}$ represent the change in heat exergy and the work and destroyed exergy respectively.

$$\Delta X_{system} = (\Delta X_{heat} - X_{work}) - X_{destroyed}. \tag{2}$$

When analyzing a particular system, it is necessary to identify the flows in and out of the system and to determine whether the system should be treated as closed or open. To do this, the system boundary and surroundings must be identified. In a closed system, mass is not permitted to pass the boundary and the mass is fixed; only energy may pass the boundary and interact with the surroundings. In an open system, both mass and exergy are permitted to flow through the system boundary. An isolated system is a closed system in which neither energy nor mass crosses the system boundary.

Calculating the exergy balance for both closed and open systems requires understanding the energy and entropy balances as the system progresses from state 1 to state 2, as shown in equations (3) and (4), where Q represents heat added or rejected, W represents work, E is energy, and S is entropy. $\int_1^2 \frac{dQ}{T} = 0$ for an adiabatic system. If the temperature of Q is not constant, then $\int_1^2 \frac{dQ}{T} = \sum \frac{Q_i}{T_i}$.

$$\Delta E_{system} = \int_1^2 dQ - W = E_2 - E_1, \tag{3}$$

$$\Delta S_{system} = \int_1^2 \frac{dQ}{T} + S_{gen} = S_2 - S_1. \tag{4}$$

The energy at each state E_1 and E_2 may be calculated from the respective internal, potential, and kinetic energies.

Equation (5) shows internal energy U based upon enthalpy H, pressure P, and volume V. Equations (6) and (7) show the exergies at each state. The exergy for each state for unit mass may be calculated by equations (8) and (9).

$$U = H - PV, \qquad (5)$$

$$E_1 = U_1 + \frac{v_1^2}{2} + gz_1, \qquad (6)$$

$$E_2 = U_2 + \frac{v_2^2}{2} + gz_2, \qquad (7)$$

$$X_1 = (U_1 - U_0) + P_0(V_1 - V_0) - T_0(S_1 - S_0) + \frac{v_1^2}{2} + gz_1, \qquad (8)$$

$$X_2 = (U_2 - U_0) + P_0(V_2 - V_0) - T_0(S_2 - S_0) + \frac{v_2^2}{2} + gz_2. \qquad (9)$$

The change in exergy for the system may then be determined from equations (5) to (9).

$$X_2 - X_1 = \int_1^2 dQ \left(1 - \frac{T_0}{T}\right) - W + P_0(V_2 - V_1) \\ - X_{destroyed}. \qquad (10)$$

2.3 Exergy flows and transfer

Exergy transfer must also be accounted for in a system. It occurs via heat, work, or mass flows. The Carnot cycle efficiency $n_{th,\text{Carnot}} = 1 - \frac{T_0}{T}$ represents the max exergy or work that can be achieved from heat transfer [7]. This heat exergy is shown in equation (11), where heat added $Q = c_p m (T_2 - T_1)$.

$$W_{heat} = W_{net,out} = Q \times n_{th,\text{Carnot}}. \qquad (11)$$

It is more realistic to use the efficiency based on the power cycle being analyzed, rather than an ideal Carnot cycle. For this case, the thermal efficiency would be $n_{th} = \frac{W_{net,out}}{Q_{in}} = \frac{Q_{in} - Q_{out}}{Q_{in}}$ [7]. A process efficiency may be calculated as the ratio of the increase in exergy over the decrease in exergy [4]. In a thermodynamic cycle, the entire cycle efficiency may be described in the same manner, representing the second law efficiency [4].

Work exergy transfer can occur via boundary work, meaning the boundary of the system changes, such as expansion/compression in a piston/cylinder system, or via mechanical/shaft or electrical work. In the former case, only a portion of the work is completely useful, while some is lost to the surroundings, as shown in equation (12). The latter exergy transfer results in entirely useful work, such that $X_{mechanical} = W$.

$$X_{boundary} = W - W_{surr} = W - P_0(V_2 - V_1). \qquad (12)$$

Mass exergy transfer occurs in an open system proportionally to the system flow rate. It has exergy, energy, and entropy, as shown in equation (13), where m represents mass. This article does not focus on mass transfer within the system analyzed.

$$X_{mass} = m\left[(h - h_0) - T_0(s - s_0) + \frac{v^2}{2} + gz\right]. \qquad (13)$$

3 Methods and materials

3.1 Nuclear renewable energy integration

This article breaks down the exergy of individual components within the Nuclear Renewable Energy Integration (NREI) system shown in Figure 1, where nuclear energy is supplementing the available wind energy through storage to meet the needs of the electrical grid. Nuclear power is also being used for the production of hydrogen via high temperature steam electrolysis.

3.2 System breakdown and assumptions

The process flows utilized in this analysis are also displayed in Figure 1. The following assumptions were made to analyze the system:

– steady state operation;
– required electrical output to grid is 245 MWe;
– wind electric production is constant at 100 MWe with 5 MWe in frictional heat losses;
– high temperature steam electrolysis requires 5 MWth and 1 MWe to produce 2780 Nl/min with a 50% thermal conversion efficiency [8];
– high temperature helium-cooled small modular reactor with 300 MWth capacity;
– compressed air energy storage is at maximum capacity of 400 MW;
– reactor outlet temperature and pressure of 850 °C and 5 MPa [8,9];
– average reactor fuel temperature of 1000 °C;
– heat loss in reactor is assumed to occur from heat transfer inefficiencies from fuel to helium;
– generator exhibits 1% heat losses [10];
– turbine isentropic efficiency of 89% [11];
– compressor isentropic efficiency of 86% [11];
– no electric losses incurred during high temperature steam electrolysis (HTSE);
– losses incurred in the low temperature heat exchanger are not considered;
– constant pressure across heat exchanger;
– 4% frictional pressure drop in turbine [11].

A closed Brayton power cycle with helium was used in this analysis, as shown in Figure 2. The overall Brayton cycle efficiency is a function of the work in and out of the system and the heat added to the system, as shown in equation (14). The turbine will turn the shaft which powers the compressor and generator. Since the compressor will use a substantial amount of the turbine work/power, it is necessary to determine fraction of work, or backwork ratio

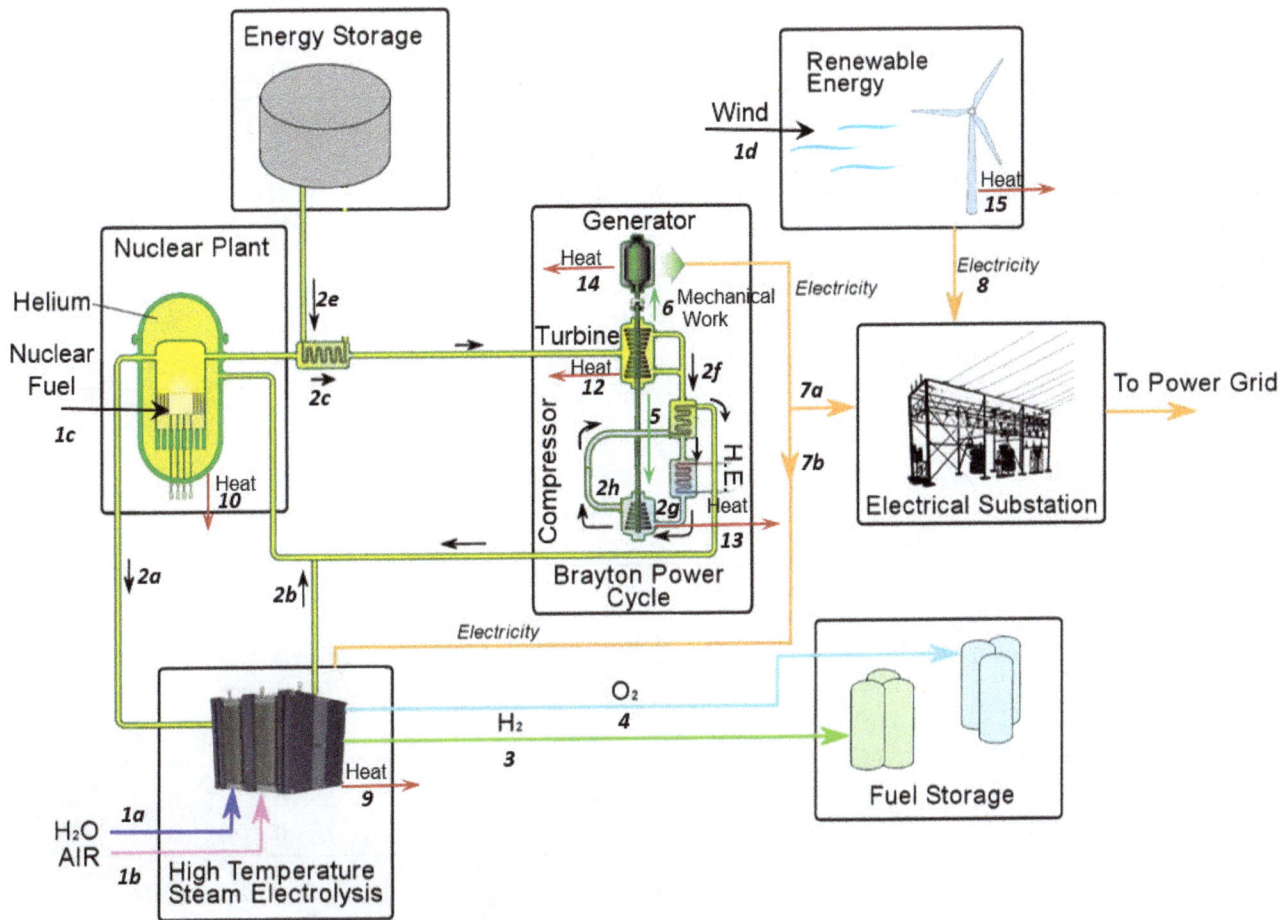

Fig. 1. Nuclear renewable energy integration schematic.

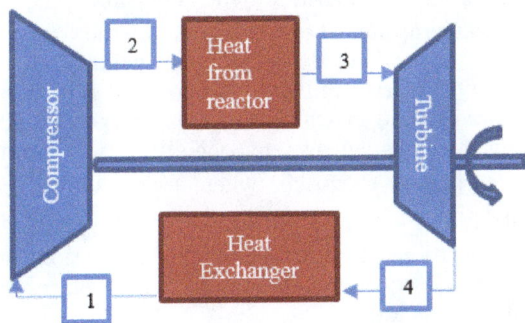

Fig. 2. NREI Brayton power cycle.

(BW), to the compressor using equation (15). The generator then sees $(1 - BW) \times Turbine\ Output$.

$$n_{BC} = \frac{w_{out} - w_{in}}{q_{in}} = \frac{(h_3 - h_4) - (h_2 - h_1)}{h_3 - h_2}, \quad (14)$$

$$BW = \frac{w_{in}}{w_{out}} = \frac{h_2 - h_1}{h_3 - h_4}. \quad (15)$$

3.3 Resources, products, and losses

To properly identify the production function for a unit and how the unit contributes to the plant function, it is necessary to define each unit's product (P), resources (F), and losses (L) and identify which flows they correspond to [3]. In other words, the production function may be determined by using the F-P-L definition, such that every flow into or out of a unit is incorporated once; every component has an overall exergy greater than or equal to zero; and the exergy balance for each unit follows the form $F - P - L = D$, where D refers to exergy destruction or unavoidable losses [3]. The total irreversibility then becomes the sum of the process losses and exergy destruction $I = F - P = L + D$. Table 1 shows the distribution of flows amongst resources, products, and losses based on unit and for the entire plant.

Once again following the F-P-L definition yields the matrix $M = M_F - M_P - M_L$, as shown in Table 2. The individual incidence matrices for resources, products, and losses were derived from Table 1, where the respective flows are shown, along with a positive or negative sign indicating whether the flow adds or subtracts from the unit's exergy.

Table 1. Unit flow distribution of resources, products, and losses.

Unit	Resources	Products	Losses
HTSE	$1a + 1b + 2a + 7b$	$2b + 3 + 4$	9
SMR	$1c + 2b + 2h$	$2a + 2c$	10
Turbine	$2c + 2e$	$5 + 6 + 2f$	12
Heat exchanger	$2f$	$2g$	–
Compressor	$2g + 5$	$2h$	13
Generator	6	$7a + 7b$	14
Wind energy	$1d$	8	15
Plant	$1a + 1b + 1c + 1d$	$3 + 4 + 7a + 8$	$9 + 10 + 12 + 13 + 14 + 15$

Table 2. *F-P-L* incidence matrix for NREI system.

	1a	1b	1c	1d	2a	2b	2c	2e	2f	2g	2h	3	4	5	6	7a	7b	8	9	10	11	12	13	14	15
1	1	1	0	0	1	−1	0	0	0	0	0	−1	−1	0	0	0	0	0	−1	0	0	0	0	0	0
2	0	0	1	0	−1	1	−1	0	0	0	1	0	0	0	0	0	0	0	0	−1	0	0	0	0	0
3	0	0	0	0	0	0	1	1	−1	0	0	0	0	−1	−1	0	0	0	0	0	0	−1	0	0	0
4	0	0	0	0	0	0	0	0	1	−1	0	0	0	0	0	0	0	0	0	0	0	0	0	0	0
5	0	0	0	0	0	0	0	0	0	1	−1	0	0	1	0	0	0	0	0	0	0	0	−1	0	0
6	0	0	0	0	0	0	0	0	0	0	0	0	0	0	1	−1	−1	0	0	0	0	0	0	−1	0
7	0	0	0	0	0	0	0	−1	0	0	0	0	0	0	0	0	0	0	0	0	−1	0	0	0	0
8	0	0	0	1	0	0	0	0	0	0	0	0	0	0	0	0	0	0	−1	0	0	0	0	0	−1

3.4 Exergetic cost of flows

The exergetic cost X_i^* (MW) may be defined as the necessary exergy for a process or flow. It provides a method for comparing units or components producing different quality products – for example, heat vs. electrical work – and also helps bridge the gap between plant thermodynamics and economics. The unit exergetic cost k_i^* is simply the ratio of the exergetic cost to the exergy X_i, or $k_i^* = \frac{X_i^*}{X_i}$ [2,3].

Several propositions have been developed based on energy/exergy relationships within a thermal system, so that the system of cost allocation equations may be determined, thereby allowing one to solve for the exergetic costs for each component and the entire plant [3,12]:

- the exergetic cost for each unit may be considered conservative;
- the exergetic cost for the plant flows may be equated with the exergy of the flow;
- exergetic cost of losses is assumed to be zero when no external assessment or influence is made;
- in the case where a flow is both an output and an input to a unit, the unit exergetic costs are considered equal. In the case where the output of a unit consists of several flows, the unit exergetic cost of each output flow is considered equal.

The first proposition states that the exergy balance for each unit must equal zero, or $M \times \boldsymbol{X}^* = 0$, where $\boldsymbol{X}^*$ is a $[23 \times 1]$ matrix containing all the unknown exergetic cost values X_i^* for each flow i. The second proposition relates the plant inputs to their respective exergies, such that the exergetic cost of each input may be equated to its exergy, $X_j^* = X_j$, where j represents plant. The third proposition makes an assumption about the value of the losses, such that $X_{k,losses} = 0$, where k represents individual subsystems or units. The fourth and final proposition refers to two separate cases, one in which the unit exergetic costs are equivalent for a subsystem yielding two products, $k_{out,1}^* = k_{out,2}^*$, and the other when a flow is both input and output for a particular subsystem, $k_{i,in}^* = k_{i,out}^*$ [2]. These propositions provide a system of 23 equations, which are presented in equations (16) to (33) with their respective proposition indicated:

- Proposition 1:

$$\text{HTSE}: \quad X_{1a}^* + X_{1b}^* + X_{2a}^* - X_{2b}^* - X_3^* - X_4^* - X_9^* = 0, \quad (16)$$

$$\text{SMR}: \quad X_{1c}^* - X_{2a}^* + X_{2b}^* - X_{2c}^* + X_{2h}^* - X_{10}^* = 0, \quad (17)$$

$$\text{Turbine}: \quad X_{2c}^* + X_{2e}^* - X_{2f}^* - X_5^* - X_6^* - X_{12}^* = 0, \quad (18)$$

$$\text{H.E.}: \quad X_{2f}^* - X_{2g}^* = 0, \quad (19)$$

$$\text{Compressor}: \quad X_{2g}^* - X_{2h}^* + X_5^* - X_{13}^* = 0, \quad (20)$$

Generator : $X_6^* - X_{7a}^* - X_{7b}^* - X_{14}^* = 0,$ (21)

Wind energy : $X_{1d}^* - X_8^* - X_{15}^* = 0.$ (22)

– Proposition 2 (Plant):

$$X_{1a}^* = X_{1a},$$ (23)

$$X_{1b}^* = X_{1b},$$ (24)

$$X_{1c}^* = X_{1c},$$ (25)

$$X_{1d}^* = X_{1d}.$$ (26)

– Proposition 3 (Losses):

$$X_9^* = X_{10}^* = X_{12}^* = X_{13}^* = X_{14}^* = X_{15}^* = 0.$$ (27)

– Proposition 4:
- product of multiple flows:

$$\frac{X_{2a}^*}{X_{2a}} = \frac{X_{2c}^*}{X_{2c}},$$ (28)

$$\frac{X_3^*}{X_3} = \frac{X_4^*}{X_4},$$ (29)

$$\frac{X_5^*}{X_5} = \frac{X_6^*}{X_6},$$ (30)

$$\frac{X_{7a}^*}{X_{7a}} = \frac{X_{7b}^*}{X_{7b}},$$ (31)

- output is also input to subsystem:

$$\frac{X_{2a}^*}{X_{2a}} = \frac{X_{2b}^*}{X_{2b}},$$ (32)

$$\frac{X_{2c}^*}{X_{2c}} = \frac{X_{2h}^*}{X_{2h}}.$$ (33)

Unit exergy consumption refers to the resource exergy consumed or used within the subsystem to yield the subsystem product [2,3]. First and second law subsystem efficiencies and unit consumption may be calculated from $\eta_{1st\ Law,i} = \frac{E_{P,i}}{E_{F,i}}$, $\eta_{2nd\ Law,i} = \frac{X_{P,i}}{X_{F,i}}$, and $k = \frac{X_{F,i}}{X_{P,i}}$, respectively, where X_F, X_F^*, and k_F^* refer to the resource exergy, exergetic cost, and unit exergetic cost and X_P, X_P^*, and k_P^* refer to the product exergy, exergetic cost, and unit exergetic cost. The total irreversibilities within each subsystem are also a function of the resources and products, $I_i = X_{F,i} - X_{P,i}$ [2,3].

3.5 Exergoeconomic cost of flows

The final stage in a thermoeconomic analysis requires incorporating economic market conditions, such as the costs associated with resources and operations. This yields the exergoeconomic cost ($/s), or rather the money required to create the previously detailed flows in the system [2]. The amortized cost rate ($/s) of installation and operations for each subsystem Z_i must be determined, such that a system of equations may be developed based on the cost balance shown in equation (34), where c_i represents the unit exergoeconomic cost of a product or resource ($/MW $\times$ s) [2,3]. Any losses are assumed to have a cost of zero. The exergoeconomic cost is then $C_i = c_i X_i$.

$$\sum_{input} c_i X_{F,i} + z = \sum_{out} c_i X_{P,i}.$$ (34)

The cost balance equations for the NREI subsystems are shown in equations (35) to (41).

$$c_{1a} X_{1a} + c_{1b} X_{1b} + c_{2a} X_{2a} + c_{7b} X_{7b} + Z_{HTSE}$$
$$= c_3 X_3 + c_4 X_4 + c_{2b} X_{2b},$$ (35)

$$c_{1c} X_{1c} + c_{2b} X_{2b} + c_{2h} X_{2h} + Z_{SMR}$$
$$= c_{2a} X_{2a} + c_{2c} X_{2c},$$ (36)

$$c_{2e} X_{2e} + c_{2c} X_{2c} + Z_{turb} = c_{2f} X_{2f} + c_5 X_5 + c_6 X_6,$$ (37)

$$c_{2f} X_{2f} + Z_{HE} = c_{2g} X_{2g},$$ (38)

$$c_{2g} X_{2g} + c_5 X_5 + Z_{Comp} = c_{2h} X_{2h},$$ (39)

$$c_6 X_6 + Z_{Gen} = c_{7a} X_{7a} + c_{7b} X_{7b},$$ (40)

$$c_{1d} X_{1d} + Z_{Wind} = c_8 X_8.$$ (41)

Most of the subsystems have more than one input and/or output, so the cost balance equations do not provide enough information to solve for the unit exergoeconomic costs. It is therefore necessary to make assumptions regarding the importance of plant products. The extraction method is used in this analysis, such that the priority is assigned to electricity production, which will bear the system's overall costs [2]. This is deemed reasonable, as only 5 MWth is utilized for hydrogen production and is negligible compared to the electric generation costs. The unit exergoeconomic cost for a subsystem's product that is not related to electricity production will equal its unit exergoeconomic cost input, or $c_{2a} = c_{1c}$. Several plant inputs (water, air, and wind) are considered free, thus $c_{1a} = c_{1b} = c_{1d} = 0$. The fuel cost may be estimated by $c_{1c} = 0.31 \times \frac{Z_{SMR}}{2}$, because fuel makes up approximately 31% of nuclear power plant operation and maintenance costs, which are assumed to account for half of the installation and operation costs [13]. The exergoeconomic cost of storage will be equivalent to the energy storage installation and operation cost rate, $c_{2e} X_{2e} = Z_{storage}$. The

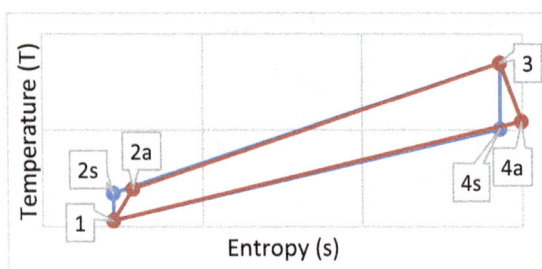

Fig. 3. *T-s* diagram (isentropic in blue and actual in red).

electricity produced will have the same unit exergoeconomic cost as will the hydrogen and oxygen flows, such that $c_{7a} = c_{7b}$ and $c_3 = c_4$.

4 Results and discussion

4.1 Brayton cycle and thermodynamic properties

The *T-s* diagram in Figure 3 illustrates the thermodynamic properties of the NREI system, which are displayed in Table 3.

4.2 Energy, exergy, and exergetic cost of flows

Table 4 displays the energy, exergy, and unit exergetic cost of the flows. There are several flows into the plant, such as air or water, which are considered limitless, so the energy is set to zero, meaning there is no exergetic cost for these products. The energy for other flows refers to the power (MW), while the exergy is the power available as a result of irreversibilities – both unavoidable heat losses and process efficiency losses. Flows 5 and 6 represent the mechanical work produced by the turbine, which must turn the shaft for the compressor and generator. A backwork ratio of 54.4% was determined from equation (15) using the required input and output work to the cycle.

The reactor losses, flow 10, were calculated from the reactor efficiency $n_{\text{reactor}} = 1 - \frac{h_{\text{max}} - h_{\text{He}}}{h_{\text{max}}}$, where h_{max} and h_{He} represent the enthalpies of helium at the fuel and reactor

outlet temperatures, respectively. The other heat losses were determined as a function of component efficiencies and cycle loss assumptions. For example, helium expands in the turbine from state 3 to state 4 in Figures 2 and 3 (or state 4s in an isentropic Brayton cycle).

The losses may be calculated from $w_{turb,loss} = h_{4a} - h_{4s}$, where $h_{4a} = h_3 - \frac{h_3 - h_{4s}}{n_{turb}}$ and n_{turb} represents the turbine isentropic efficiency. Similarly, the compressor losses are $w_{comp,loss} = h_{2a} - h_{2s}$. The exergy of all heat losses is zero. The exergy of hydrogen and oxygen were also set to 0.5 MW or half of the resultant energy (50% efficient process) [8]. The energy values for the helium flows were all determined based on what was available from the reactor after hydrogen production, respective component efficiencies, and the energy demand from the substation. In general, the exergy is then simple to determine for heat and work flows, $X_j = E_j \times \left(1 - \frac{T_0}{T_j}\right)$, where E_j is the flow energy, T_0 is the reference temperature of the surroundings (25 °C), and T_j is the flow temperature [3].

4.3 Subsystem resource exergy consumption, efficiencies, and irreversibilities

The exergy of the resources and products of individual subsystems may be determined from the exergy of each flow, as shown in Table 5. The resource/product exergy and exergetic cost were calculated by summing the exergy and exergetic cost flows, respectively, from Table 4 in the manner shown in Table 1. The first law or energy efficiency is also included in Table 5, to better demonstrate where improvements could be made to enhance the system performance [2,3]. Systems with very high second law or exergy efficiencies have limited room for improvement, as the majority of irreversibilities are unavoidable losses due to exergy destruction. This is not always clear from the first law efficiency, as in the case of the SMR, where the second law efficiency is quite high at 93.3%, while the first law efficiency is 81.5%. Similarly, the difference between the first and second law efficiencies for the HTSE demonstrates how there is less work potential than anticipated from the first law efficiency. Some systems, like the heat exchanger, see a very high first law efficiency and a low second law efficiency, illustrating how the magnitude of energy lost is very small compared to the input energy, but the lost energy is useful work potential not captured by the turbine.

Table 3. Thermodynamic properties of Brayton cycle.

State	Temperature (°C)	Pressure (MPa)	Enthalpy (kJ/kg)
1	32.8	1.9	46.62
2s	174.3	5	790.6
2a	197.6	5	911.8
3	850	4.8	4297
4s	507.2	1.9	2510
4a	545.1	1.9	2706

Table 4. NREI component conditions (MW).

Flow #	Flow	Power (MW)	Exergy (MW)	Unit exergetic cost (MW/MW)
1a	Water	0	0	1
1b	Air	0	0	1
1c	Fuel	300	254	1
1d	Wind	100	100	1
2a	Helium 725 psi, 850 °C	5	4.853	1.075
2b	Helium 725 psi, 197.6 °C	3.853	3.365	1.075
2c	Helium 725 psi, 850 °C	248.1	240.8	1.075
2e	Helium 725 psi, 850 °C	127.6	123.9	1
2f	Helium 280 psi, 545.1 °C	30.28	28.89	15.31
2g	Helium 280 psi, 32.8 °C	28.89	6.855	64.50
2h	Helium 725 psi, 197.6 °C	6.855	5.988	1.075
3	Hydrogen	1	0.5	1.599
4	Oxygen	1	0.5	1.599
5	Mechanical work	181.8	181.8	2.467
6	Mechanical work	152.5	152.5	2.467
7a	Electrical work	150	150	2.492
7b	Electrical work	1	1	2.492
8	Electrical work	95	95	1.053
9	Heat	2.5	0	0
10	Heat	46.02	0	0
12	Heat	35.08	0	0
13	Heat	16.69	0	0
14	Heat	16.78	0	0
15	Heat	5	0	0

Table 5. Subsystem resource and product energy and exergy (MW), consumption, irreversibilities (MW), and efficiencies.

	Resource energy E_F	Resource exergy X_F	Product energy E_P	Product exergy X_P	Irreversibility I	Consumption k	$\eta_{1st\ Law}$	$\eta_{2nd\ Law}$
HTSE	6	5.853	5.853	4.853	1	1.206	0.976	0.829
SMR	310.708	263.353	253.1	245.653	17.7	1.072	0.815	0.933
Turbine	375.7	364.7	364.58	363.19	30.4	1.004	0.970	0.996
Heat exchanger	30.28	28.89	28.89	6.855	22.035	4.214	0.954	0.237
Compressor	210.69	188.655	6.855	5.988	182.667	31.506	0.033	0.032
Generator	152.5	152.5	151	151	1.5	1.010	0.990	0.990
Wind energy	100	100	95	95	5	1.053	0.950	0.950

4.4 Exergoeconomics and optimization

Using the following example installation/operation cost rates and equations (35) to (41) yields the unit exergoeconomic costs for the product flows shown in Table 6: $Z_{HTSE} = \$0.06/s$, $Z_{SMR} = \$0.30/s$, $Z_{turb} = \$0.06/s$, $Z_{HE} = \$0.003/s$, $Z_{comp} = \$0.04/s$, $Z_{Gen} = \$0.003/s$, $Z_{Wind} = \$0.20/s$, and $c_{2e} = Z_{storage} = \$0.03/s$.

An optimized thermoeconomic analysis of the NREI system would require setting additional constraints and determining an appropriate function to optimize [2]. To optimize costs, or more specifically the levelized cost to create the products, the objective function may follow the form, min $C_0 = c_e X_e + \sum_i Z_i$, where c_e and X_e are the cost and exergy for resources external to the system [2]. Incorporating possible subsystem efficiency improvements and mathematically iterating would provide a method of optimizing the exergoeconomic costs for NREI.

Tsatsaronis and Moran described how complex systems, such as NREI, may be difficult to optimize using typical mathematical models, due to lack of information on individual components – particularly if they are purchased,

Table 6. Unit exergoeconomic costs.

Flow	Unit exergoeconomic cost ($/s)
$1a$, $1b$, $1d$	0
$1c$	0.018
$2a$, $2b$	0.018
$2c$	0.453
$2e$	0.024
$2f$	0.453
$2g$	1.954
$2h$	12.45
3, 4	6.347
5, 6	0.314
$7a$, $7b$	0.319
8	0.211

computational time, and possible changes to the overall system structure that may be much more profitable than the initial design [14]. A mathematical model would not be able to consciously make these structural changes. Manually iterating based on possible changes to temperature, turbine/compressor pressure ratios and efficiencies, etc. provides a better understanding of the system at hand, and for increasingly complex systems, may be a more effective manner of balancing exergy and costs [14].

5 Conclusion

A thermoeconomic study of an energy system provides significant information on how the system operates on both a technical – efficiencies and losses – and economic level. The coupling of these two fields through exergy and exergoeconomic analyses offers insight into methods of improving the economic competitiveness of complex systems incorporating advanced small modular reactors, energy storage, hydrogen production, and renewable energy technologies, all of which serve to meet escalating energy demands in a sustainable manner. This paper details relevant exergy concepts, explores how exergy and exergoeconomic analyses could be beneficial in assessing the Nuclear Renewable Energy Integrated system, and provides theoretical first and second law efficiencies, exergetic costs, and exergoeconomic costs for this example system.

We would like to thank the DOE for its support through the Nuclear Energy University Program Graduate Fellowship. Any opinions, findings, conclusions or recommendations expressed are those of the authors and do not necessarily reflect the view of the Department of Energy Office of Nuclear Energy or Idaho National Laboratory.

Nomenclature

BC	Brayton cycle
BW	backwork ratio
η	efficiency
E	energy/power (MW)
h	enthalpy (kJ/kg)
s	entropy
S_{gen}	entropy generated
X^*	exergetic cost (MW)
C	exergoeconomic cost ($/s)
X	exergy (MW)
D, $X_{destroyed}$	exergy destruction (MW)
q, Q	heat added
Z	installation and operation cost rate ($/s)
H.E.	heat exchanger
HTSE	high temperature steam electrolysis
I	irreversibility (MW)
L	losses (MW)
MJ, MW	mega-Joule, mega-Watt
NREI	nuclear renewable energy integration
P, P_0	pressure, reference pressure (101.4 kPa)
P, F	product or resource (subscript)
k	resource consumption
SMR	small modular reactor
T, T_0	temperature, reference temperature (25 °C)
W	work
$k*$	unit exergetic cost (MW/MW)
c	unit exergoeconomic cost ($/MJ)

References

1. T.J. Kotas, *The exergy method of thermal plant analysis* (Department of Mechanical Engineering, Queen Mary and Westfield College, University of London, UK, 1995)

2. A. Valero, C. Torres, Thermoeconomic analysis, Center of Research for Energy Resources and Consumption, Centro Politecnico Superior, Universidad de Zaragoza, Spain

3. M.A. Lozano, A. Valero, Theory of the exergetic cost, Energy **18**, 939 (1993)

4. D. Abata, Exergy, in *The Concept of Exergy* (2011), Chap. 8

5. P. Le Goff, *Énergétique industrielle. Tome 1 : Analyse thermodynamique et mécanique des économies d'énergie* (Technique et Documentation, Paris, France, 1979)

6. A. Valero, Bases termoeconómicas des ahorro de energía, in *2ª Conferencia national sobre ahorro energético y alternativas energéticas, Zaragoza, Spain* (1982)

7. D. Kaminski, M. Jensen, *Introduction to thermal and fluids engineering* (Wiley Publishing, 2011)

8. High-temperature electrolysis unlocking hydrogen's potential with nuclear energy, Idaho National Laboratory, Document 08-GA50044-06, 2005

9. Q. Zhang, H. Yoshikawa, H. Ishii, H. Shimoda, Thermodynamic and economic analyses of HTGR cogeneration system performance at various operating conditions for proposing optimized deployment scenarios, J. Nucl. Sci. Technol. **45**, 1316 (2008)

10. Efficiency in electricity generation, EURELECTRIC preservation of resources working group's upstream sub-group in collaboration with VGB, 2003

11. Gas turbines and jet engines, University of Tulsa, 2000, www. personal.utulsa.edu/~kenneth-weston/chapter5.pdf

12. A. Valero, M.A. Lozano, M. Muñoz, A general theory of exergy saving: Part I. On the exergetic cost, Part II. On the thermoeconomic cost, Part III. Exergy saving and thermoeconomics, Comput. Aided Eng. Energy Syst. **3**, 1 (1986)

13. Nuclear Energy Institute, Fuel as a percent of production costs, 2013, available at: http://www.nei.org/Knowledge-Center/Nuclear-Statistics/Costs-Fuel,-Operation,-Waste-Disposal-Life-Cycle/Fuel-as-a-Percent-of-Production-Costs

14. G. Tsatsaronis, M. Moran, Exergy-aided cost minimization, Energy Convers. Mgmt **38**, 1535 (1997)

6

A new MV bus transfer scheme for nuclear power plants

Choong-Koo Chang[*]

Department of NPP Engineering, KEPCO International Nuclear Graduate School (KINGS), Ulsan, Korea

Abstract. Fast bus transfer method is the most popular and residual voltage transfer method that is used as a backup in medium voltage buses in general. The use of the advanced technology like open circuit voltage prediction and digital signal processing algorithms can improve the reliability of fast transfer scheme. However, according to the survey results of the recent operation records in nuclear power plants, there were many instances where the fast transfer scheme has failed. To assure bus transfer in any conditions and circumstances, uninterruptible bus transfer scheme utilizing the state of the art medium voltage UPS is discussed and elaborated.

1 Introduction

The auxiliary power system of many generating stations consists of offsite power supply system and onsite power supply system, including emergency diesel generators (EDG) to provide secure power to auxiliary loads. If a normal power supply fails to supply power, then the power source is transferred to a standby power supply. In the case of nuclear power plants (NPP), the unit auxiliary transformer (UAT) and standby auxiliary transformer (SAT) – or station service transformer – are installed and powered from two offsite power circuits to meet regulatory requirements (10 CFR 50, App. A). Figure 1 is a simplified single line diagram of the auxiliary power system for APR 1400 type nuclear power plant. The transfer methods of a motor bus from a normal source to a standby source used in power generating stations are fast bus transfer, in-phase transfer, or residual transfer. Three important parameters, which are crucial from a bus transfer point of view, are the magnitude of the residual voltage, decay time, and the associated phase angle of the residual voltage [1]. The problem is, if the parameters do not meet the transfer conditions, the bus transfer will fail. Therefore, a new MV (medium voltage) bus transfer scheme is presented in this paper. It guarantees a successful bus transfer regardless of the bus condition.

2 MV bus transfer requirements in NPP

2.1 Reactor coolant pump motor buses

Both of the reactor coolant loops should be available and two (2) reactor coolant pumps of each loop should be in operation for the normal operation of the reactor in pressurized water reactors. If any one (1) of four (4) pumps is inoperable, then the operation mode should be changed to the hot standby mode from normal operation mode. It means that the fast transfer is essential to the RCP motor buses (Divisions I and II) to maintain reactor operation. Typical transfer time of the fast transfer is 4 to 9 cycles. If both of the RCP motor buses fail to transfer, the reactor must go into the hot shutdown and cold shutdown mode successively. If a reactor is shut down, it takes a long time to restart generation. The time required to reach full power from the hot-tripped condition is around 4 to 6 hours. And, the minimum time required for starting up large LWRs (light water reactors) from the cold tripped condition may be around 20 hours [2].

2.2 Medium voltage safety (class 1E) bus

The preferred power supply (PPS) to the safety power systems is from the grid. During power operation, the power supply is normally from the main generator connected to the grid. The offsite power should be supplied by two or more physically independent offsite supplies designed and located in order to minimize the likelihood of their simultaneous failure. A minimum of one offsite circuit should be designed to be automatically available to provide power to its associated safety divisions within a few seconds, following a design basis accident to meet the accident analysis requirements. A second offsite circuit should be designed to be available within a short time period [3].

The load capacity of the safety bus varies with the operation mode of the plant. Therefore, the most practical bus transfer method of MV class 1E buses is residual voltage

*e-mail: ckchang@kings.ac.kr

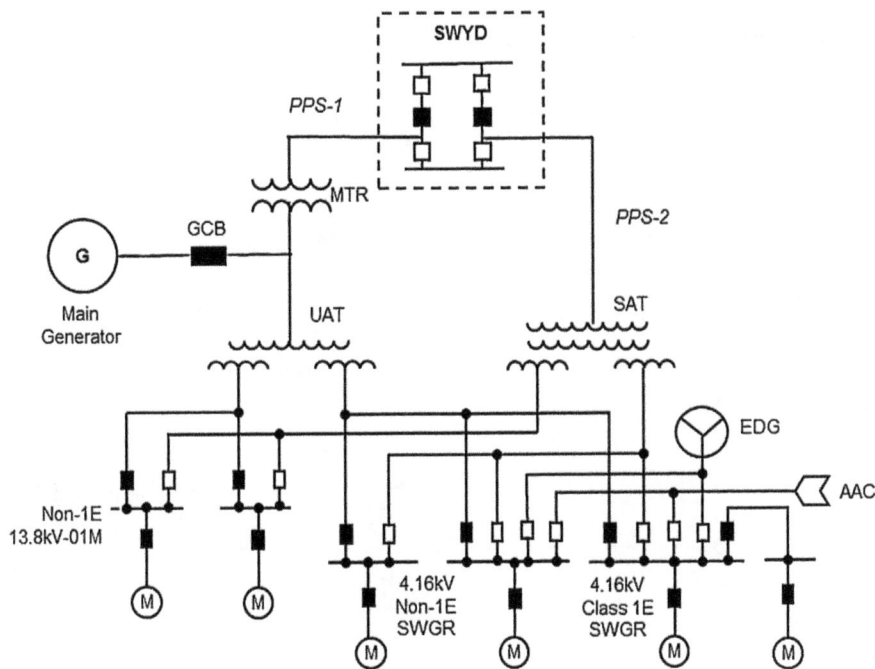

Fig. 1. Simplified MV bus single line diagram (Division I).

transfer. Furthermore, to avoid the alternative source overloading in the case of motors' low voltage restarting, it is required to implement low voltage load-shedding function before the residual voltage transfer. Typical transfer time of residual transfer is 0.5 to 3 seconds.

Each division of the MV class 1E power system is supplied with emergency standby power from an independent EDG. In the case of APR 1400, if the loss of voltage is detected by four time delay type undervoltage relays, the EDG is started to attain rated voltage and frequency within 20 seconds after receipt of a start signal, then supply power to 4.16 kV AC class 1E bus within 22 seconds.

3 Bus transfer methods in general

There are several bus transfer methods used to transfer a MV motor bus from a normal source to a standby source. The detail features and functions of a bus transfer device are dependent on manufacturers. However, basic concept is the same and a typical MV bus transfer scheme is as follows [4,5].

3.1 Fast transfer

In the auxiliary power system of power stations and industrial plants, lots of asynchronous motors are connected. In the case of the main source interruption, the residual voltage will be induced on the busbar by connected asynchronous motors. Studies to determine the magnitude of the transient current and torque are recognized to be complex and require detailed knowledge of the motor, the driven equipment, and the power supply.

3.1.1 Sequential transfer

Under the sequential transfer, a bus transfer device will firstly issue an open command to the running source CB (circuit breaker) after the device gets the starting request command. Sequential transfer can only issue close command after the running source CB is opened. The switching sequence of the sequential transfer scheme is illustrated in Figure 2 [5]. This approach provides increased security because the bus has been disconnected from the main source prior to the standby source breaker closing. A bus dead time of 5 ~ 7 cycles is normally encountered with this type of transfer [6,7].

3.1.2 Simultaneous transfer

If two sources are not allowed to work on a busbar in parallel, the simultaneous sequence can be used for the power supply transfer [7].

Under the simultaneous sequence, a bus transfer device will firstly issue an open command to the running source CB after the device gets the starting command. Meanwhile, the device will issue a close command to the standby source CB if criteria are met [5]. This type of transfer has the shortest dead time of 1 ~ 2 cycles. This type of transfer may not be possible when the main source is lost due to a close-in electrical fault or abnormal condition that causes the phase angle to move instantaneously. Also, a breaker failure scheme is required for the simultaneous transfer so that if the main source breaker fails to open during the transfer, then the standby source breaker is tripped to avoid paralleling of two sources [6].

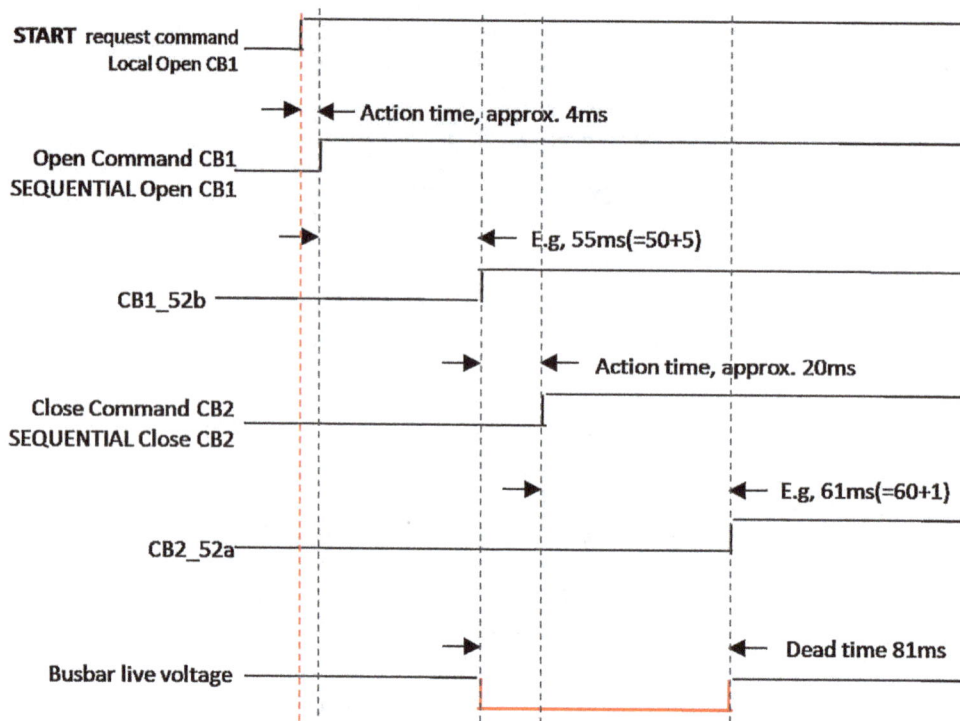

Fig. 2. Switching sequence illustration of sequential transfer.

3.2 In-phase transfer

In the in-phase transfer scheme, the standby source breaker is closed when the phase angle of the bus voltage is in-phase with the phase angle of the standby source voltage. The bus transfer device estimates the phase angle difference $d\varphi$ at the instance of CB closing based on real-time slipping rate and the settable "CB Closing Time". If all the quantity of predicted $d\varphi$, the real-time df, and residual voltage U_{res} meet the defined criteria, the device will immediately issue the close command to the alternative source CB [5,7].

3.3 Residual voltage transfer

If the above-mentioned transfer modes fail, the transfer can still go on with residual voltage transfer mode. When the residual voltage U_{res} under-shots the settable parameter "U_{res} Threshold", the residual voltage transfer mode will perform and the device will immediately issue the close command to the standby source CB. The typical setting could be 30% of rated bus voltage U_n [5]. Typically, residual voltage transfers are done at $25\% \sim 30\%$ of the rated voltage, irrespective of the phase angle of the motor bus. As the residual transfer is slow, process interruption is likely to take place. Also, in the majority of cases, motors cannot be reaccelerated simultaneously following such a transfer as their speeds have fallen so low that inrush currents approach motor locked-rotor values, and stalling would occur due to depressed voltage [8].

4 Challenges in MV bus transfer system

Research and development on the MV bus transfer system has been continuing since 1950s to meet the necessity of highly reliable power system in the industry and power plants [8]. Especially, in a nuclear power plant, reliable power supply is essential for the safe operation of the power plant and the prevention of the release of radioactive fission products. However, successful bus transfer was not assured until now; although, there were remarkable achievements in the development of improved fast bus transfer system.

4.1 Typical recent research results and achievements

To ensure the success of fast bus transfer, many research activities had been conducted. Typical recent research results and achievements are as follows.

4.1.1 Prediction of open circuit voltage

The open circuit characteristic of a motor bus auxiliary is influenced by the type of motors selected, breakdown torque of induction motors, load characteristics, and motor inertia. The type of bus transfer – such as sequential fast transfer or in-phase transfer of the motor bus – is also influenced by the above factors. The estimation of the open circuit time constant of the motor bus is therefore critical to accurately predict how the voltage and frequency of the motor bus is going to decay. A successful bus transfer depends on the thorough understanding of the process and a proper analysis of the auxiliary system using proper simulation tools. A robust bus transfer technique is presented in the Balamourougan's paper which determines, in approximately one cycle after the motor bus has been interrupted, whether a sequential fast bus transfer is possible or not by estimating the rate of decay of the motor bus residual voltage, magnitude of residual voltage, and the phase angle [1].

4.1.2 Digital signal processing algorithms

The design of a digital high-speed bus transfer system has been presented in the Yalla's and Sidhu's paper [6]. Digital signal processing algorithms that calculate the magnitude and phase angle of a voltage signal over a wide frequency range have been presented. Also presented is an algorithm to predict the phase coincidence between the motor bus voltage and the standby source voltage using delta frequency, the rate of change of delta frequency, and breaker closing time, which results in a very accurate prediction of phase coincidence. The algorithm has also calculated the bus voltage magnitude very accurately down to 4 Hz, which is important for the residual transfer method.

4.1.3 Advanced modeling and digital simulation

Presently, advanced modeling and digital simulation tools help in modeling and simulation of complex fast transfer scheme. They are used to analyze whether the conditions are met to adopt fast transfer with the realistic loads. The preliminary assessment assures successful fast transfer for the estimated transfer time and the inertia constants based on the preliminary data sheets from the vendors and the generic modeling of turbine generator excitation and governing system [8].

4.1.4 Implementation of IEC 61850 standard

The new communications technology and the newly developed International Electrotechnical Commission (IEC) standard IEC 61850, for generic object-oriented substation events (GOOSE), bring many advantages to industrial protection and control applications. Some of the applications benefiting the most are those associated with the bus-transfer and load-shedding schemes, together with more beneficial communication-assisted schemes, like zone interlocking, fast bus trip, and arc-flash reduction. A fast bus trip scheme using GOOSE messaging is performed by the relays from the main breakers without the need of adding a bus differential relay. In such cases, relays from the main breakers are connected via fiber optic or copper twisted pair Ethernet cables to the entire feeder relays to exchange GOOSE data [10].

4.2 Causes of the failure of bus transfer schemes in nuclear power plants

Fast bus transfer method is the most popular and residual voltage transfer method that is used as a backup in nuclear power plants of Korea. During the last 10 years (2005 ~ 2014), 30 cases of reactor shutdown events caused by electrical faults had been reported by Korea Institute of Nuclear Safety [11]. Among them, 20 cases are the events for which fast bus transfer (or residual voltage transfer) is required. However, there was only one successful bus transfer. In the other 12 cases, the bus transfer had failed. No information is available for the remaining seven cases. Major reasons of bus transfer failure are as follows and summarized in Table 1.

4.2.1 Malfunction of circuit breaker

The malfunction of circuit breaker closing device caused the failure of the bus transfer. Accordingly, RCPs were stopped and followed by reactor shutdown.

4.2.2 Ground fault on the normal source

The ground fault occurring on the low voltage side of the Main Transformer made low bus voltage. Accordingly, fast bus transfer to alternative source was unsuccessful. In the other case, the ground fault occurred on the secondary side of the Startup Transformer which caused loss of class 1E bus voltage. Then EDG was started automatically.

4.2.3 Perturbation of grid voltage

Bus transfer was initiated when a grid lost load-following capability. It caused voltage dip and phase angle change. Accordingly, bus transfer was unsuccessful.

4.2.4 Other unidentified failures

Causes of some bus transfer failures were not verified clearly. But it is assumed that fast bus transfer or residual bus transfer requirements were not satisfied by any reason.

Table 1. Types of bus transfer failure.

Cause	Number	Remark
CB malfunction	4	Failure of closing
Ground fault	2	Low bus voltage
Perturbation of grid voltage	1	Low bus voltage
Others	5	Dissatisfaction of bus transfer conditions
Total	12	

5 Uninterruptible MV bus transfer scheme

To guarantee the safety of nuclear power plants, interruption of power supply shall be prevented. For the purpose of satisfying such requirements, emergency power systems are provided including EDGs, DC batteries, and vital AC power supplies. Nowadays, it is possible to reach almost 100% availability of power supply for low voltage systems. However, there is still a gap when facing applying these emergency power systems in medium voltage systems, for several reasons: investment, space, long-term energy losses cost, high temperatures or dirtiness, and regenerative loads. To resolve this problem, uninterruptible MV bus transfer scheme is proposed by combination of existing bus transfer system and industrial UPS (uninterruptible power supply). In general, the fast transfer of 13.8 kV bus is more successful than 4.16 kV bus because the RCP motors fed from 13.8 kV bus have high inertia and are large size. Furthermore, 4.16 kV class 1E loads are more critical than 13.8 kV loads in terms of safety. All reactor coolant pumps may be de-energized for up to ≤1 hour per 8-hour period, provided: (a) no operations are permitted that would cause reduction of the RCS boron concentration; and (b) core outlet temperature is maintained at least 5.6 °C (10 °F) below saturation temperature [12]. Therefore, uninterruptible MV transfer system is proposed for 4.16 kV class 1E bus. And, the available capacity of energy storage device also has been considered.

5.1 Existing class 1E 4.16 kV buses operation scheme

Each division of the 4.16 kV AC class 1E power system is supplied with emergency standby power from an independent EDG. The EDG is designed and sized with sufficient capacity to operate all the needed emergency shutdown loads powered from its respective class 1E buses. Each EDG is designed to attain rated voltage and frequency within

20 seconds after receipt of a start signal, supply power to 4.16 kV AC class 1E bus within 22 seconds [12].

Degraded voltage is detected by the time delay type relay of which the setting is higher than the setting value of the undervoltage relay for loss of voltage and lower than the required minimum operating voltage. A detection signal is provided to ESF-CCS (Engineered Safety Feature-Component Control System). The EDG is started on ESF actuation signal, and ready for operation. However, the EDG is not connected to 4.16 kV class 1E bus when normal or standby power is available but remains in standby. The class 1E loads are powered from the normal or standby power source.

When the loss of voltage is detected, if the standby power is available, the 4.16 kV class 1E buses are transferred to standby power source by using residual voltage transfer scheme with the shedding of non-class 1E loads. In case of bus transfer failure, the EDG is started by undervoltage on the bus, all breakers for 4.16 kV motor feeders are tripped and the load sequencer is reset. Upon detection of the EDG rated speed and rated voltage, the EDG circuit breakers on class 1E 4.6 kV bus can be closed and the load sequence logic starts automatically, sequencing the safety related loads on the EDG. The required safety related loads are connected to the bus in the preselected interval time. Thus, the EDG can be operated stable and minimize motor acceleration time.

Each EDG and the automatic sequencers necessary for generator loading are designed such that flow is delivered to the reactor vessel within a maximum of 40 seconds after an SIAS set point is reached [12].

5.2 Implementation of MV UPS on the class 1E 4.16 kV buses in a NPP

By applying a MV UPS on the class 1E switchgear incoming feeder as shown in Figure 3, seamless transfer can be achieved without shedding of loads. When loss of voltage is

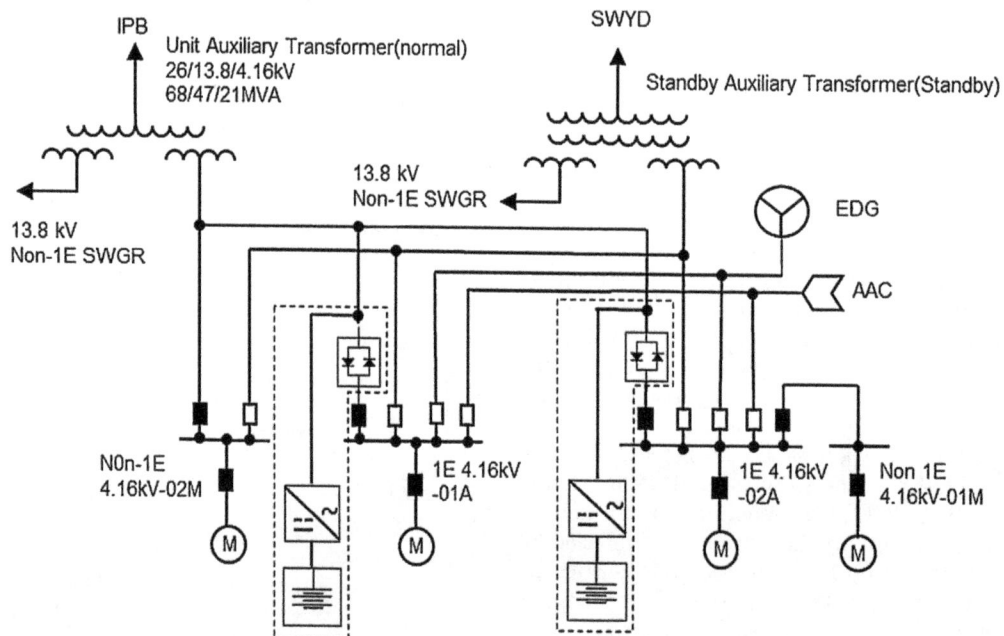

Fig. 3. 4.16 kV buses of a nuclear power plant (Division I only).

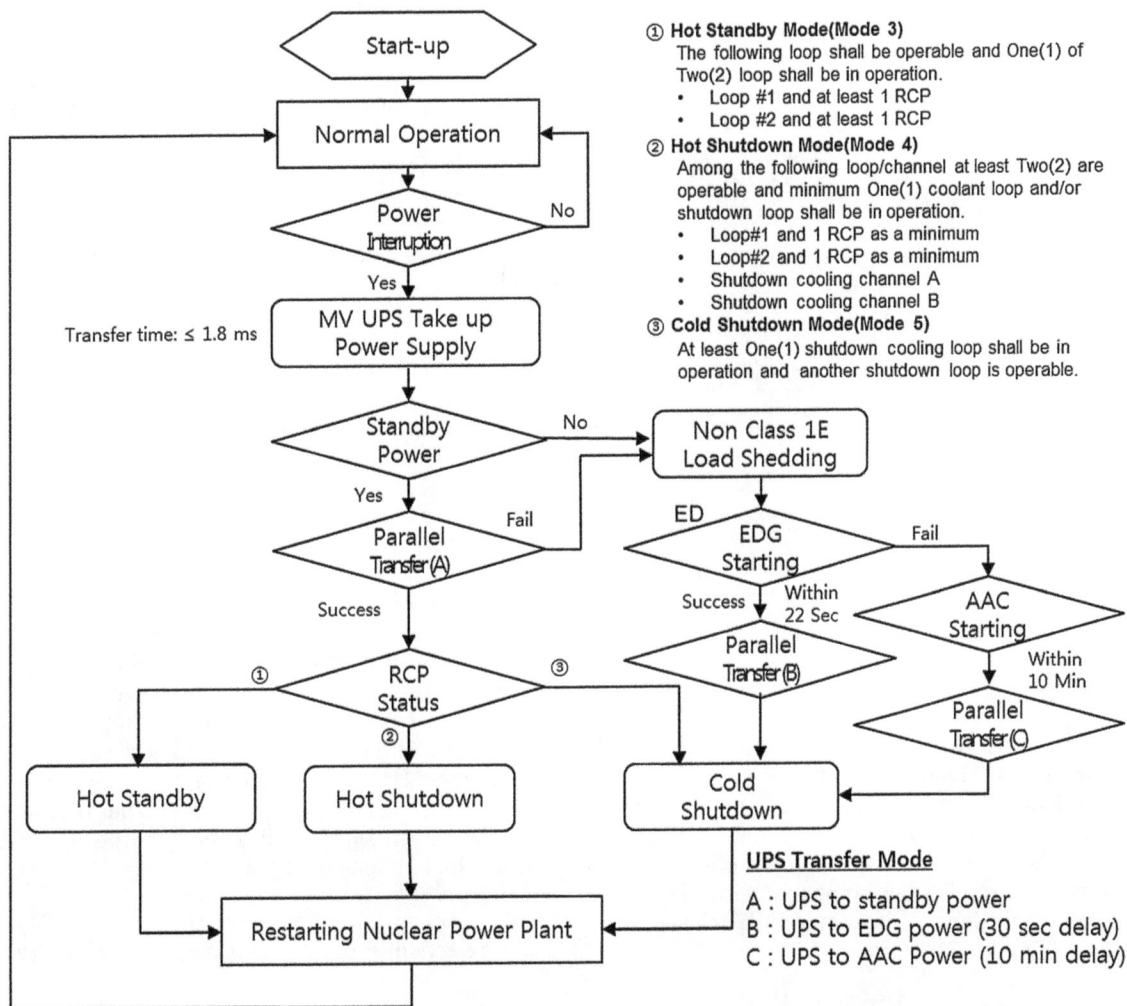

Fig. 4. Uninterruptible MV bus transfer system functional flow diagram.

occurred in the normal source, if the standby power source is available, the class 1E bus is transferred to the standby power source with the momentary backup of the UPS without interruption of power supply to the bus.

The flow of uninterruptible bus transfer scheme is shown in Figure 4.

When the loss of both normal and standby source is detected, the UPS supply power to the class 1E bus until EDG is initiated and supplies power to the class 1E bus for about 30 seconds. Therefore, the power of class 1E 4.16 kV bus is not interrupted.

If the EDG also fails to start, then the Alternative AC (AAC) generator is started manually within 10 minutes. Until the AAC generator starts, the UPS supplies power to the 4.16 kV class 1E buses. As a result, power supply to nuclear safety systems is not interrupted all the time. It means that motors are not needed to stop and restart during the bus transfer.

5.3 Medium voltage UPS

In case of power supply loss, the MV UPS operates disconnected from the normal power but continues to supply loads (Fig. 5). Then, the MV UPS synchronizes the islanded bus to the standby source. Estimated capacity of the MV UPS for the Division I, class 1E buses is approximately 10 MVA each. The MV UPS supports the load when the voltage is outside a user set window. Targeted protections are voltage sags, voltage swells, and short and long outages [9,13].

5.3.1 Converter module

MV UPS uses the LV power modules which employ IGBT's and integrated sinusoidal filters. Multiple modules are connected in parallel to provide higher power. The modules are current rated and the available power output depends on the AC coupling voltage. The AC coupling voltage is further defined by the lowest possible DC link voltage. This is the minimum operational or discharging voltage of the storage.

5.3.2 Energy storage system

The MV UPS is designed by using energy storage device. The most common are super capacitors, lithium-ion

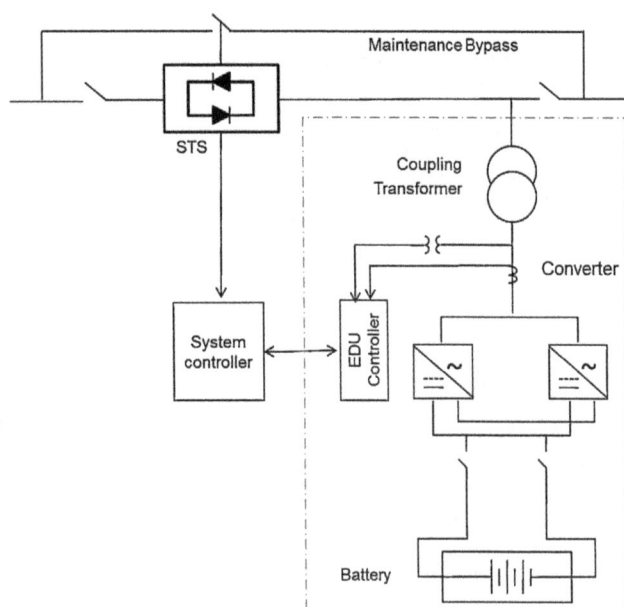

Fig. 5. Schematic diagram of MV UPS.

5.3.4 Control

The MV UPS controls its own voltage and frequency, enabling it to create a micro or islanded grid. When the motor bus is disconnected from the utility, the MV UPS will support the MV bus loads with minimal disturbance. The monitoring and indicating of a normal source failure can be done externally or by internal supervision based on frequency/voltage monitoring. After confirming the standby source, the MV UPS can synchronize the MV bus to the standby source.

5.3.5 Technical specifications and rating of MV UPS

Table 2 shows the typical technical specifications of the 4.16 kV UPS for the Division I.

The power converter and energy storage remain at low voltage, with a transformer coupling these to medium voltage. Also at the medium voltage level is a thyristor-based disconnection switch which prevents backfeed into the grid in the event of power loss or voltage sag. The MV UPS is compatible with a wide range of energy storage depending on the duration of protection required. Super capacitors and flywheels provide high-density coverage for seconds while batteries can be used for longer backup times for up to 15 minutes. The 10 MVA MV UPS is recommended for the 4.16 kV class 1E buses considering the maximum loads safety margin. As shown in Figure 3, class 1E bus 02A and Non-1E bus 01M are interconnected. Therefore, the MV UPS shall be sized to include bus 01M load.

batteries, and high-discharge sealed lead-acid batteries. It is expected that super capacitors will be widely used in industrial applications due to their long life and compact size. For longer-autonomy applications, lithium-ion batteries similar to those used in electric cars offer reduced footprint and increased life when compared with the lowest-cost, lead-acid solutions. Lithium-ion batteries have excellent cycle life characteristics.

5.3.3 Coupling transformer

The AC connection voltage of the MV UPS depends on the batteries used. Therefore, usually a coupling transformer is needed to obtain required bus voltage.

5.3.6 Selection of battery type and sizing capacity

Energy storage devices applicable to MV UPS are super capacitors, lead-acid batteries, lithium-ion batteries, and flywheels. Characteristics of each solution are as specified in Table 3. Among them, lithium-ion battery is preferred to be used for the MV UPS. The energy density of lithium-ion battery is three to five times that of the lead-acid stationary

Table 2. Technical specifications of MV UPS.

Item	Specification
Nominal voltage	4.16 kV ± 10%
Rating	10 MW
Efficiency (full load)	99.5%
Autonomy	Up to 15 min
Displacement power FACTOR	0.7 lagging to 0.9 leading
Typical transfer time	≤1.8 ms (typical)
UPS footprint	54 m^2 (without storage)
System DC nominal voltage	750 V DC (812 ~ 554 V DC)

The MV UPS may be equipped with one of three storage devices

Energy storage	Volt/Cell	AH	Autonomy	Footprint
Super capacitor	–	–	1 s	10 m^2
Lead-acid	12 V	50	30 s	25.6 m^2
Lithium-ion	3.7 V	60	15 min	42 m^2

Table 3. Comparisons of energy storage devices.

Item	Super capacitor	Lead-acid battery	Lithium-ion battery	Flywheel
Energy density	High	Standard	High	–
Design life	15 years	10 years	20 years	Long with maintenance
Autonomy	1 s	30 s ~ 15 min	30 s ~ 15 min	10 ~ 15 s
Ambient temperature	25 °C	25 °C	40 °C	–
Etc.	High discharge	–	–	Very high cycle life

Table 4. Load capacities at normal operation.

Description	Non-1E 4.16 kV-01M	Non-1E 4.16 kV-02M	Class 1E 4.16 kV-01A	Class 1E 4.16 kV-02A	02A + 01M
Apparent (MVA)	1.97	5.85	5.17	1.38	3.35
Active (MW)	1.74	5.04	4.62	1.19	2.93
Reactive (Mvar)	0.92	2.96	2.32	0.70	1.62

Table 5. Load capacities at LOCA mode.

Description	Non-1E 4.16 kV-01M	Non-1E 4.16 kV-02M	Class 1E 4.16 kV-01A	Class 1E 4.16 kV-02A	02A + 01M
Apparent (MVA)	0.54	5.85	6.79	2.82	3.36
Active (MW)	0.46	5.04	6.08	2.49	2.95
Reactive (Mvar)	0.28	2.96	3.03	1.33	1.41

battery. Furthermore, lithium-ion battery is a low mainte-nance battery, an advantage that most other chemistries cannot claim. There is no memory and no scheduled cycling is required to prolong the battery's life. Of course, despite its overall advantage, lithium-ion battery has its draw-backs. It is fragile and requires a protection circuit to maintain safe operation [14]. Due to insufficient backup time, the super capacitor and flywheel are not applicable to the MV UPS.

5.4 MV UPS transfer performance

Tables 4 and 5 are load lists of each bus at normal operation and LOCA mode. The load capacity of the 01A bus at the LOCA mode is largest.

Accordingly, test on the LOCA mode operation is required. Typical transfer performance of a MV UPS complying with the IEC 62040-3 is as shown in Figure 6 [13,15].

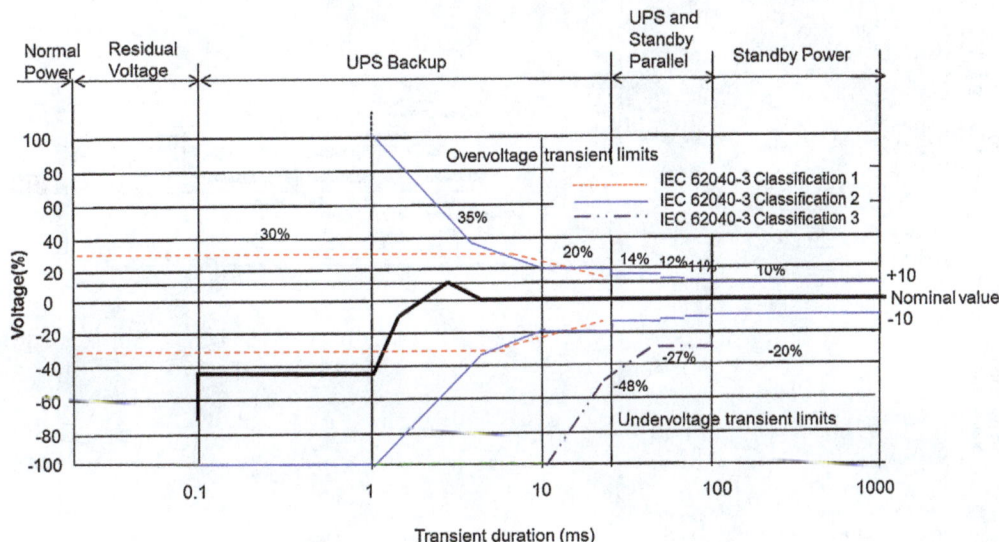

Fig. 6. Typical transfer performance of a MV UPS.

Table 6. Test criteria.

Item	Specification
Transfer time (from normal to inverter)	Less than 1.8 ms
Output voltage setting time (to within ±10% of set point)	Less than 5.0 ms

Table 7. Test facilities.

Item	Description
Test date	April 2012
Device under test	180A 400 V 3PH + N Battery UPS-I
Software revision	R1E2
Measurement method	Multi Channel Hioki 8861 Memory HiCORDER
Sag generator	IGBT instantaneous sag generator

Up until now, 10 MVA MV UPS has been developed and its test is ongoing. As an alternative, test results of the same type LV UPS are presented below. In the MV UPS system, the power converter and energy storage remain at LV with a transformer coupling these to MV. Therefore, it is expected that the MV UPS has the same performance as the same type LV UPS.

5.4.1 Test configuration

The following test results are typical performance of PCS-100 UPS-I product when a normal power failure event occurs. Results are shown for unity and inductive power factors. The two performance characteristics are presented in Tables 6 and 7 [16].

5.4.2 Test results with inductive load (PF 0.5)

Transfer characteristics under a 0.5 lagging power factor load are shown below. Even with 0.5 lagging power factor the UPS-I can transfer within 1.8 ms (actual transfer time is 1.48 ms) (Fig. 7).

The Figure 8 is the same event as above, but processed in Excel to calculate the vector magnitude of the three-phase overvoltage (purple line). The dashed lines indicate the ±10% levels.

$$\text{Vector magnitude} = \sqrt{2} \times \text{maximum voltage.}$$

The output voltage has settled to within ±10% of nominal within 5.0 ms and is greater than 90% voltage after 3.0 ms.

The MV UPS comply with the dynamic output performance classification of the IEC 62030-3 uninterruptible power systems (UPS) - Part 3: Method of specifying the performance and test requirements. Class 1 and Class 2 are applied to overvoltage and undervoltage respectively.

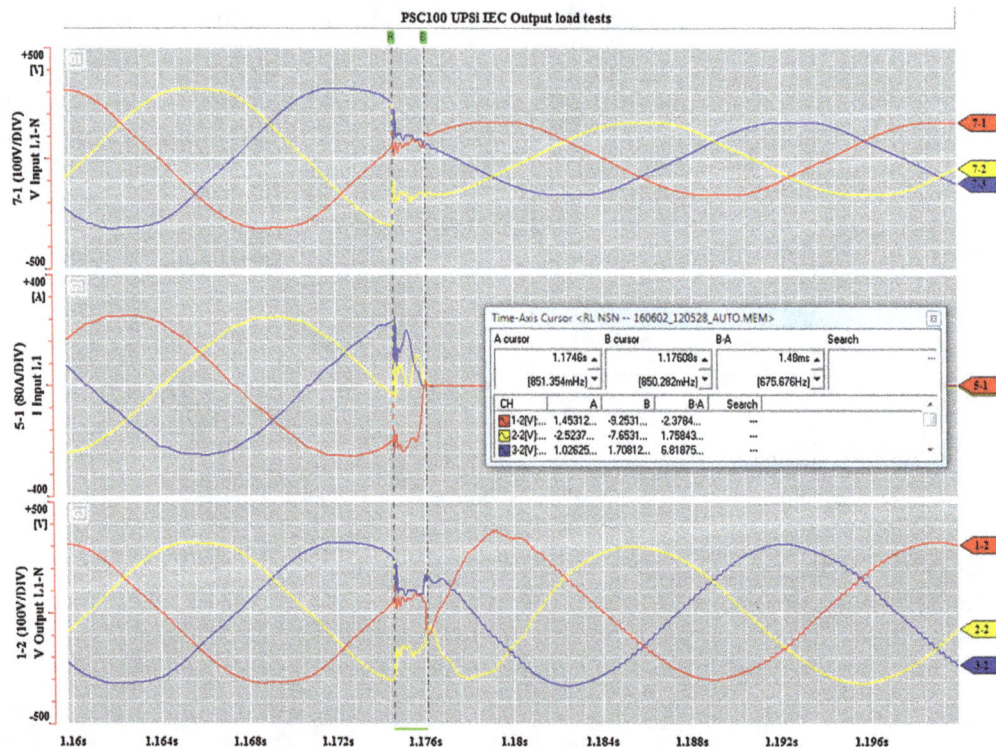

Fig. 7. UPS transfer characteristic curve. Top: input voltage; Mid: input current; Bottom: output voltage.

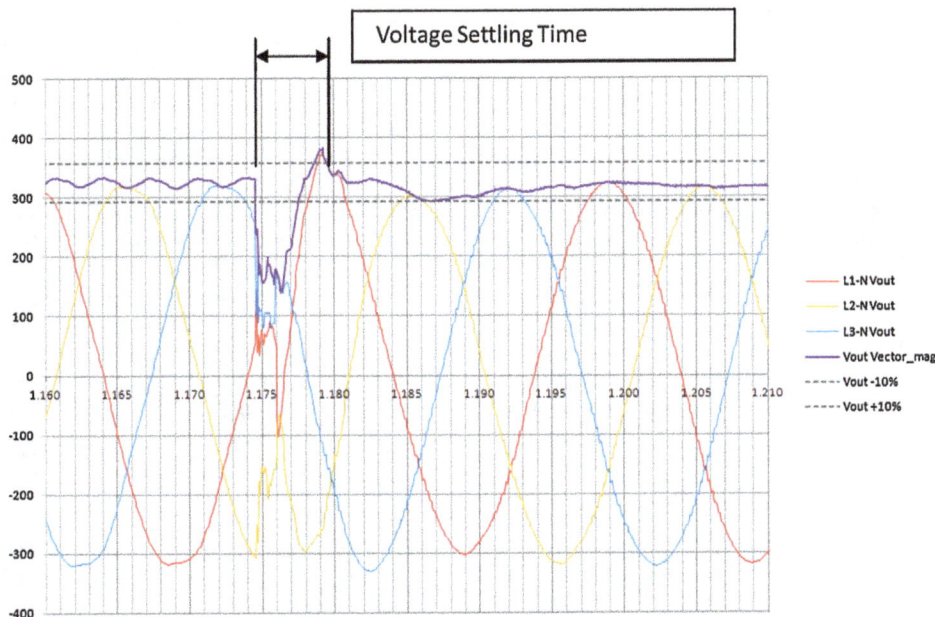

Fig. 8. Voltage settling time.

6 Conclusion

It takes 140 ms to transfer from normal to standby source after receiving starting request command in sequential fast transfer system. If the fast transfer has failed, residual transfer follows. Typical transfer time of residual transfer is 0.5 to 3 seconds. On the other hand, the transfer time of MV UPS is only 1.8 ms and output voltage setting time to within ±10% of set point is 5.0 ms.

Due to the various reasons specified in Table 1, fast bus transfer failure occurs occasionally in the existing system. However, the MV UPS can prevent such kind of failures except the failures caused by CB malfunction. In addition, the MV UPS makes the class 1E bus to transfer from normal source to the EDG seamlessly. It means that no load shedding and sequential loading are required to start up the EDG. During the severe accident condition when the LOOP and LOCA occur at the same time, the class 1E loads can operate continuously without interruption. Accordingly, the operator can shorten the countermeasure action time for accident with simplified operation procedure.

As a result, the uninterruptible MV bus transfer scheme improves the availability of the power of the safety bus and also safety of the nuclear power plant.

The size and price of the MV UPS varies with the autonomy time. According to the study, lead-acid battery is preferred for the purpose of short time backup (30 s; EDG starting time) and lithium-ion battery is preferred for the long time backup (10 min; AAC DG starting time). The latter one requires about 65% more space than the former one as shown in the Table 2.

The effect of the safety enhancement, especially in the nuclear power plant, is scarcely converted to price. On the other hand, economic effect of the operation loss reduction in the specific power plant can be estimated based on the operation records. Further study will be conducted to evaluate the economic effect resulted from the proposed system before the deployment of the first uninterruptible MV bus transfer scheme to a nuclear power plant.

Nomenclature

AAC	alternative AC
CB	circuit breaker
EDG	emergency diesel generator
LV	low voltage
LOCA	loss of coolant accident
LOOP	loss of offsite power
MV	medium voltage
RCP	reactor coolant pump
SAT	standby auxiliary transformer
UAT	unit auxiliary transformer
UPS	uninterruptible power supply

References

1. V. Nalamourougan, B. Kasztenny, A new high speed bus transfer relay design, implementation and testing, in *Proceedings of the 2006 IEEE Power India Conference* (2006)
2. IAEA Technical Report TR-271, Introducing nuclear power plant into electrical power systems of limited capacity: problems and remedial measures, 1987
3. International Atomic Energy Agency, Design of electric power systems for nuclear power plants, Draft Safety Guide DS-430, April 2012
4. R.D. Pettigrew, P. Powel, Motor bus transfer, IEEE Trans. Power Deliv. **8**, 1747 (1993)
5. Siemens, 7VU683 High Speed Busbar Transfer Device, Chapter for the Catalog SIP, Edition No. 7, March 2014
6. M.V.V.S. Yalla, Design of a high-speed motor bus transfer system, IEEE Trans. Ind. Appl. **46**, 612 (2010)

7. A. Raje, A. Raje, J. McCall, A. Chaudhary, Bus transfer systems: requirements, implementation, and experiences, IEEE Trans. Ind. Appl. **39**, 34 (2003)

8. M. Prasad, N. Theivarian, Normal power supply system of a nuclear power plant-modelling and simulation studies for fast bus transfer, in *Proceedings of 2011 Electrical Energy System conference* (IEEE, 2011), p. 294

9. M. Paliwal, R. Chandra Verma, S. Rastogi, Voltage sag compensation using dynamic voltage restorer, Adv. Electron. Electr. Eng. **4**, 645 (2014)

10. L. Sevov, T.W. Zhao, I. Voloh, The power of IEC 61850, bus-transfer and load-shedding applications, IEEE Ind. Appl. Mag. **19**, 60 (2013)

11. Korea Institute of Nuclear Safety (KINS), Operation performance information system for nuclear power plant, accidents and events list, http://opis.kins.re.kr/index.jsp?Lan=KR, Oct. 17, 2014

12. KHNP, NPP Systems, Auxiliary Power System, Text Book, No. BNP-FU-COM-SYS-TB-APS, Korea Hydro & Nuclear Power Co. Rev. 2, p. 18

13. ABB, PCS100 Industrial Medium Voltage UPS Technical Proposal, No. 2UCD130469-T, Rev. A, 2013

14. Cadex Electronics Inc., Lithium-ion based batteries, http://batteryuniversity.com/learn/article/lithium_based_bateries, Aug. 19, 2015

15. Uninterruptible Power Systems (UPS). Part 3: Method of specifying the performance and test requirements, International Electrotechnical Commission, IEC 62040-3, 2011

16. ABB, ABB Technical Document, PCS100 UPS-I Transfer performance, Doc. No. 2UCD120000E023, 2014

7

Flexblue® core design: optimisation of fuel poisoning for a soluble boron free core with full or half core refuelling

Jean-Jacques Ingremeau*,** and Maxence Cordiez*

DCNS, France, 143 bis, avenue de Verdun, 92442 Issy-les-Moulineaux, France

Abstract. Flexblue® is a 160 MWe, transportable and subsea-based nuclear power unit, operating up to 100 m depth, several kilometers away from the shore. If being underwater has significant safety advantages, especially using passive safety systems, it leads to two main challenges for core design. The first one is to control reactivity in operation without soluble boron because of its prohibitive drawbacks for a submerged reactor (system size, maintenance, effluents, and safety considerations). The second one is to achieve a long cycle in order to maximise the availability of the reactor, because Flexblue® refuelling and maintenance will be performed in a shared support facility away from the production site. In this paper, these two topics are dealt with, from a neutronic point of view. Firstly, an overview of the main challenges of operating without soluble boron is proposed (cold shutdown, reactivity swing during cycle, load following, xenon stability). Secondly, an economic optimisation of the Flexblue® core size and cycle length is performed, using the QUABOX/CUBBOX code. Thirdly, the fuel enrichment and poisoning using gadolinium oxide are optimized for full core or half core refuelling, with the DRAGON code. For the specific case of the full core refuelling, an innovative heterogeneous configuration of gadolinium is used. This specific configuration is computed using a properly adapted state-of-the-art calculation scheme within the above-mentioned lattice code. The results in this specific configuration allow a reactivity curve very close to the core leakage one during the whole cycle.

1 Introduction

Flexblue® is a Small Modular Reactor (SMR) delivering 160 MWe to the grid. The power plant is subsea-based (up to 100 m depth and a few kilometers away from the shore) and transportable (Tab. 1). It is entirely manufactured in shipyard and requires neither levelling nor civil engineering work, making the final cost of the output energy competitive. Thanks to these characteristics and its small electrical output, Flexblue® makes the nuclear energy more accessible for countries where regular large land-based nuclear plants are not adapted, and where fossil-fuelled units currently prevail on low-carbon solutions. Immersion provides the reactor with an infinite heat sink – the ocean – around the containment boundary, which is a cylindrical metallic hull hosting the nuclear steam supply systems.

Several modules can be gathered into a single seabed production farm and operate simultaneously (Fig. 1). The reactor is meant to operate only when moored on the seabed. Every 3 years, production stops and the module is emerged and transported back to a coastal refuelling facility which hosts the fuel pool. This facility can be shared between several Flexblue® modules and farms. During operation, each module is monitored and possibly controlled from an onshore control center. Redundant submarine cables convey both information and electricity output to the shore. A complete description of the Flexblue® concept, including market analysis, regulation and public acceptance, security and environmental aspects can be found in reference [1]. A more detailed description of the PWR reactor design and the thermal-hydraulic accident analysis can also be found in reference [2].

The purpose of this paper is to present a suitable design of the Flexblue® core, taking into account the specificities of the reactor. The first major option of this reactor is a soluble boron free control, which is analyzed in Section 2. The second main core characteristic is a three-year-long cycle. This duration together with the core size, enrichment and the refuelling scheme are justified, using an economic analysis, in Section 3. In the last part, an optimization of the burnable poison (gadolinium [Gd]) in the fuel assembly is performed, using an innovative heterogeneous configuration.

---

*Present address: IRSN, 31, avenue Division Leclerc, 92260 Fontenay-aux-Roses, France

**e-mail: jjingremeau@gmail.com

Table 1. Flexblue® module main characteristics.

Parameter	Value
Unit power rating	160 MWe
Length	150 m
Diameter	14 m
Immersion depth	100 m
Lifetime	60 years

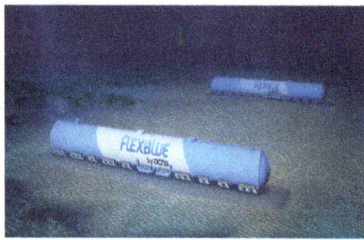

Fig. 1. Artist view of a Flexblue® farm.

2 Operating without soluble boron

2.1 Motivations

The use of soluble boron in the primary coolant is very common in large electricity generator PWR, such as French EDF or American ones. It is used there for three main purposes:

- cold shutdown: in these reactors soluble boron is the only system able to provide sufficient negative reactivity to achieve cold shutdown;
- reactivity swing during cycle: the use of soluble boron enables to mitigate the high reactivity of fresh fuel and to control the reactivity during the fuel depletion;
- load following: soluble boron is a convenient manner to control reactivity during short and limited variation of reactivity (load following, xenon transient).

Moreover, soluble boron has the advantage to be homogeneously distributed in the core, which is favorable to flatten the power distribution in order to reduce the power peak.

But, in the Flexblue® case, it has significant drawbacks. First of all, the use of soluble boron requires voluminous recycling systems, that cannot be afforded in the limited space available in an underwater reactor. Furthermore, these systems require frequent maintenance, which is hardly suitable for Flexblue®. Finally, operating without soluble boron also eliminates all the boron dilution accidents. This point is particularly important for severe accidents, if the flooding of the reactor compartment by seawater is considered; in such a case, if soluble boron is required to achieve cold shutdown, criticality may occur. This last point, even if associated to a very unlikely accident, prohibits the use of soluble boron for Flexblue®.

A soluble boron free reactor also has significant safety advantages, such as less primary corrosion, an increased

moderator coefficient (in absolute value) which is favorable for several accidents (uncontrolled control rod withdraw, unprotected loss of flow accident...),[1] and no criticality in case of main steam line break (in such an accident, the core cooling could be sufficient to make the core critical even with all the control rods inserted in reactors using soluble boron).

The manners to solve the cold shutdown and load following issues in a soluble boron free reactor are presented below, together with a consideration about the shutdown system redundancy. The way to solve the reactivity swing during cycle is analyzed in Section 4.

2.2 Cold shutdown

Due to moderator effect, the reactivity strongly increases between the hot and cold shutdown (around 5,000 pcm), and a safety margin of negative reactivity of 5,000 pcm is also required [3,4]. In order to provide this negative reactivity without soluble boron, the only manner is to increase the control rod worth. Several ways can be investigated:

- use of particularly absorbing materials, such as enriched boron or Hafnium [5]; control rod using B_4C with 90% of ^{10}B worth 40% more than with natural boron[2] in an infinite medium;
- an increased number of control rod pins: the use of 36 pins (compared to the classical 24 for 17×17 fuel assembly) can increase the control rod worth of 70%[2] (250%[2] using enriched B^{10}) in an infinite medium. But these attractive results are not directly applicable, because in a real core, the control rods only cover a fraction of the core, and a space-shielding effect in controlled fuel assemblies strongly limits the negative reactivity of those solutions. For example, a 97 standard 17×17 fuel assembly's core, with half of them controlled with 24-pins control rods using enriched boron, with optimized poisoning (Sect. 4.2), does not achieve cold shutdown. With the most reactive rod stuck above the core, the reactivity is positive, around 2000 pcm;
- an increased number of control rods; another way to avoid this space-shielding effect is to increase the number of rodded fuel assemblies, above 50% possibly up to 100%

---

[1] A high moderator coefficient (in absolute value) is however unfavorable to overcooling accidents, such as main steam line break. But, this is not a drawback for soluble boron free reactor compared to reactor using soluble boron; indeed both have the same maximum moderator coefficient (in absolute value), at end of cycle when both do not have soluble boron in their primary coolant. The impact to be soluble boron free only reduces the moderator coefficient variation from approximately -40 pcm/°C in begin of cycle to -60 pcm/°C at end of cycle compared to ~ 0 pcm/°C to -60 pcm/°C for soluble boron reactors. As safety studies only consider the maximum coefficient, it has no impact. Moreover, concerning the main steam line break accident, as there is no criticality after automatic shutdown (thanks to the increased control rod worth) in soluble free reactor, it is much less an issue.
[2] These results have been obtained with QUABOX/CUBBOX, using cross-section libraries generated with DRAGON (Sect. 4).

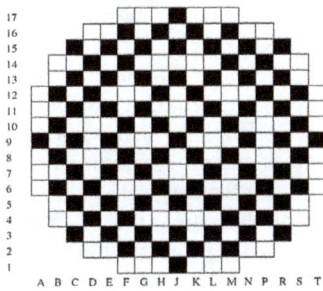

Fig. 2. Control rod position in the EPR Core.

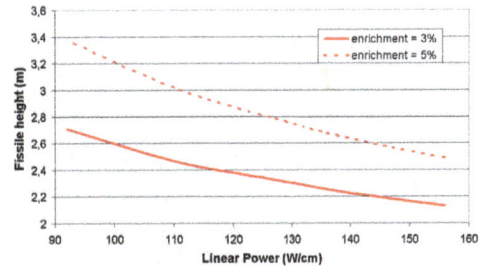

Fig. 3. Limit fissile height for axial xenon stability.

of the core. Calculations performed for this paper show that for 100% of fuel assemblies rodded, even with 24 pins of natural B_4C or AIC, cold shutdown is easily achieved (−7,000 pcm for the above-mentioned core, with the most reactive rod stuck).[3] But, this last solution has a major limitation: the current size of Control Drive rod Mechanism (CDM), ~30 cm, is larger than the fuel assembly size (21.5 cm). That is why in current French PWR, the fraction of rodded fuel assemblies is always below 50% and diagonally spread in the core[4] (Fig. 2).

A way to solve this issue is to insert the control rod mechanism inside the reactor primary vessel; this exempts them to stand the pressure difference (155 bars − 1 bar), and enables to make them more compact. For example, Babcock and Wilcox have chosen this solution for the mPower integral reactor.[5] But it does not really suit the Flexblue® reactor, which is a loop type reactor. Furthermore, the development of such immerged CDM could be long, risky and costly.

For the Flexblue® case, another option has been preferred: it consists in using more compact external CDM. Indeed, the CDM size is mainly imposed by the control rod weight and the primary pressure. If immerged CDM use the lack of pressure to reduce CDM size, reducing the weight can also be an option. For a SMR concept such as Flexblue®, the reduced height (2.15 m of fissile height compared to around 4 m for large land-based PWR) automatically divides by two the control rod weight. The power required in the CDM is therefore also divided by two, and, for a constant height the CDM radial side could be reduced by approximately $\sqrt{2}$, and fitted with the fuel assembly dimensions. This solution, requiring less development than immerged CDM, is the reference for Flexblue® reactor.

Another way to cover 100% of fuel assemblies with classical CDM is to use bigger control rods, recovering several fuel assemblies. That is the idea developed by DCNS in four patent applications [6–9];

– adapted fuel assemblies: another way to cover 100% of the core with control rods is to use larger assemblies, in order

to have the same size for fuel assemblies and CDM. For example, 21 × 21 fuel assemblies, with a moderator ratio of 3 (compared to 2 for current standard PWR, meaning more water between the pins) have a size of around 30 cm. But the change of a standard 17 × 17 fuel assembly to a 21 × 21 larger fuel assembly would have a significant impact on all the fuel facilities, and may raise criticality issues. That is why the reduced size CDM is preferred to this solution.

Several solutions can be combined in order to achieve cold shutdown. For example, reference [10] uses 28 control rods pins for a 17 × 17 lattice, and a fraction of rodded fuel assemblies of 62%. Reference [5] uses Hf as absorber, an increased number of control rods pins, an increased moderator ratio (2.5) and a fraction of rodded fuel assemblies below 50%.

In conclusion, several ways are possible to achieve cold shutdown without soluble boron. Without significant modifications in the fuel assembly, a fraction of rodded fuel assemblies above 50% is required. The reference solution for the Flexblue® project is to keep a standard (but shortened) 17 × 17 fuel assembly, and to adapt the design of the CDM in order to be able to insert a control rod in every assembly in the core. This solution has been chosen for its minimal required developments.

2.3 Xenon stability and load following

In large current PWR, boron is used in order to limit the control rod displacement and avoid the risk of axial xenon instabilities (increase of axial power oscillations due to xenon). For Flexblue®, the limited fissile height (2.15 m) is very favourable in terms of stability. In order to analyse the risk of xenon axial instability, a simplified conservative analytical model based on reference [11] has been used. The model estimates the maximum fissile height for which xenon oscillations are stable, for an axially uniform power profile (conservative hypothesis for stability), as a function of linear power (assuming a standard 17 × 17 fuel assembly) and enrichment.[6] The results, presented in Figure 3, show that for a 5% enrichment, and a linear power below 125 W/cm (Flexblue® core), the stability limit is estimated above

---

[3]It is not necessary to have 100% of rodded fuel assemblies; a fraction around 70–80% seems to be enough (function of the enrichment, size of the core and fuel refuelling strategy). But, this point should be more deeply studied.

[4]The CDM lattice is diagonally oriented compare to the fuel assembly lattice.

[5]www.generationmpower.com.

[6]The xenon density, and neutronic worth are directly dependent on the fission rate (by producing ^{135}I), i.e. the linear power. The xenon density is also function of the neutron flux (for captures), which is strongly dependent on the enrichment for a given power density.

2.8 m. Accordingly, despite the uncertainty of the model, it can be assumed that xenon oscillations are stable in the Flexblue® core. Oscillations may occur consequently to a load follow, or a significant control rod movement, but they will decrease, without leading to safety concerns. Reference [10] even claims to be able to design a soluble boron free core, stable with 3.8 m of fissile height using axially heterogeneous poisoning.

Soluble boron is also currently used to manage significant reactivity variations due to xenon poisoning during a load follow. The way to manage it in a soluble boron free core has been well studied in references [12,13]. The idea is to adapt the average coolant temperature during a transient, in order to use moderator effect to balance the xenon variations. Control rod movements are also required during such a transient, in order to limit the temperature variations, but the study in reference [12] concludes that they are small enough to keep an acceptable form factor.

2.4 Other safety considerations

The soluble boron suppression also raises some other safety considerations. Firstly, a safety requirement of the European Utility requirements [3], similar to a requirement of the NRC in reference [14], is: 'The control of the core reactivity shall be accomplished by means of at least two independent and diverse systems for the shutdown'. Usually, boron and control rods are these diverse shutdown systems. That is why, even if the reactor is soluble boron free in normal operation, the Flexblue® auxiliary systems include an emergency boron injection system, similar to VVER ones [15]. It consists of two tanks, full of borated water at the primary pressure, connected to the primary pumps (Fig. 4).

In case of an Anticipated Transient Without Scram (ATWS), the pump inertia provides the passive injection in the cold leg. After such an injection, the reactor must be transported back to a coastal maintenance facility in order to remove the boron.

Secondly, the reduced weight of control rods has also an impact on their falling time, which is expected to be slightly increased. The impact of this increase cannot be evaluated at this project phase but has to be carefully considered for future detailed transient studies.

2.5 Control rod ejection

For SMR reactors, the control rod ejection is much more problematic compared to large PWR. Indeed, due to the small core size, the neutronic worth of each control rod is strongly increased, reaching 5,000 pcm for a 24 pins, natural B_4C control rod, in a 77 assemblies core. This value has to be compared to approximately 600 pcm for the same control rod in a large PWR, which is high enough to lead to prompt-criticality, a power excursion up to 10 times the nominal power and an energy release of 75% of the safety related criterion of 200 cal/g [16]. Considering that the energy release is roughly proportional to ρ_{CR}-β (where ρ_{CR} is the control rod worth and β the delayed neutron fraction)[7], it is clear that the safety criterion cannot be respected with such insertion of reactivity (up to 30 times the criterion). Even with a control rod of 2000 pcm, the criterion is 10 times exceeded.

Furthermore, for Flexblue®, this point is emphasized by the soluble boron free conception; the control rods are inserted deeper and for a longer time in the core, for long-term reactivity variation and Axial Offset regulation. This makes the control rod ejection accident more likely and even more problematic. Additionally, a control rod ejection may deteriorate the third containment barrier (the module hull), if a dedicated protection is not added above the reactor. However, this place is very critical in terms of component arrangement, due to the module compactness. All these reasons make the control rod ejection a potential issue for safety. That is why, within the Flexblue® project, the strategy is to eliminate the possibility of a control rod ejection. This is achievable using anti-ejection devices, such as described in CEA or Combustion Engineering patents in references [17–19]. Many patents on preventing control rod ejection devices can be found, some associated with "nut screw" CDM, others with "pawl-push" ones. There are too many to be all listed and described here.

This problem is another reason why a re-design of a specific CDM is required for Flexblue®, taking into account two major issues: to be sufficiently compact to achieve one CDM by fuel assembly (to reach cold shutdown), and to eliminate the control rod ejection accident.

2.6 Conclusion

In conclusion, one of the main challenges to operate without soluble boron is achieving cold shutdown. In addition, one of the main challenges of designing a SMR core, especially a soluble boron free one, is control rod ejection accident. These two issues can be solved, keeping a standard 17 × 17 fuel assembly, by using an adapted CDM, more compact, in order to be able to insert one control rod per assembly, and integrating an anti-control-rod-ejection device. The following assumes that such CDM is achieved. The reduced fissile height of the core ensures the stability of axial xenon oscillation, and the load follow can be managed by adapting the coolant average temperature. In order to fit safety requirements, a passive emergency boron injection is added.

The last main challenge for operating a Flexblue® without soluble boron is to manage the reactivity swing

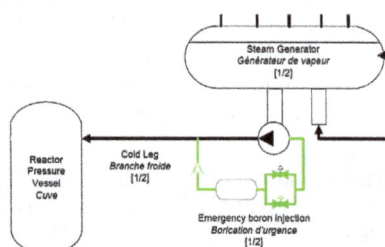

Fig. 4. Scheme of the Emergency Boron Injection.

---

[7]www.cea.fr/energie/la-neutronique/ (in French).

during cycle. This last point will be presented in Section 4. Meanwhile the next part describes the core design strategy and results.

3 Core design

3.1 QUABOX/CUBBOX calculations and control rods regulation

QUABOX/CUBBOX is a diffusion 3D code, developed by GRS (in German "Gesellshaft für Anlagen- und Reaktorsicherheit"). It is integrated in all the GRS reactor physics chain, and especially coupled to ATHLET code for neutronic/thermal-hydraulic transients. It has been validated by benchmark (see for example Refs. [20,21]).

In this study, QUABOX/CUBBOX uses library cross-sections generated by DRAGON (Sect. 4.3). The coupling between the two codes has been developed by DCNS in Python. A validation of this new calculation chain has been performed on standard and Cyclades refuelling strategies on 900 MWe French PWR, with a few percents of discrepancy on burn-up and cycle length.

Cycle calculations have been performed with imposed temperature profile and moderator density (no thermal-hydraulic feedback). For the soluble boron free operation, the current version of the code uses a very simplified control rod regulation; all groups are inserted or withdrawn at the same time, keeping a constant relative distance. These simplifications have a quite small impact on the cycle length, but strongly limit the ability of the current version to estimate precise form factors. Despite these limitations, some optimizations of the refuelling scheme have been performed, and some 3D form factors are presented below, in order to evaluate the performance of poisoning optimization. These values are not very accurate, but give a good idea of what kind of performance can be achieved.

In order to control the Axial Offset, a fuel with heterogeneity has been used, considering a layer of 21.5 cm for two-batch cycle and 18 cm for single-batch without Gd at the top of the core.

3.2 Methodology

Considering that the transportation, between the production site and the refuelling facility, might have an impact on the average availability, the focus has been placed on the following features. Firstly, the conception of the module and the maintenance planning are optimized to shorten the maintenance duration, especially using standard exchange for some components. Secondly, and that is this paper's objective, the core has been designed to optimize the cycle length in order to minimize the Levelized Cost of Energy (LCOE).

The optimized cycle is a compromise between the availability (which is improved by increasing the cycle length), the fuel cost (which is dependent on the enrichment and the refuelling strategy: single or two-batch) and the core size (to increase the reactor vessel size increases the reactor investment).

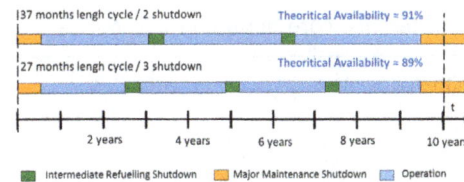

Fig. 5. Examples of 10 years Flexblue® cycles.

One major parameter is the refuelling strategy. Indeed, a single-batch refuelling (100% of fresh fuel at each refuelling) enables to reach a long duration cycle, but misuses the fuel with typical burn-up below 30 GWd/t_{UO2} for 5% enrichment. On the other hand, a two-batch refuelling reduces the cycle length by approximately one third, compared to a complete refuelling, but increases the final fuel burn-up by one third,[8] reducing the fuel total cost.

Another key parameter is the core size. Indeed, a bigger core reduces the power density, and linear power. As a result, it increases the cycle length (thus the availability) for a given burn-up. But it also increases the reactor-vessel cost, and the initial investment to build a module. Taking into account the financial aspect of this investment, with an 8% actualization rate, it has an impact on the LCOE. The linear power is also limited by safety considerations, especially for a soluble boron free core, in which the form factor is expected to increase (Sect. 3.3).

Considering a major shutdown for maintenance of several months every 10 years adds another aspect to take into consideration, because the fuel cycle length should be close to a fraction of this 10-year cycle. It is worthless to achieve a 32-month cycle, because it is not long enough to have only two intermediate refuelling shutdowns, and 27 months are sufficient to have three intermediate shutdowns (Fig. 5). A margin is useful to provide flexibility for the shutdown operation date (function of electricity consumption) but is already provided by stretching possibilities and burn-up economy realized during load following.

All these parameters have been included in a general economic model in order to evaluate the LCOE of several Flexblue® farms. This model takes into account some operation hypotheses (maintenance and transportation durations), cost evaluation (module, fuel, transportation, decommissioning, maintenance facility cost including its own investment and cost strategy), and models for a progressive development and investment in each farm, all the financial fluxes, planned shutdowns and electricity production. In order to evaluate the maximum cycle length for a given core size, enrichment and refuelling strategy (single, two or three batches), polynomial interpolations sets on several hundreds of QUABOX/CUBBOX calculations are used. These calculations are performed assuming a standard 17 × 17 fuel assembly, with a fissile height of 2.15 m. The average quadratic discrepancy between the interpolations and the calculation is 2%. The model also optimizes the core enrichment in order to adapt the cycle length to the number of refuelling shutdowns required, and

---

[8]Using the well-known approximation $Bu(n) = \frac{2n}{n+1} Bu$ (1), where n is the refuelling strategy (1 for single batch, 2 for two batch), and Bu the burn-up [22].

Table 2. Reference core main characteristics.

General characteristics	Number of fuel assemblies	77
	Fissile height	2.15 m
	Equivalent diameter	2.13 m
	Height/Diameter ratio	1.01
	Thermal power	550 MW[a]
	Average linear power	126 W/cm
	Internal/External vessel diameter	3.2/3.5 m
	Reflector	Iron
Single batch refuelling (reference cycle)	Enrichment	4.95%
	Gadolinium content	44 pins/8%w
	Radial assembly form factor	1.24
	3D core form factor	3.0
	Maximum linear power	< 465 W/cm
	Cycle length	38 months
	Burn-up	27 GWd/t$_{UO2}$
Two-batch refuelling	Enrichment	4.95%
	Gadolinium content	32 pins/9%w
	Radial assembly form factor	1.16
	3D core form factor	2.2
	Maximum linear power	< 322 W/cm
	Cycle length	27 months
	Burn-up	38 GWd/t$_{UO2}$
	ΔLCOE (compared to single batch)	+2 €/MWh

[a]The thermal power is, in this study, considered to be 550 MWth, including 20 MWth of margin (conservative on the cycle length and DNBR), compared to the reference 530 MWth [2].

reduces fuel costs. A 5% maximum enrichment is imposed, mainly because most industrial enrichment capabilities cannot reach higher values.

This paper does not deal with the evaluation of the LCOE of Flexblue®, and only focuses on the impact of core design on the LCOE, in order to guide core pre-conception. The results presented below are subsequently only relative between themselves, and to a certain extent functions of the economical hypotheses, and distance between the production site and maintenance facility.

3.3 Core design results

One of the first results obtained with this model is the fact that, achieving a very long cycle of 55 months (4 years and half), in order to maximise the availability, by doing only one refuelling shutdown, is not the economic optimum. Indeed, despite the fact that it would be quite difficult from a maintenance point of view, the gain in terms of availability is annihilated by the fuel misuse and increased investment due to the pressure vessel size.

Another result concerns the decision to use a heavy reflector (iron, like the EPR®) or water like current French PWR. The model shows, with the considered hypotheses, that the use of a heavy reflector is always interesting. The iron reflector reduces the neutron leakage (by ≈1800 pcm compared to a water reflector, for such small cores), therefore

the enrichment (≈0.2–0.3% for a given cycle length) and fuel cost, enough to compensate for its own cost. It also reduces the vessel neutronic damages and flattens the radial-core-power distribution. Consequently, a heavy reflector is today the reference design option for the Flexblue® project.

Every core size (between 69 and 121 fuel assemblies[9]) and refuelling strategy has been evaluated, and among all the results two particular core designs have been selected. More detailed neutronic calculations have been performed for these ones.

The first one, which is the reference core for the Flexblue® project, is a 77 fuel assemblies core whose main characteristics are presented in Table 2.

One of the main characteristics of this core is its quite low average linear power (126 W/cm, compared to 175 W/cm for current French PWR). Such a low power density has been chosen in order to ensure enough safety margins, especially on the DNBR, despite the degraded power distribution in the core due to soluble boron free operation (more control rods inserted) and the scale effect on DNBR (Sect. 3.4). The H/D ratio is almost 1, corresponding to minimum leakage.

This core can be used with a single batch refuelling. It enables to reach cycles longer than 37 months and only two intermediate refuelling shutdowns (Fig. 5), increasing the availability. This is the current reference cycle. On the

---

[9]Odd geometries only, for better symmetry of the control rod scheme.

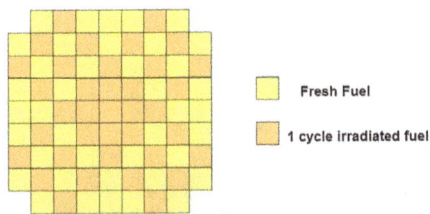

Fig. 6. Two-batch refuelling scheme of the reference core.

contrary, with a two-batch refuelling (Fig. 6) the cycle length is limited to 27 months, requiring three intermediate refuelling shutdowns. In this specific situation, the LCOE increase due to reduced availability is bigger than the fuel economy, indeed LCOE is increased by about 2 €/MWh. But, even with a slightly increased LCOE, this cycle is quite attractive, because it has an increased burn-up, meaning less waste management. In addition, if for regulatory or maintenance reasons, the 37-month cycle could not be achieved at the beginning, it offers an interesting alternative.

Moreover, a two-batch refuelling has significant advantages in terms of reactivity management (Sect. 4.2) and power distribution flattening, with a very low 3D form factor (2.2 compared to 3.0 for single-batch). Especially, it has to be noticed that current single-batch duration cycle is limited by the ability to control the power distribution (the axial offset and form factor increase up to +35% and 3.5 after 38 months), while two-batch cycles are only limited by fuel depletion.

The second selected core is a 97-fuel-assemblies core (Fig. 7) whose main characteristics are presented in Table 3. Its very low linear power (100 W/cm) enables to reach a 37-month cycle, with only two intermediate refuellings, with a two-batch cycle. With the best availability, and a quite good fuel economy (burn-up of 40 GWd/t_{UO2}), this core has a reduced LCOE of 2 €/MWh compared to the reference one, and is the cheapest among all studied cores. Moreover, the very low linear power provides significant safety margins. Its main drawback is its size, but such a primary vessel could still be integrated in Flexblue® containment.

3.4 Concerning the maximal linear power and DNBR

Low linear powers are required for Flexblue® core, because of two main safety reasons. First, the soluble boron free operation increases the power-distribution heterogeneity in the core, due to an increased use of neutron poison and control rods, especially for a single batch refuelling. Reference [12] shows that it can be managed for a 3-batch

Table 3. Ninety-seven fuel assemblies core main characteristics.

General characteristics	Number of fuel assemblies	97
	Fissile height	2.15 m
	Equivalent diameter	2.39 m
	Height/Diameter ratio	0.9
	Thermal power	550 MW[a]
	Average linear power	100 W/cm
	Internal/External vessel diameter	3.4/3.7 m
	Reflector	Iron
Two-batch refuelling	Enrichment	4.95%
	Gadolinium content	32 pins/9%w
	Radial assembly form factor	1.16
	3D core form factor	2.4
	Maximum linear power	< 278 W/cm
	Cycle length	38 months
	Burn-up	40 GWd/t_{UO2}
	ΔLCOE (compared to single batch reference core)	− 2 €/MWh

[a] The thermal power is, in this study, considered to be 550 MWth, including 20 MWth of margin (conservative on the cycle length and DNBR), compared to the reference 530 MWth [2].

refuelling, but it is quite different from a single-batch, with longer cycle, higher reactivity swing and initially uniform fuel assemblies. In Tables 2 and 3, the maximum linear power is below 330 W/cm for two-batch cycles, which is acceptable for normal operation compared to 430 W/cm for current PWR, but slightly above the large reactor reference for a single-batch operation (465 W/cm). This last cycle still requires an optimized power distribution.

Secondly, one of the main safety core parameters is the Departure from Nucleate Boiling Ratio. Considering it, SMR are slightly disadvantaged, compared to large PWR. Indeed, with an approximately half fissile length, for similar core-coolant temperature variation, the flow is roughly divided by two. This has a significant impact on flow turbulence and DNBR, and this single effect approximately reduces by 30% the DNBR (for Flexblue® case compared to a 1300 MWe PWR), and so the acceptable maximum linear power (Fig. 8).

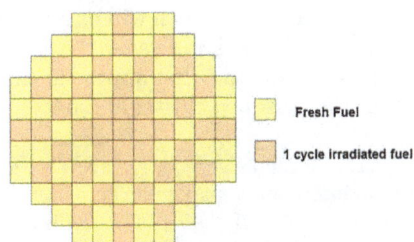

Fig. 7. Two-batch refuelling scheme of the 97 fuel assemblies core.

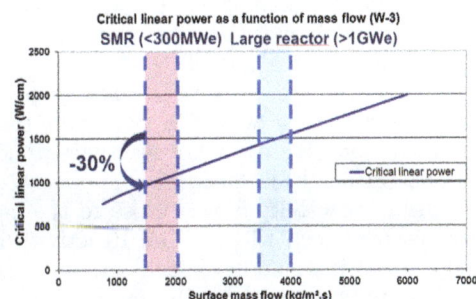

Fig. 8. Scale effect on the DNBR.

Other parameters must be taken into account such as outlet temperature, but this scale effect implies a roughly 30% lower linear power for SMRs. And, as shown in Tables 2 and 3, the above-mentioned two-batch cycles, with unoptimized core power distributions, show a minimum 25% reduction of maximal linear power, which is considered sufficient to have an acceptable DNBR. It is not the case for the single batch cycle, which requires a better power distribution optimization. For future work, this will have to be verified by dedicated thermal-hydraulic codes.

3.5 Conclusion

An economic study has explored the most suited core for Flexblue®, and two particular cores of 77 and 97 fuel assemblies have been selected. The reference one (77) shows good economic characteristics for a single batch but unsatisfactory 3D form factor and has to be optimized with a better control rods regulation. This core also shows acceptable economic features for a back-up 27-month-cycle with enough safety margins. The 97 one has a reduced LCOE of 2 €/MWh and improved safety characteristics, but its size is a limitation.

This study highlights the fact that with different economic features (particularly a longer shutdown for refuelling), the optimum refuelling strategies are very different from current French PWR (3-batch and 4-batch ones). Lower linear power is also required, not only in order to achieve long cycles, but also for safety reasons.

4 Fuel optimisation for 1- and 2-batch refuelling

In this last part, the fuel optimisation for the above-mentioned cycles is performed, considering a standard 17×17 fuel assembly and Gd homogeneously dispersed in fuel pellets as neutron burnable poison. The optimisation aims to obtain a reactivity curve close to the leakage in order to minimise the control rods insertion.

4.1 Influence of gadolinium on reactivity

Gadolinium is a burnable neutron poison. It means that once it has absorbed a neutron, it becomes almost transparent to them. Therefore, it brings negative reactivity at the beginning of irradiation, which decreases with Gd depletion. An example of the influence of Gd on reactivity, in rods homogeneously distributed in an assembly, is displayed on Figure 9.

Two major parts are present in this graph. First, reactivity increases along with Gd depletion to reach, around 18 GWd/t on Figure 9, a "Gd reactivity peak" when almost all the Gd has been burnt. In a second part, the linear decrease of reactivity is very close to the one that would have been noticed without Gd. Reactivity is then slightly below what it would have been without Gd, owing to the lower uranium enrichment in poisoned rods and to the residual absorption of Gd.

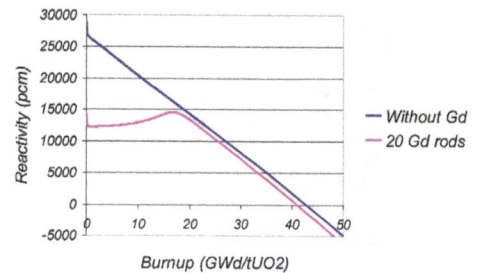

Fig. 9. Influence of gadolinium poisoning on fuel reactivity.

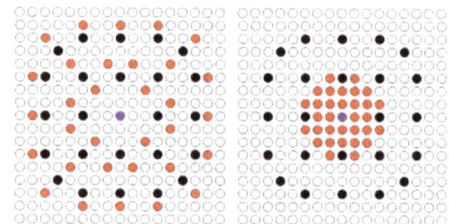

Fig. 10. Fuel bundles containing 32 poisoned rods homogeneously distributed (left) and gathered in a cluster (right). White: fuel, red: poisoned fuel, black: control rods, purple: instrumental rod.

In a homogeneous core of PWR containing one type of assembly made of homogeneous pellets, there are mostly three ways to modify the reactivity using Gd. The first option is to change the number of poisoned rods, keeping the Gd ratio in the poisoned rods the same. Since the effective surface is modified, negative reactivity at the beginning of irradiation is accordingly modified. But the ratio being the same, the burn-up of the Gd peak does not change. The second option is to increase the Gd ratio in the fuel. Since more Gd is available with the same effective surface, it remains longer in the fuel and the Gd peak is shifted to higher burn-up values. Actually, this ratio is limited because gadolinium oxide strongly lowers the fuel thermal conductivity and it could become a safety issue if Gd were to be used in too high proportions. Indeed, when Gd is burnt, what stays is a rod with fresh uranium, almost no fission products to mitigate reactivity and deteriorated thermal conductivity. Hence it increases the risk of fuel melt once Gd has been consumed. This phenomenon explains why the enrichment of poisoned rods has to be lower than the one of the other rods.[10] Industrially in France, gadolinium oxide has never exceeded 9%w, with an 8% standard value. The last option is to modify the poisoned rods distribution in the assembly to make clusters of them (Fig. 10). This way the inter-rods spatial self-shielding protects Gd of the inner rods which will be available later in the cycle. Hence, the Gd peak is delayed. At the same time, the effective surface being in a first approach the one of the external layer of the poisoned cluster and not the sum of all

---

[10]In this paper, an interpolated maximum enrichment in poisoned fuel pellets is used: $e_{Gd} (\%) = 1.8965 + 0.6207e - 0.25r_{Gd}$. Where e is the enrichment in other pellets (%), and r_{Gd} the gadolinium content (%w). This interpolation has been set on open data on irradiated fuel.

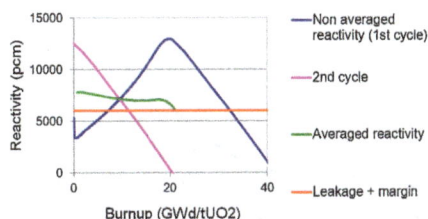

Fig. 11. Reactivity profile of first- and second-cycle fuel, along with the averaged reactivity.

the poisoned rods surfaces, initial negative reactivity is much lower than the one of a homogeneous distribution of the poisoned rods. This kind of configuration has been presented by Soldatov [23], but with questionable parameters (8% enriched uranium even in poisoned rods). What is presented in the present paper (Sect. 4.3) is an improved application of this interesting idea.

4.2 Fuel optimisation for 2-batch fuel management

In order to optimise the fuel for a half refuelling management, the averaged reactivity of the first cycle and second cycle fuel has been studied. This assumption has proven being a good approximation for a core where first- and second-cycle fuel bundles are positioned alternatively.

If the refuelling happens during the Gd peak, namely around 20 GWd/t_{UO2} with 8%w of Gd, the reactivity of first-cycle fuel increases (burn-up before the Gd peak) and the reactivity of the second-cycle fuel decreases (normal reactivity decrease with fuel consumption). With an optimisation of the number of homogeneously distributed poisoned rods and the ratio of Gd in them, it was possible to obtain a flat reactivity curve up to 21 GWd/t_{UO2}[11] above the leakage (considered as constant around 3,000 pcm) and margins for operations (Fig. 11). This assembly, enriched up to 4.95% and containing 32 homogeneously distributed rods poisoned with 9%w of Gd, is the one used in the two-batch cycle of Section 3.

4.3 Fuel optimisation for full refuelling management

In case of full refuelling management, no assembly shows a reactivity decrease to compensate for the reactivity increase linked to Gd consumption. Hence, fuel reactivity has to follow the leakage without being averaged on the core.

For this kind of fuel management, a phenomenon usually negligible for 3- or 4-batch fuel management must also be considered. The fuel burns at different speeds according to its location in the core (faster in the middle than on its edges). To neutron leakage (around 3,000 pcm), it adds a reactivity penalty (up to 4,000 pcm at 25 GWd/t_{UO2}) compared to reactivity in an infinite medium ρ_{inf}, due to the fact that the fuel assemblies which have the most impact on core reactivity are at higher burn-up than the average burn-up of the core. To optimise ρ_{inf}, these two phenomena have to be considered. Therefore, the reactivity

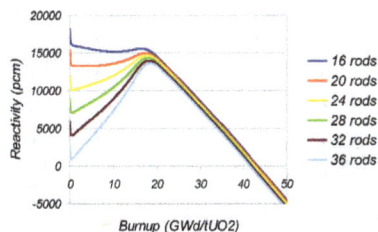

Fig. 12. Influence of the number of poisoned rods on reactivity (U enrichment: 4.95%, Gd ratio: 8%).

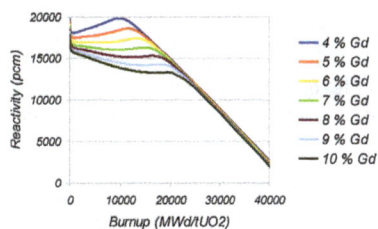

Fig. 13. Influence of gadolinium in poisoned rods on reactivity (U enrichment: 4.95%, 16 poisoned rods).

of the optimised fuel should increase with the burn-up in order to follow this evolution (Fig. 16).

For the considered cycle, Gd peak cannot be compensated at the core scale if the study includes only homogeneous poisoned rods distributions. It would imply an important resort to control rods to mitigate the reactivity and this would penalise the core form factor. And from that observation, two contradictory parameters had to be considered. First, the initial reactivity should not to be too high, so the poisoned rods should be numerous enough (to increase the effective surface of the poison), as displayed on Figure 12. Secondly, the reactivity when all the Gd is burnt (during the Gd peak) should not be too high. Using only homogeneous distributions, one would have to increase the Gd ratio in poisoned rods (Fig. 13). However, as evoked previously, this resort to the Gd ratio is limited to 9% percent of gadolinium oxide, so that all the poison is burnt at around 20 GWd/t_{UO2} and reactivity is still very high (Fig. 13). In other words, to control that peak, another way to preserve Gd negative reactivity after 20 GWd/t_{UO2} has to be found.

A solution to answer that question and keep negative reactivity after 20 GWd/t_{UO2} is to group the poisoned rods in clusters, as described in Soldatov's Ph.D. thesis. In this kind of configuration called heterogeneous distribution, Gd

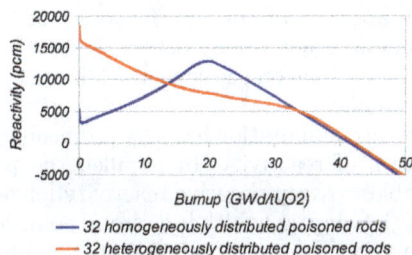

Fig. 14. Comparison of reactivity of two assemblies with 32 poisoned rods (U enrichment: 4.95%, Gd ratio: 8%).

---

[11] Cycle length, which means a final burn-up of 42 GWd/t_{UO2}.

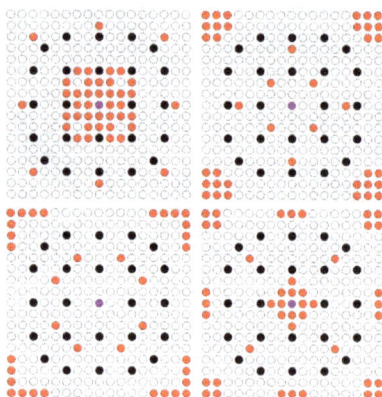

Fig. 15. Examples of mixed assemblies (white: fuel, red: poisoned fuel, black: control rods, purple: instrumental rod).

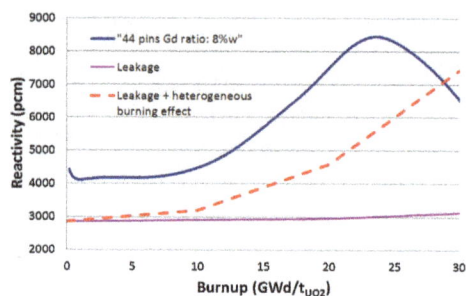

Fig. 16. Reactivity of a mixed configuration made of 44 poisoned rods: 40 in 9 small clusters and 4 isolated.

is burnt from the edge of the cluster to the center. The effective surface is reduced: it is no more the summation of all the poisoned rods' areas but the area of the rods constituting the external layer of the cluster. Because of that, initial negative reactivity is much lower than the one of a homogeneous distribution with the same number of poisoned rods. At the same time, this inter-rods spatial self-shielding enables to use more poisoned rods. Owing to these two aspects (the protection given by the cluster and the increased number of poisoned rods), Gd is preserved after $20\,\mathrm{GWd/t_{UO2}}$, with Gd ratios lower than or equal to 8%. Reactivity is then mitigated below 10,000 pcm and there is no more Gd peak. The problem of such a configuration, as displayed on Figure 14 for the fuel displayed on Figure 10, is that initial reactivity is too high. This is due to the reduced amount of poisoned rods initially subjected to the neutron flux. Besides, the radial form factor in the assembly can be relatively high (1.2–1.3).

To summarise the last two paragraphs, homogeneous distributions enable an interesting reactivity mitigation at the beginning of irradiation but with the price of an intense Gd peak afterwards ($> 10{,}000\,\mathrm{pcm}$), when heterogeneous ones display high reactivity at first, mitigated in a second time owing to the absorption of initially protected Gd.

Another solution, called "mixed" was considered. Inspired of both, it has been designed to ensure reactivity mitigation all the irradiation long. In this idea, the assembly contains cluster(s) of poisoned rods and isolated poisoned rods homogeneously distributed amongst the other rods (see some examples on Figure 15, from the DCNS patent [24]). The distribution of the isolated rods has to take into account the presence of clusters. In addition, considering that an assembly is not alone in a core, it should be kept in mind that the clusters on the edge are adjacent to the ones of other assemblies.

This kind of configuration has very promising properties on the control of reactivity. In parallel, the presence of clusters enables to keep negative reactivity longer without using too high Gd ratio, which is a strong point for the fuel safety. A last interesting aspect is that if the radial form factor in an assembly is higher than the one of a homogeneous configuration, it is still lower than the one

of a heterogeneous distribution, owing to the isolated poisoned rods which mitigate the fluxes in the areas where it is the highest in heterogeneous configurations.

The reactivity at the beginning of irradiation can be set by the number of isolated poisoned rods and the peak reactivity by the size of the cluster (or the number of clusters). Therefore, by adjusting the number of poisoned rods, their Gd ratio and their repartition, the fuel designer can model reactivity at any time of the fuel's life.

To study a mixed configuration, a 44 poisoned rods assembly, with a distribution of poisoned rods shown on bottom right on Figure 15, with a 4.95% enrichment and 8%w Gd, has been selected. Its reactivity curve is displayed on Figure 16. This fuel assembly has been used to evaluate the reference single-batch cycle of the reference core.[12] Compared to homogeneous or heterogeneous configurations, reactivity is indeed much closer to the summation of the leakage and heterogeneous burning effect, from the beginning of irradiation to the end (around $30\,\mathrm{GWd/t_{UO2}}$). This leads to the insertion of fewer control rods in the core, and the ability to a better optimisation of the core power distribution. However, due to the simple control rods regulation and the lack of thermal-hydraulic feedback, current calculations do not achieve to manage the core power distribution in this particular case, and the maximum 3D form factor reaches 3.0. Furthermore, it is the increase of the axial offset and the form factor at the end of cycle which requires limiting the cycle length at 38 months. For future work, an optimisation of the control rods regulation, including thermal-hydraulic feedback, is required in order to confirm the interest of this solution.

A drawback of such a configuration is an increased assembly form factor. For the considered 44 poisoned rods it reaches 1.24 at the beginning of cycle, compared to typically 1.06 for an assembly without poison, and 1.18 for the 36 poisoned rods used for the two-batch cycles. But such an increase may be acceptable if it enables a significant reduction of the core form factor. Moreover, this form factor decreases with irradiation and is about 1.10 at the end of cycle. Another issue might be the thermal difference between the rods inside and outside the poisoned cluster.

---

[12]The fact that reactivity is at the beginning very close to the leakage (low margin) is to be compensated by the non-poisoned layer of fuel on the top of the core.

4.4 Calculation strategy for heterogeneous and mixed assemblies

Calculations were led with DRAGON, a freely available calculation code developed by École Polytechnique de Montréal. DRAGON has been designed to solve the neutron transport equation considering every prominent physical phenomenon [25,26]. Evolution calculations have been performed using the interface current tracking method and the JEFF-3.1.2 cross-sections library, with 281 energy groups. The energy self-shielding is done by the subgroups method on the isotopes of Zr, U, Pu, Am and Gd.

For homogeneous fuel configurations, each rod (being fuel or poisoned fuel) evolves in a quasi-isotropic medium regarding neutron flux; hence a discretisation of the rods in concentric rings is sufficient (4 rings for fuel and 11 for poisoned fuel owing to the importance of spatial self-shielding).

For heterogeneous or mixed configurations, this isotropy hypothesis is not valid anymore. Actually, the interest of that kind of configuration lies in its anisotropy that guarantees the inter-rods self-shielding. For that reason, a pellet discretisation according to its diagonals is required to describe adequately the fuel (cf. example on Fig. 17).

With this new calculation scheme, each quarter of ring, corresponding to one side of the pellet, is calculated with his own flux, and own isotopic depletion (energy self-shielding being the same). Consequently, a single Gd pellet requires 44 nodes, compared to 11 for a homogeneous configuration.

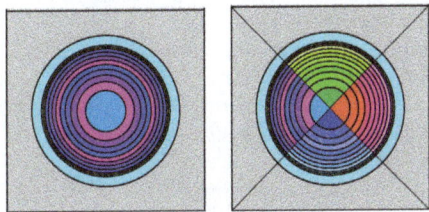

Fig. 19. Example of assembly discretisation for "mixed" distributions.

Furthermore, on the assembly scale, the same issue is raised; in a homogeneous distribution (Fig. 18) every fuel pellet without Gd can be assumed to have the same isotopic depletion. The same hypothesis can be done (and is done in standard calculation scheme) for the Gd pellet. The total number of fuel nodes is then: $11 + 4 = 15$ nodes.

But for a heterogeneous or mixed distribution, such assumptions cannot be done, and each Gd pellets, with surrounding fuel pellets, has to be differentiated. In Figure 19, corresponding to the previously mentioned 44 poisoned rods assembly, 88 depletion nodes are required for fuel pellets and 275 for Gd pellets. Such calculation scheme drastically increases the computation time and memory.

Comparisons between calculations, with and without azimuthal discretisation, have been performed for a 20-poisoned-pins homogeneous configuration (Fig. 20) and the 44-poisoned-pins mixed configuration mentioned above (Fig. 21). The results show a discrepancy up to 260 pcm for the homogeneous one, and around 1,000 pcm for the mixed one. In order to analyse precisely the impact of azimuthal discretization, according to heterogeneous or homogeneous configurations, another calculation has been performed for a heterogeneous configuration with 20 poisoned rods in a central single cluster; the discrepancy is about 425 pcm, to be compared to the 260 pcm for the homogeneous one.

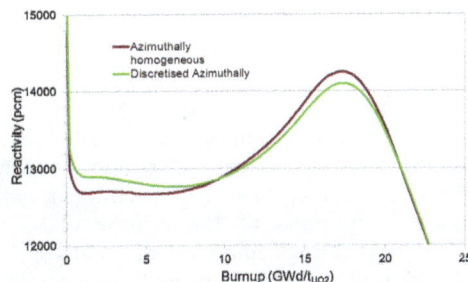

Fig. 17. Poisoned rod discretisation. On the left: without azimuthal discretisation, acceptable description for homogeneous distributions, and on the right: with azimuthal discretisation adapted to heterogeneous or mixed distributions.

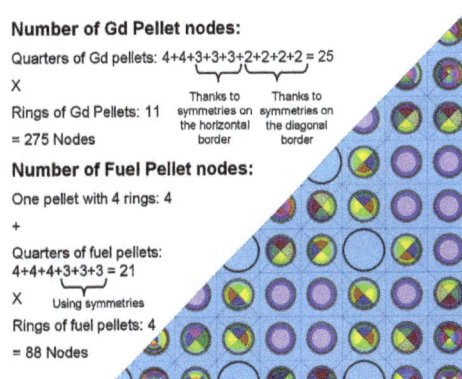

Fig. 18. Example of a standard assembly discretisation for homogeneous distributions.

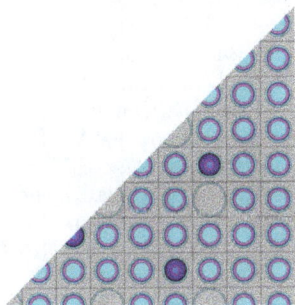

Fig. 20. Reactivity with or without azimuthal discretisation for a homogeneous configuration with 20 rods poisoned with 8%w of Gd.

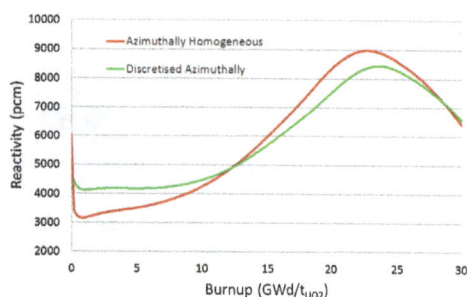

Fig. 21. Reactivity with or without azimuthal discretisation for a mixed configuration with 44 rods poisoned with 8%w of Gd.

In each case, without the azimuthal discretisation the reactivity is under-estimated at beginning of cycle (due to an artificial over-consumption of Gd) and over-estimated during the Gd peak.

Therefore, for a preliminary optimisation of fuel assemblies, if the azimuthal discretisation is not strictly necessary for homogeneous configurations, it is clearly required, and used in this paper, to study heterogeneous and mixed configurations with such number of Gd rods.

The use of the interface current tracking method is also a discrepancy source, especially for heterogeneous and mixed geometries, because this method is not very accurate for treating the strong flux gradients of such configurations. Reference [27] shows an impact of around 300 pcm for a homogeneous configuration. It is expected that the impact is increased for a heterogeneous configuration. However, it is assumed that the impact is lower than the azimuthal discretisation one.

4.5 Conclusion

For two-batch cycle, the classic homogeneous configurations of Gd poisoning are well suited and enable to achieve quite good safety performances.

For the specific case of single-batch long-cycle core, new heterogeneous configurations of Gd inside the fuel assembly are proposed. They offer a powerful manner to control reactivity during the whole cycle. But, in order to adequately model such geometries, the calculation scheme has to be adapted to take into account an azimuthal discretisation.

5 Conclusions

Within the Flexblue® project, a soluble boron free core has been designed. Several manners to achieve cold shutdown have been explored and a compact CDM, including an anti-ejection device has been selected. An economic optimisation of core design and fuel has been performed, with two selected cores, functions of the size available in the containment. For the specific case of single-batch long-cycle, a new kind of heterogeneous fuel-assembly poisoning is proposed, which may enable an improved reactivity regulation, and so, power distribution in the core. These

configurations require adapted calculation scheme including an azimuthal discretisation of poisoned rods.

The authors would like to thank warmly Pr. Alain Hébert and Vivian Salino from IRSN for their precious help with the use of DRAGON, and Yann Périn from GRS, for his help with the use of QUABOX/CUBBOX.

References

1. G. Haratyk, C. Lecomte, F.X. Briffod, Flexblue®: a subsea and transportable small modular power plant, in *Proceedings of the ICAPP 2014, Charlotte, USA* (2014)

2. G. Haratyk, V. Gourmel, Preliminary accident analyses of Flexblue® underwater reactor, in *Proceedings of the ICAPP 2015, Nice, France* (2015) Paper 15149

3. *European utility requirements for LWR Nuclear Power Plants*, Revision D (Oct. 2012), Vol. 2, Chap. 8: Functional Requirements: Systems & Processes, § 2.1.1.5

4. EPRI, *Advanced Light Water Reactor Utility requirements document* (1999), Vol. III.

5. P. Thomet, Feasibility studies of a soluble boron-free 900-MW (electric) PWR. Core physics I: motivations, assembly design, and core control, Nucl. Technol. **127**, 259 (1999)

6. Patent FR1302763, DCNS, 2015

7. Patent FR1302764, DCNS, 2015

8. Patent FR1302765, DCNS, 2015

9. Patent FR1302766, DCNS, 2015

10. S.Y. Kim, J.K. Kim, Conceptual core design of 1300-MWe reactor for soluble boron free operation using a new fuel concept, J. Korean Nucl. Soc. **31**, 391 (1999)

11. P. Reuss, *Exercices de Neutronique* (EDP Sciences, 2014)

12. J.P. Defain, P. Alexandre, P. Thomet, Feasibility studies of a soluble boron-free 900-MW (electric) PWR. Core physics II: control rod follow, load follow and reactivity-initiated accident linked to RCCAs, Nucl. Technol. **127**, 267 (1999)

13. G.L. Fiorini, G.M. Gautier, Feasibility studies of a soluble boron-free 900-MW (electric) PWR. Safety systems: consequences of the partial or total elimination of soluble boron on plant safety and plant systems architecture, Nucl. Technol. **127**, 239 (1999)

14. NRC 10CFR50 Appendix A to Part 50 "Domestic licensing of production and utilization facilities – General Design Criteria for Nuclear Power Plants", 2007

15. *Development of IRSN-GRS Safety Approach in view of the EPR detailed Design* (IRSN/GRS, 2003), Vol. 5

16. B. Tarride, *Physique, fonctionnement et sûreté des REP* (Collection génie atomique, EDP Sciences, 2013)

17. Patent FR2432197A1, CEA, 1978

18. Patent FR2075928A7, Combustion Engineering, 1971

19. Patent FR2728098, CEA, 1994

20. H. Finnemann, H. Bauer, A. Galati, R. Martinelli, Results of LWR core transient benchmarks, NEA/NS/DO(93)25, 1993

21. R. Fraikin, PWR benchmark on uncontrolled rods withdrawal at zero power, Final report, NEA/NSC/DOC(96)20, 1997

22. P. Reuss, *Précis de neutronique* (EDP Sciences, 2003)

23. A.I. Soldatov, Design and analysis of a nuclear reactor core for innovative small light water reactors, Ph.D. Thesis, 2009

24. Patent Application filled before the INPI (Institut National de la Propriété Industrielle), FR1402827, by DCNS, December 2014

25. G. Marleau, R. Roy, A. Hébert, DRAGON: a collision probability transport code for cell and supercell calculations,

Report IGE-157, Institut de génie nucléaire, École Polytechnique de Montréal, Montréal, Québec, 1994

26. G. Marleau, A. Hébert, R. Roy, A user guide for DRAGON, Report IGE-294, Institut de génie nucléaire, École Polytechnique de Montréal, Montréal, Québec, 2013

27. D. Calic, A. Trkov, M. Kromar, Use of Lattice code DRAGON in reactor calculations, in *Proceedings of the 22nd International Conference Nuclear Energy for New Europe, September, 2013* (2013)

8

Beyond designed functional margins in CANDU type NPP. Radioactive nuclei assessment in an LOCA type accident

Andrei Razvan Budu and Gabriel Lazaro Pavel[*]

University Politehnica of Bucharest, Faculty of Power Engineering, Splaiul Independentei No. 313, Sector 6, Bucharest, 060042, Romania

Abstract. European Union's energy roadmap up to year 2050 states that in order to have an efficient and sustainable economy, with minimum or decreasing greenhouse gas emissions, along with use of renewable resources, each constituent state has the option for nuclear energy production as one desirable option. Every scenario considered for tackling climate change issues, along with security of supply positions the nuclear energy as a recommended option, an option that is highly competitive with respect to others. Nuclear energy, along with other renewable power sources are considered to be the main pillars in the energy sector for greenhouse gas emission mitigation at European level. European Union considers that nuclear energy must be treated as a highly recommended option since it can contribute to security of energy supply. Romania showed excellent track-records in operating in a safe and economically sound manner of Cernavoda NPP Units 1&2. Both Units are in top 10 worldwide in terms of capacity factor. Due to Romania's need to ensure the security of electricity supply, to meet the environmental targets and to move to low carbon generation technologies, Cernavoda Units 3&4 Project appears as a must. This Project was started in 2010 and it is expected to have the Units running by 2025. Cost effective and safety operation of a Nuclear Power Plant is made taking into consideration functional limits of its equipment. As common practice, every nuclear reactor type (technology used) is tested according to the worse credible accident or equipment failure that can occur. For CANDU type reactor, this is a Loss of Cooling Accident (LOCA). In a LOCA type accident in a CANDU NPP, using RELAP/SCDAP code for fuel bundle damage assessment the radioactive nuclei are to be quantified. Recently, CANDU type NPP accidents are studied using the RELAP/SCDAP code only. The code formerly developed for PWR type reactors was adapted for the CANDU geometry and can assess the accident progression consequences up to a certain point. The code assesses the fuel bundle damage progression, but cannot assess further core damage for a CANDU type core, and starting from these data the amount of damaged fuel can be calculated. The radio nuclei present in the damaged fuel are supposed to be released into the main heat transport system and after that into the containment building in the worst case scenario. Assessing the radioactive nuclei maximum release is the purpose of the present paper. The radioactive nuclei release is needed for the accident management plan, limiting the environmental and population impact of the supposed accident, and furthermore for a later site remediation plan that can be put in action after the complete mitigation of the accident consequences. The maximum quantity of radio nuclei released during the accident calculated in this paper is a worst case scenario evaluation that can lead to better preparedness in an accident scenario.

1 Introduction

Nuclear power is today among the non-CO_2 emitting energy sources and nuclear fuel reserves are surpassing the fossil fuel reserves in terms of potential energy production.

Although there are many reactor years of experience in the design and operation field of nuclear power plants, events through the years have shown that there is no certainty to safe nuclear power operation and nuclear risk arises from even the most mundane operation activities.

Thus, even though best estimate evaluations of nuclear safety are performed for every type of operating nuclear power plant, the worst case scenario can lead to innovating new solutions for future nuclear power plants.

This paper proposes new values for release factors for fission products resulting from a severe accident, starting from the fuel bundle damage occurring in a LOCA/LOECC

*e-mail: gabriel.pavel@gmail.com

(Loss of Cooling Accident/Loss of Emergency Core Cooling) accident in a CANada Deuterium Uranium (CANDU) type Nuclear Power Plant (NPP).

The paper presents beforehand the main steps in using the SCDAP/RELAP5 code for CANDU type NPP severe analysis, and modifying the code to suit that type of power plants characteristics, and a severe accident transient to evaluate the fuel bundle and fuel pins damage occurred.

The fuel damage occurred leads to the release factor calculated and proposed for use in future environmental impact assessment done for a CANDU type NPP.

2 SCDAP/RELAP5 use in CANDU type NPP accident analysis

RELAP5 is a Light Water Reactor (LWR) transient analysis code developed initially for the US NRC at the Idaho National Laboratory as a base for nuclear power plant analysis, operating manual review, licensing calculations auditing and nuclear power regulation. It has a mono-dimensional transitory hydrodynamic model, with two-phase flow of water-steam mixture that may contain non-condensable components in the steam phase and a soluble component in the liquid phase.

The SCDAP/RELAP5 coupled code was developed for best-estimate simulation of light water reactors during severe accidents. The code models behavior of the main reactors cooling system coupled with that of the core and radioactive fission product release during a severe accident. This is the result of the unification of the RELAP5 used for

thermal-hydraulic analysis, the study of control systems interaction, reactor kinetics and non-condensable gases transport with the SCDAP code that models the core behavior during severe accidents. The result is a flexible tool due to its generic approach to modeling that allows the modeling of specific systems according with the demand and is used consequently in the study of a large transient collection for power stations, research reactors and experiments in small installations.

Due to the moderator and cooling agent separation and horizontal flow in the fuel channels in the CANDU core, direct use of the detailed core degradation models of the existing system codes as SCDAP/RELAP5, MELCOR, ICARE/CATHARE or ATHLET-CD cannot be done. But, due to the flexibility of SCDAP/RELAP5 code and validation results for other reactor system analysis, the early phase modeling of some severe CANDU6 type accident was done. Furthermore, based on studies linked to those simulations, basic evaluation of the code aptitudes was conducted along with its development and adaptation needs due to the special conditions and phenomena in severe accidents for CANDU systems.

The SCDAP/RELAP5 code is adaptable to CANDU power plants systems due to heavy water library use and horizontal flow modeling capabilities. From the beginning of SCDAP/RELAP5 use in Romania, the code was added to, modified and improved to meet CANDU specifications.

The first step in the early stages was the use of the SCDAP/RELAP5 code in modeling a severe accident in a CANDU type coolant loop. Figure 1 shows an early complete mapping used to analyze a LOCA type accident in

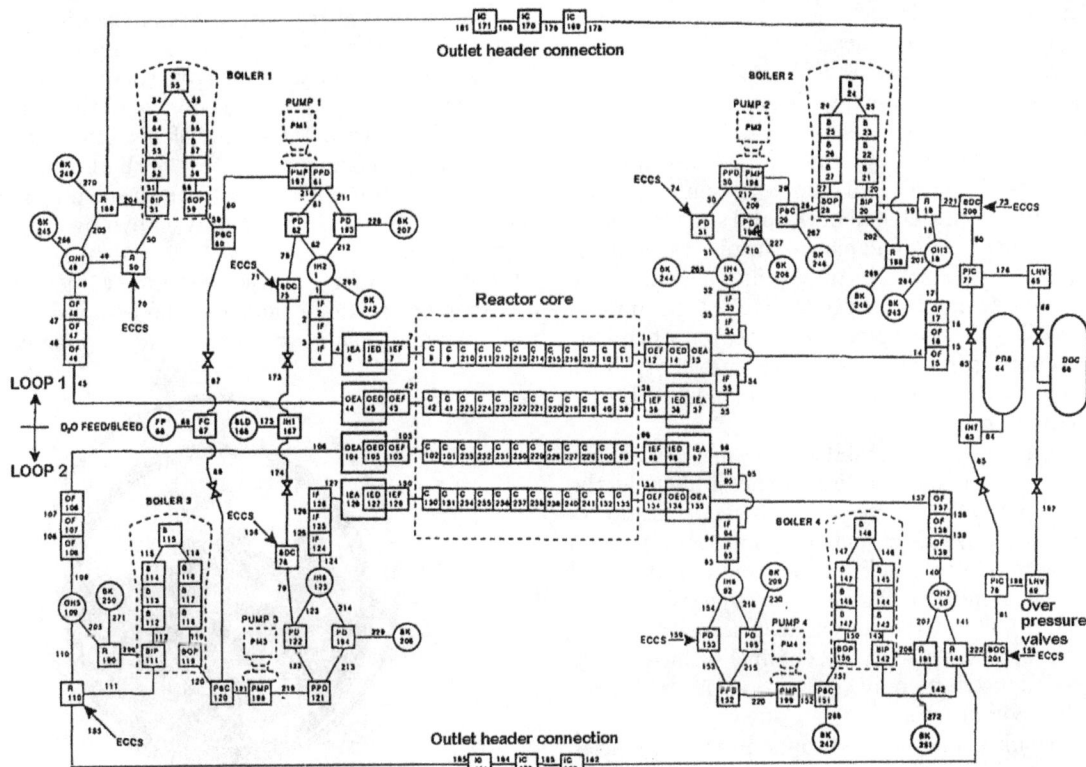

Fig. 1. CANDU coolant main circuit mapping [1].

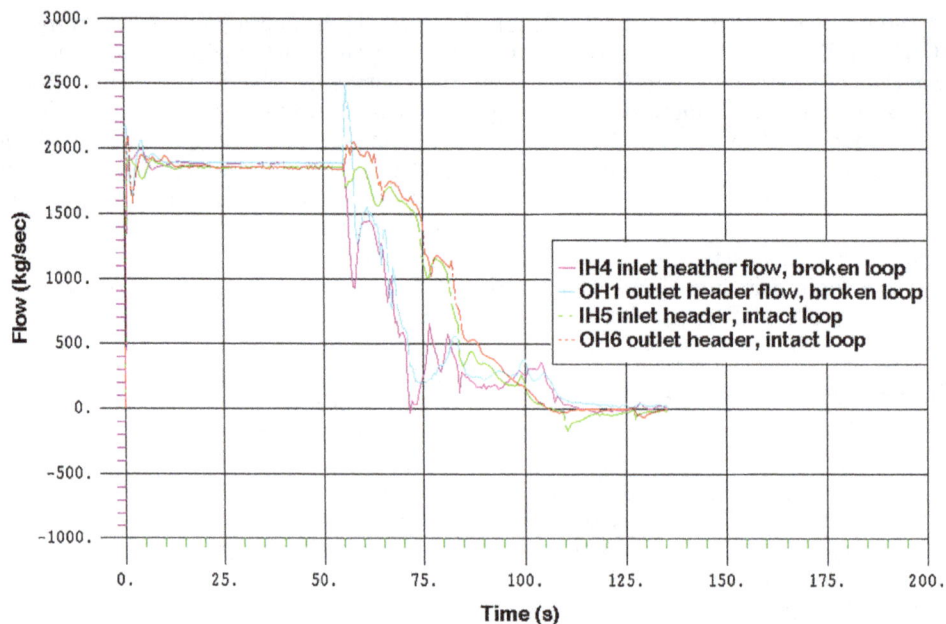

Fig. 2. Flow variation in LOCA accident with 100% break of inlet header [1].

a CANDU NPP using SCDAP/RELAP5. The accident presumes complete failure of a reactor coolant inlet header that deprives the reactor of the decay heat removal even after reactor emergency shutdown. Figure 2 shows flow through the system as the accident progresses. These results are the outcome of a PhD thesis defended in University Politehnica of Bucharest by Negut Gheorghe.

Another important step in the adaptation of the SCDAP/RELAP5 code for the CANDU NPP severe accident analysis was modifying it for analysis of the early phases of a loss of coolant accident with loss of the emergency core cooling system LOCA/LOECC.

In this accident the coolant loss leads to the fuel pins heat up and internal structure loss for the horizontal fuel bundle. The initial vertical typical PWR fuel bundle is losing its structural integrity in a completely different way than the CANDU bundle. Due to horizontal stacking of the fuel pins and the fuel bundle end plates, the pins sag and the bundles collapse to the bottom of the fuel channel. The collapse of the fuel bundle, added to the lack of coolant can lead to poor cooling for some pins and better for others due to steam flow rerouting, as shown in Figure 3 [2]. This configuration for the fuel bundle was used in the SCDAP/RELAP5 modified model.

This configuration was calculated by a new restart file at the moment that a temperature reaches 1400 K in the fuel bundle and the cladding loses its mechanical resistance. Beyond this point, a bypass flow channel was introduced to account for the modified conditions surrounding the fuel pins. After the bundle collapses, the bypass channel occupies around 48% of the initial heat transfer surface of the channel; meanwhile, the total channel surface for the fuel pins was reduced to only 52% worsening the heat transfer to the coolant.

After this result came the need to modify the way that material relocates during the melting of the fuel pins. In the original PWR model, molten droplets move axially along

the fuel pin due to the vertical position of the bundle. The horizontal position of the CANDU bundle means that molten droplets move along the pins circumference and pool at the bottom of the flow channel.

A new horizontal geometry model was created by the collaboration of a group from University Politehnica of Bucharest and the Nuclear Research Institute of Pitesti and contained modifications of the LIQSOL module included in the original SCDAP/RELAP5 code. A presumption for the new module is that there is material relocation between pins that implies a new possibility: pins that are not melting and have an intact oxide layer or have solidified drops on them can receive molten material from the melting pins surrounding them. A fraction of a molten droplet out of a fuel pin can come into contact with another pin or even the pressure tube. After that the droplets cannot change their axial position. They only can move along the pin circumference.

Figures 4 and 5 show the intact fuel bundle with different power rated pins and the collapsed bundle with the different coolant availability and cooling conditions.

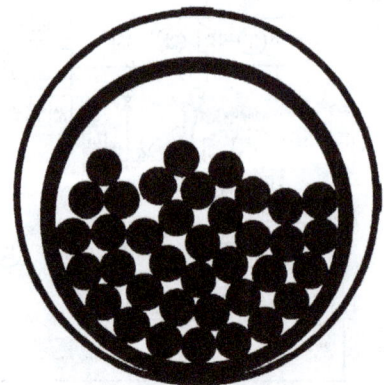

Fig. 3. Flow rerouting due to bundle collapse [3].

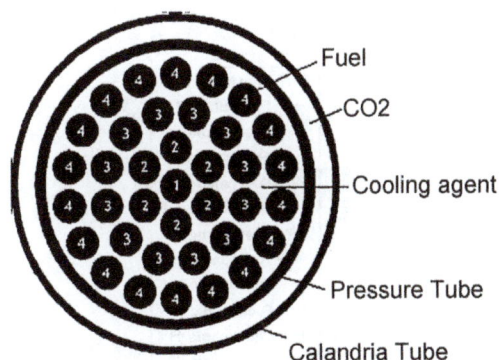

Fig. 4. Intact fuel bundle [4].

In Figure 4 [4] we can observe the four different types of pins used to model the CANDU fuel bundle. The intact bundle has four types of pins according to the different power rating of the pins. Thermal neutrons have the moderator as their source, thus the most outside ring of fuel pins receiving the highest neutron flux and producing more power than the inner ring pins. The pins in Figure 5 [4] are numbered according to the different cooling conditions. Due to coolant depravations, the pins at the bottom of the fuel channel receive less steam than the ones at the top of the collapsed bundle due to thermal stacking of the fluid left in the fuel channel.

Model used implies that the droplets are released at the point that the temperature reaches the point of initial oxide layer breakage. This means that the melted material is available to relocate at the set temperature independent from the oxidation status. This temperature may be even between 2098 and 2125 K (or the beta Zr melting margin). The temperature at which the droplets continue to relocate is set 50 degrees over the temperature at which the intact shield starts to flow in order to avoid mixing the droplets from the intact shield with the ones melted after solidifying, although the model permits the existence of both relocation pathways. The physical motive is the increase of melting temperature for the droplets compared to the intact shield due to hydrogen addition.

Modifying the LIQSOL module was the work of M. Mladin as part of his PhD thesis, the results of which were published, some of them being listed in the references section for this paper [2,3].

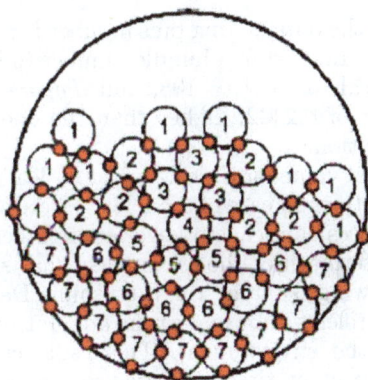

Fig. 5. Collapsed fuel bundle [4].

Although the SCDAP/RELAP5 code is suitable for CANDU type severe accident analysis, modifications to the code in order to better use it on this type of reactor were performed only in Romania by M. Mladin.

3 Fuel degradation analysis in a LOCA/ LOECC accident

This section analyses the parameter evolution on the maximum power channel for a CANDU 600 reactor during a LOCA type accident. The analysis implies the loss of moderator cooling (considered a heat sink during CANDU accident sequences) due to moderator pump failure. In addition, for the worst case scenario calculation, there is a loss of emergency core cooling system during the hole sequence [5].

The aim is to determine the extent of fuel degradation during the accident. The case study illustrated below presumes loss of coolant circulation through the pressure tube in a 100,000 seconds transient, the first 1000 seconds modeling a stationary, normal operation status. Coolant flow starts from 24 kg/s in normal operation, decreasing to 5 kg/s between the 1000 and the 1002 seconds and stabilized at 5 g/s during the whole transient.

At the start of the accident, the reactor is shut down, decay and oxidation heat being the sources for the fuel bundle heat-up and melt. The 2000 seconds mark the loss of moderator cooling.

The radioactive nuclei possibly released out of the containment depend on the amount of fuel bundles/pins destroyed during a transient. The worst case scenario is the one in which all of the radioactive nuclei inventory is released and assessing the release is closely linked to the amount of fuel bundles or pins damaged during the transient.

In the conditions listed above, the fuel bundles defects were evaluated by the SCDAP/RELAP5 code between 0 (undamaged pin) up to 1 (totally damaged pin). In the model fuel pins have different power ratings, the ones in the outside ring in the bundle receiving the higher rating and the central pin the lowest, so damage occurs in the outside ring pins rather that in the central pin.

Figure 6 shows damage progression for the outside ring pins and as it is depicted six central fuel bundles suffer total damage during the transient, the other six being only slightly damaged.

The fuel pins on the internal fuel bundle rings were almost undamaged due to low power operation, so we can conclude that the release of radioactive nuclei is mainly due to the outside ring pins.

4 Radioactive nuclei evaluation

In June 2014, the Canadian Nuclear Safety Commission released a draft report, "Study of Consequences of a Hypothetical Severe Nuclear Accident and Effectiveness of Mitigation Measures". This study lists the fractions for equilibrium core inventory of radionuclide contained in the fuel released to the environment as can be observed below.

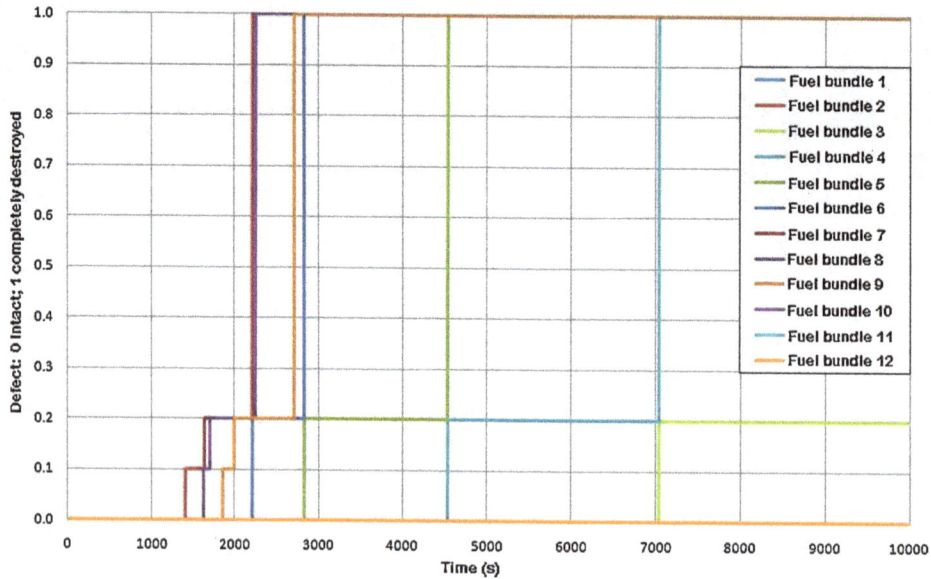

Fig. 6. Fuel degradation for the outside fuel pin ring in the CANDU bundle [5].

These release fractions can be used to assess the environmental impact of a severe accident and, furthermore, to develop the mitigation actions to be carried by the authorities after a postulated event. The amount of radioactive material released is thus important to the plans and costs for the mitigation measures.

In the previous section we have shown results that indicate the damage of the outer ring fuel pins for six fuel bundles in a CANDU channel during a LOCA/LOECC accident. The outer ring contains 18 fuel pins that are supposed to be totally damaged during the transient.

For a worst case scenario assumption, we are proposing to modify the source term used in environmental impact assessment in order to accommodate for a larger release as the one considered in Table 1.

The evaluation takes into account the number of bundles affected by the accident, the pin rings that are the most affected by the accident, and the total release of the inventory present in the damaged fuel pins, regardless of their position in the fuel channel or in the fuel bundle.

Table 1. Fission product groupings of the generic large release.

Fission product group	Release fraction
Noble gases	0.412
Halogens	0.00152
Alkali metals	0.00152
Alkaline earths	2.3×10^{-8}
Refractory metals	0.000253
Lanthanides	8.51×10^{-9}
Actinides	5.16×10^{-8}
Barium	1.68×10^{-7}

Source term CNSC-Study [6].

Given the number of bundles affected by the accident and the number of pins from each bundle, we can estimate a different release fraction, unified for all the groupings.

$$Rf = Ap/Tnp, \qquad (1)$$

where Rf is the release fraction, Ap is the number of affected pins in the channel, Tnp is the total number of pins in the channel.

We can assume that the entire inventory of the damaged pins is released, in the worst case scenario due to transportation in the containment and unforeseen events that lead to containment failure (and the Fukushima event gives the means for this assumption).

Thus:

$$Tnp = Pn \times Nfb, \qquad (2)$$

where Pn is the total pins number per bundle (37 for CANDU 600), and Nfb the number of fuel bundles in a channel (12 for CANDU 600).

And:

$$Ap = Orp \times Dfn, \qquad (3)$$

where Orp is the outside ring pins number (18 in this case), and Dfn the damaged fuel bundle number (6 in this case).

We can calculate $Ap = 108$, and $Tnp = 444$, giving a release factor of 0.2432, higher than the one used in the CNSC evaluation.

This higher value for the release factor leads to different mitigation actions in case of a nuclear severe accident, and proper measures can lead to lower environmental impact.

In the aftermath of the earthquake that shook Japan, and the following tsunami, the Fukushima Daiichi nuclear power plant released an important amount of radioactive material in the environment. This radioactive material must be, at present time, collected and accounted for in order to reduce the consequences of the accident and to restore the normal living conditions in the area affected by the accident.

The initial state restoration activities after a severe nuclear accident are directly linked to the released amount of radioactive material, and proper planning for these activities is of utmost importance.

After the Chernobyl disaster, the clean-up operations were carried out by the military and all the cost was engulfed in the communist economic system, all military or civilian personnel using state-provided equipment and all costs being neglected.

In our days, with the majority of nuclear power plant located in non-communist countries, the mitigation cost for a severe nuclear power plant accident must be provided for either by the utility owner or by the authority through special funds, and the necessary funds ready at any time.

These funds must be very well spent in order to optimize the cost/effects ratio. A well-evaluated released quantity of radioactive nuclei leads to a well thought plan of action, depending of course on weather conditions. A larger release needs a bigger effort and that effort may be well coordinated if the quantities are well evaluated from the start.

For example, a lower quantity released, along with the improper evaluation leads to over-evaluating the personnel and equipment needed to mitigate the consequences, and mobilizing a large number of unnecessary people and equipment is not cost effective.

Another example is the case in which the evaluation is below the released quantities, and the under-evaluation leads to poor mitigation results.

This means that a proper evaluation of the source term for the environmental release of radio nuclides is very important for the costs and people and material resources mobilized in the mitigation actions.

5 Conclusions

European Union (EU) through its legislation and directives set an action plan up to year 2050 (SET Plan[1]) which describes the pathway the energy sector has to follow in order to be at the forefront of EU citizen's needs. In order to reduce EU dependency on primary energy it is recommended to each member state to produce and follow a valid strategy with respect to energy. This strategy should be based on security of supply, on ensuring a comfortable energy mix and to reduce the losses. EU leaves the option to each member state to decide the best suitable energy mix for respective country. Along with renewable sources of energy, the nuclear industry is perceived as a major player in greenhouse gas reduction and in safe and secure provider for power with predictable and affordable prices.

Safe operation of all types of NPPs has been observed all over Europe. One key element on defining safety for a nuclear power plant is analyzing its functional margins, its safe limits for operation. R&D has played a major role and brought a huge contribution to this situation.

SCDAP/RELAP5 code can be used for severe accident analysis for the CANDU 6 type reactor due to heavy water properties library and horizontal flow calculations capabilities. Starting from these premises, the code was adapted and used for CANDU type NPP severe accident analysis and it can be used to assess the fuel bundle and fuel pin damage in a LOCA/LOECC accident with loss of moderator cooling.

The code was used to model different systems for the CANDU type reactor and power plant in other countries other than Romania but it has been modified to take into account the horizontal geometry of the fuel bundles during severe accident analysis.

The results of the analysis done using SCDAP/RELAP5 are used to estimate the release factor for fission products present in the fuel as a worst case scenario evaluation. These factors are higher than the ones calculated and used in source term evaluation in a recently release study performed by the CNSC study.

This evaluation is meant as a starting point towards a better assessment of the radioactive contamination following a severe nuclear power plant accident.

Better assessment of release factors lead to better preparedness of the authorities and of the involved institutions, utility owner or government, better planning and thus in better use of funds and human resources in a severe nuclear power plant accident event.

We are proposing a new evaluation of the source term for the radioactive nuclei release during a severe accident, evaluation that may lead to a better preparedness and more flexible planning from the authority and the utility owners that means better fund, material and human resources usage. Knowing that an under-evaluation of the radioactive release leads to inefficient mitigation actions under low resources allocated to the mitigation teams and an over-evaluation leads sometimes to waste of resources by the mitigating teams, the proper evaluation means an economy for resources, both human and material.

The work has been partly funded by the Sectoral Operational Programme Human Resources Development 2007-2013 of the Ministry of European Funds through the Financial Agreement POSDRU/159/1.5/S/132395 and partly by the Sectoral Operational Programme Human Resources Development 2007-2013 of the Ministry of European Funds through the Financial Agreement POSDRU/159/1.5/S/134398.

References

1. G. Negut, Contributii la studiul dinamicii proceselor termohidraulice tranzitorii din reactoarele centralei nuclearoelectrice de la Cernavodă, PhD thesis, Universitatea Politehnica Bucuresti, Bucharest, 2006

2. O. Akalin, C. Blahnik, B. Phan, F. Rance, Fuel temperature escalation in severe accidents, in *Canadian Nuclear Society 6th Annual CNS Conference Ottawa, Canada, June 3–4, 1985* (1985)

3. M. Mladin, D. Dupleac, I. Prisecaru, SCDAP/RELAP5 application to CANDU6 fuel channel analysis under postulated LLOCA/LOECC conditions, Nucl. Eng. Design **239**, 353 (2008)

---

[1]http://ec.europa.eu/energy/en/topics/technology-and-innovation/strategic-energy-technology-plan, as of 22.2.2015.

4. M. Mladin, D. Dupleac, I. Prisecaru, Modifications in SCDAP code for early phase degradation in a CANDU fuel channel, Ann. Nucl. Energy **36**, 634 (2009)

5. A. Budu, Contributii la studiul accidentelor din centralele nuclearoelectrice CANDU, PhD thesis, Universitatea Politehnica Bucuresti, Bucharest, 2011

6. Canadian Nuclear Safety Commission, Study of consequences of a hypothetical severe nuclear accident and effectiveness of mitigation measures - Draft report, June 2014, e-Doc 4160563 (WORD), e-Doc 4449079 (PDF), p. 25, http://www.opg.com/about/safety/nuclear-safety/Documents/CNSC_Study.pdf

9

Sustainability of thorium-uranium in pebble-bed fluoride salt-cooled high temperature reactor

Guifeng Zhu[1,3], Yang Zou[1,2], and Hongjie Xu[1,2*]

[1] Shanghai Institute of Applied Physics, Chinese Academy of Sciences, Jialuo Road 2019#, Jiading District, 201800 Shanghai, P.R. China
[2] Key Laboratory of Nuclear Radiation and Nuclear Energy Technology, Chinese Academy of Sciences, Jialuo Road 2019#, Jiading District, Shanghai, P.R. China
[3] University of Chinese Academy of Sciences, No. 19A Yuquan Road, Beijing, P.R. China

Abstract. Sustainability of thorium fuel in a Pebble-Bed Fluoride salt-cooled High temperature Reactor (PB-FHR) is investigated to find the feasible region of high discharge burnup and negative Flibe ($2LiF-BeF_2$) salt Temperature Reactivity Coefficient (TRC). Dispersion fuel or pellet fuel with SiC cladding and SiC matrix is used to replace the tristructural-isotropic (TRISO) coated particle system for increasing fuel loading and decreasing excessive moderation. To analyze the neutronic characteristics, an equilibrium calculation method of thorium fuel self-sustainability is developed. We have compared two refueling schemes (mixing flow pattern and directional flow pattern) and two kinds of reflector materials (SiC and graphite). This method found that the feasible region of breeding and negative Flibe TRC is between 20 vol% and 62 vol% fuel loading in the fuel. A discharge burnup could be achieved up to about 200 MWd/kgHM. The case with directional flow pattern and SiC reflector showed superior burnup characteristics but the worst radial power peak factor, while the case with mixing flow pattern and SiC reflector, which was the best tradeoff between discharge burnup and radial power peak factor, could provide burnup of 140 MWd/kgHM and about 1.4 radial power peak factor with 50 vol% dispersion fuel. In addition, Flibe salt displays good neutron properties as a coolant of quasi-fast reactors due to the strong $^9Be(n,2n)$ reaction and low neutron absorption of ^6Li (even at 1000 ppm) in fast spectrum. Preliminary thermal hydraulic calculation shows good safety margin. The greatest challenge of this reactor may be the decades irradiation time of the pebble fuel.

1 Introduction

The sustainability of nuclear energy resources has aroused great interest and attention since the Generation IV International Forum. A reactor system with breeding capability is very essential to extend the sustainability of nuclear fuel resources. Liquid metal-cooled fast reactor is the preferred choice to achieve a high breeding ratio. However, it has some obstacles due to safety concerns associated with a positive void reactivity.

Thorium seems an attractive option of nuclear resources mainly due to its abundance, the opportunity to reduce the need for enrichment in the fuel cycle, the high conversion ratios (to ^{233}U) achievable in a thermal neutron spectrum, and also due to other neutron and thermal physical properties studied early in the development of nuclear power [1]. Due to the high effective number of neutrons for each ^{233}U fission in a thermal and epithermal neutron spectrum, thorium breeding is feasible in most existing and prospective reactor designs (including LWRs [2,3], HWRs [4–8], HTGRs [9] and molten salt reactors [10–12]), and it can provide the negative void reactivity coefficient due to the softer neutron spectrum than that of fast reactor. However, the thorium breeding gain in these reactors is far lower than fast reactor's. From an economical view, it is better to maintain fissile self-sustainability and to improve burnup for decreasing reprocessing mass per electricity.

This work focuses on sustainability of thorium in a Pebble-Bed Fluoride salt-cooled High temperature Reactor [13–16] (PB-FHR), to find its feasible region of high burnup and negative void reactivity coefficient. Expectant advantages of Flibe salt ($2LiF-BeF_2$) as breeder reactor coolant [17] are that heat-carrying capacity and boiling point are both high; weak neutron slowing-down power will allow more coolant volume ratio than HWRs; and it may provide more negative temperature reactivity coefficient due to strong (n,2n) reaction of ^9Be in the fast spectrum.

* e-mail: xuhongjie@sinap.ac.cn

Furthermore, PB-FHR is neutron saving with refueling online, and [233]Pa has the chance to decay away when thorium fuel is periodically removed from the core. However, one disadvantage is that Flibe salt in a flowing pebble bed will occupy about 40% volume of core, which enhances the moderation of Flibe and decreases the fuel inventory, as a result, a critical design should be required in fuel system of breeder PB-FHR.

The system of tristructural-isotropic (TRISO) coated fuel particles embedded in massive graphite matrix in thermal spectrum PB-FHRs is not adaptable to breeder reactor concepts due to its low fissile loading, the high irradiation swelling behavior of graphite in a quasi-fast spectrum, and the excessive moderation due to the large graphite/fuel ratio. Two kinds of fuel system [18] are developed for gas-cooled fast reactor (GFR) in order to increase fuel loading and improve radiation resistivity, which could be applied to PB-FHRs. One is pin-type GFR fuel with refractory cladding material (Fig. 1a); another one is dispersion fuel (or composite fuel or sphere-pac fuel) consisting of a distribution of discrete fuel particles embedded in a non-fuel matrix (Fig. 1b). Usually, fuel loading in dispersion fuel can reach 50 vol%, and in pin-type fuel is beyond 75 vol% [19]. Buffer layers are both designed in pin-type fuel and dispersion fuel to provide volume for

fission gas and provide volume for fuel particle swelling. SiC is a good candidate cladding material or matrix material because of the good irradiation swelling behavior of SiC [20–25], the large irradiation behavior database, and the experience in use of SiC as a component in TRISO fuel. In addition, SiC has excellent oxidation resistance due to rapid formation of a dense, adherent SiO_2 surface scale on exposure to air at elevated temperature, which offers protection from further oxidation. SiC is effective for retention of the solid fission products [26], but the migration of Ag in polycrystalline SiC can occur. Middle metallic liner designed in pin-type fuel and SiC matrix in dispersion fuel ensure such fission product confinement within the fuel system.

For a preliminary concept design, the fuel system of thorium fuel self-sustainability in PB-FHRs is considered as dispersed fuel particle filled in a sphere cladding. In order to simplify neutron calculation, an equivalent fuel system with only thorium fuel region and SiC region (Fig. 1c) is used because the weak moderation of SiC makes the space self-shielding effect insignificant. Oxy-carbide thorium fuel is chosen in this work due to stable fission products bound by oxygen, low internal pressure for low product of free oxygen and compatibility with SiC material.

In order to simplify refueling scheme, homogeneous system with one kind of $^{233}U/^{232}Th$ pebble is carried out, in which mixing flow pattern and directional flow pattern are both performed. For neutronic analysis of thorium fuel self-sustainability, neutron spectrum is adjusted by fuel loading variable V_f, which is defined as fuel volume dividing the volume of fuel system. In addition, graphite reflector is compared with SiC reflector to evaluate the moderation effect of reflector. Reactor model and refueling scheme are introduced in Section 2. Equilibrium calculation method of fissile self-sustainability is represented in Section 2.2. In Section 3, we show the results and discussions, in which achievable burnup of thorium fuel self-sustainability, temperature reactivity coefficient of Flibe, radial power distribution and preliminary thermal hydraulics are analyzed. Conclusions are drawn in Section 4.

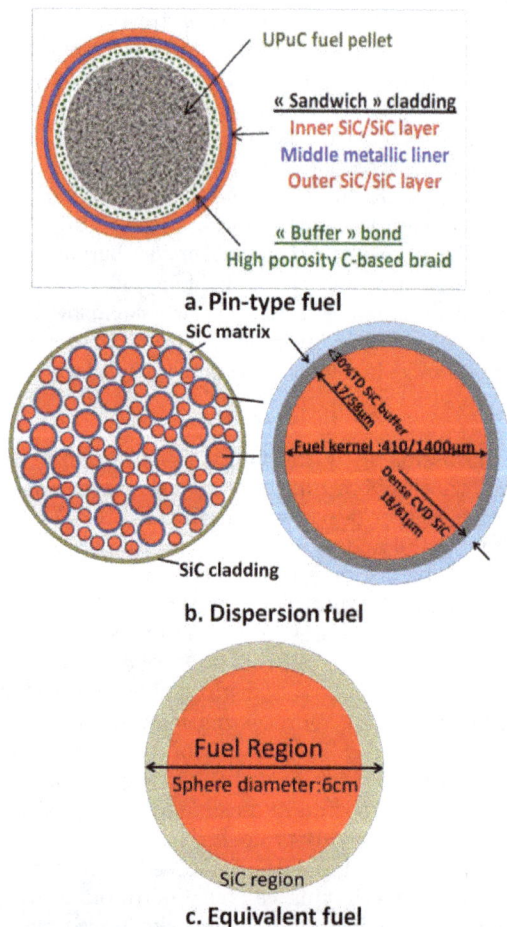

Fig. 1. Fuel system: a. pin-type fuel with SiC/SiC cladding; b. dispersion fuel filled with two kinds of coated fuel particle; c. equivalent fuel used for neutron calculation in this work.

2 Model and calculation method

2.1 Reactor model and refueling scheme

Reactor model is simplified to a cylinder (Fig. 2). The core is divided into five radial annular flow channels with the same cross-sectional area. Each channel is uniformly segmented into seven axial layers. In all, 35 burnup regions are used for neutronic calculation. Graphite or SiC is chosen as the material of both the axial and the radial reflector. Vacuum boundary condition is assumed outside the reflector. The layout of control rods and the B_4C shielding layer are outside the scope of this article, and are omitted in the equilibrium calculation. Dimensions of reactor are shown in Table 1. The diameter of pebble is chosen as 6 cm, but it may be changed for thermal hydraulic considerations. $^{233}U/^{232}Th$ pebbles are loaded in the core with a volumetric filling fraction of 0.6.

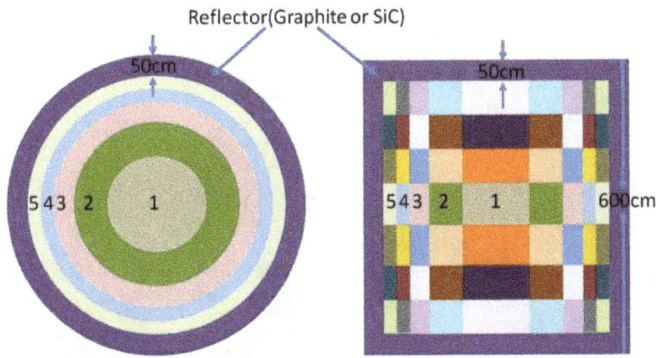

Fig. 2. Schematic view of the reactor geometry used during the neutronics calculation. On the left is the vertical view of the middle layer of the right horizontal view. Arabic numbers represent radial channels.

Table 2. Material properties of reactor.

Material	Temperature (K)	Density (g/cm^3)
$Th_2CO_3/^{233}U_2CO_3$	1050	9.86
SiC Matrix and cladding	1000	3.2
Flibe salt	920	1.96
Reflector (Graphite/SiC)	880	1.74/3.2

Table 3. Core cases.

Case 1	Mixing Flow, Graphite Reflector
Case 2	Directional Flow, Graphite Reflector
Case 3	Mixing Flow, SiC Reflector
Case 4	Directional Flow, SiC Reflector

Vf in pebble is varied by changing the packing factor. Usually, packing factor for the binary size particles is higher than unary size particle, in this paper, the packing factor in pin filling model is 0.73 calculated by equation from literature [27] (in sphere filling model, it will be lower than 0.73), fuel loading in particle could reach 78%, thus, the limiting Vf is $0.73 \times 0.78 = 0.57$. However, for neutronic analysis, Vf beyond 0.57 is also performed. Material properties of reactor are listed in Table 2. ^6Li in Flibe salt is assumed to be 22 ppm referred to literature [28], while the equilibrium concentration of ^6Li will be analyzed in the following section. Fresh $^{233}U/^{232}Th$ ratio (UTR) is automatically adjusted in the equilibrium calculation for fissile self-sustainability.

Multiple-passage-through-the-core (ten passage chosen in this work) with two kinds of flow patterns is simulated to flatten the axial power distribution. The mixing flow pattern is defined as that where pebbles, unloaded from each channel and not reached the limit of discharge burnup, are mixed with a batch of fresh pebbles and then are randomly recycled into five channels. The directional flow pattern is defined as that where a batch of fresh pebbles is

recycled 10 times in channel 1, and then 10 times in channel 2, and so forth until discharged from channel 5. It is noteworthy that the radial position of pebbles in the core is determined by their inlet position [29], which implies that the directional flow could be easily achieved by only setting four baffles in the inlet. The out-pile residence time of pebble is supposed to be equal to in-pile residence time.

For the reprocessing of discharge fuel, only ^{233}U and ^{232}Th are extracted, while other uranium isotopes such as ^{234}U, ^{235}U, ^{236}U, are omitted in the calculation due to the long equilibrium cycle. ^{233}Pa from discharge fuel is regarded as ^{233}U, and will be returned to core. In the general model, average power density is $10 MW/m^3$ (corresponding to 980 MW total power), which will be changed in the analysis of ^{233}Pa effect. According to the refueling scheme and reflector material, 4 cases are analyzed, as shown in Table 3.

2.2 Equilibrium calculation method of thorium fuel self-sustainability

Equilibrium calculation of thorium fuel self-sustainability involves searching the fuel feed rate (or in-pile residence time) and UTR to keep k_{eff} convergent to 1 and to keep the ^{233}U fed into the core equivalent to ^{233}U from discharge fuel under different energy spectra. Convergence methods are analyzed below.

Ignoring the chain of ^{233}Pa and ^{233}Th, the evolution equations of ^{232}Th and ^{233}U can be shown as:

$$\frac{dN_{Th}}{dt} = -A_{Th}N_{Th}, \qquad (1)$$

$$\frac{dN_{U3}}{dt} = -A_{U3}N_{U3} + A_{Th}N_{Th}. \qquad (2)$$

N_{Th} is the concentration of ^{232}Th, and A_{Th} is a function of fluxes and one-group capture cross-sections of ^{232}Th in

Table 1. Dimensions of reactor.

Parameter	Dimension (cm)
Outer radius of Channel 1	107.33
Outer radius of Channel 2	151.79
Outer radius of Channel 3	185.9
Outer radius of Channel 4	214.66
Outer radius of Channel 5	240
Height of active core	500
Thickness of axial and radial reflector	50
Diameter of equivalent fuel pebble	6
Kernel diameters of fuel particle	0.0410/0.1400
Buffer thicknesses of fuel particle	0.0017/0.0058
SiC thicknesses of fuel particle	0.0018/0.0061

different regions. N_{U3} is the concentration of ^{233}U, and A_{U3} is a function as fluxes and one-group absorption cross-sections of ^{233}U in different regions. After in-pile residence time T, the concentration of ^{233}U can be solved as follows:

$$N_{U3}(T) = \left(N_{U30} - \frac{\overline{A}_{Th}\cdot N_{Th0}}{\overline{A}_{U3} - \overline{A}_{Th}}\right)\exp(-\overline{A}_{U3}T)$$
$$+ \frac{\overline{A}_{Th}\cdot N_{Th0}}{\overline{A}_{U3} - \overline{A}_{Th}}\exp(-\overline{A}_{Th}T). \quad (3)$$

N_{U30} is the fresh concentration of ^{233}U, and N_{Th0} is the fresh concentration of ^{232}Th. $\overline{A}$ is time-averaged A. For fissile self-sustainability, $N_{U3}(T) = N_{U30}$. It can be deduced that:

$$UTR = \frac{N_{U30}}{N_{Th0}} = \frac{\overline{A}_{Th}}{\overline{A}_{U3} - \overline{A}_{Th}}\left(\exp(-\overline{A}_{Th}T) - \exp(-\overline{A}_{U3}T)\right)/\left(1 - \exp(-\overline{A}_{U3}T)\right). \quad (4)$$

UTR always can be determined by in-pile residence time under specific A which is affected by neutron energy spectrum and can be adjusted by Vf.

In addition, for simplified analysis, an equation can be established to connect k_{eff} with T for fissile self-sustainability.

$$k_{eff} = \frac{1}{2/\eta + L + Abs_{fp}\cdot T}, \quad (5)$$

η is the effective number of ^{233}U fission neutrons, usually about 2.25 in epithermal spectrum. L is the sum of neutron absorption rate of structure material and leakage rate of core. Abs_{fp} is the equivalent capture absorption rate of fission products and transuranic elements. For the differential equation (5),

$$\frac{dk_{eff}}{k_{eff}} = \frac{-Abs_{fp}\cdot T}{2 + L + Abs_{fp}\cdot T}\frac{dT}{T}$$
$$= \frac{Abs_{fp}\cdot T}{2 + L + Abs_{fp}\cdot T}\frac{dV}{V} \left(\text{for } \frac{dT}{T} = -\frac{dV}{V}\right). \quad (6)$$

V is feed rate of fresh fuel. Equation (6) can be changed into:

$$\frac{dV}{V} = \left(1 + \frac{2 + L}{Abs_{fp}\cdot T}\right)\frac{dk_{eff}}{k_{eff}}. \quad (7)$$

Supposing L is equal to 2%, and k_{eff} is 1, $Abs_{fp}\cdot T$ can be obtained from equation (5). Equation (7) is changed into:

$$\frac{dV}{V} \approx 10\frac{dk_{eff}}{k_{eff}}. \quad (8)$$

Equation (8) describes a positive correlation between feed rate of fresh fuel and k_{eff}, and is used to modify feed rate of fresh fuel with previous k_{eff}. Constant 10 in equation (8) does not affect the accuracy but determines the rate of convergence.

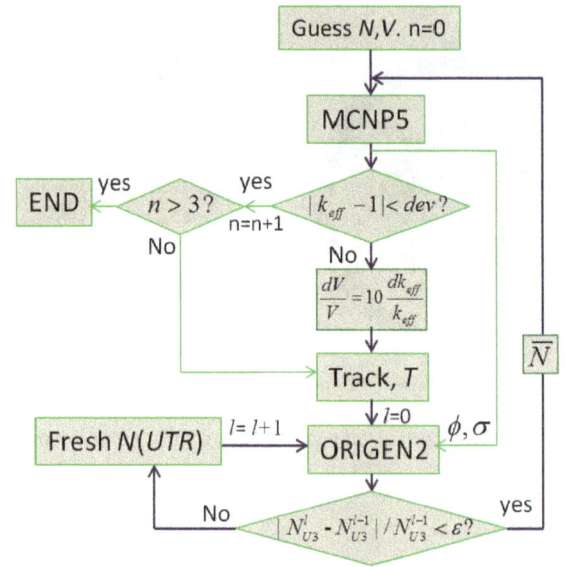

Fig. 3. Flow chart of PBRE with thorium self-sustainability module.

Therefore, two loops are necessary for equilibrium calculation of thorium fuel self-sustainability. The outer loop modifies the feed rate of fresh fuel or in-pile residence time to make k_{eff} convergent, and the inner loop changes the UTR for fissile self-sustainability. It is notable that the neutron transportation calculation is only performed in the outer loop, which can obviously save computing time.

Equilibrium calculation of thorium fuel self-sustainability has been achieved in PBRE code [30], which is accurately verified by VSOP [31] code with the HTR-10 model. PBRE is an equilibrium state searching code directly skipping the initial state and intermediate state. Method of PBRE is similar to literature [15,32,33]. The flow chart of PBRE with thorium self-sustainability is depicted in Figure 3. Guessing an equilibrium concentration of nuclides, MCNP code calculates equilibrium fluxes and one-group cross-sections of different regions. With the refueling scheme, equilibrium residence time in each region and pebble tracks are determined. Therefore, ORIGEN2 can give average concentrations in different regions and discharge concentrations. By modifying the UTR, fissile self-sustainability can be realized. Iteratively, average concentrations return to MCNP code until the k_{eff} and concentrations are convergent. If the outer loop is diverging, it means that there does not exist the condition to meet fissile self-sustainability and reactor criticality.

Pebble tracks not only give the calculating order of different regions, but also contain the decay calculation when pebbles are unloaded from each channel. In addition, for mixing flow pattern, a mixing treatment for the same batch from different channels is performed.

3 Results and discussions

Discharge burnup for thorium fuel self-sustainability and Temperature Reactivity Coefficient (TRC) of Flibe varied

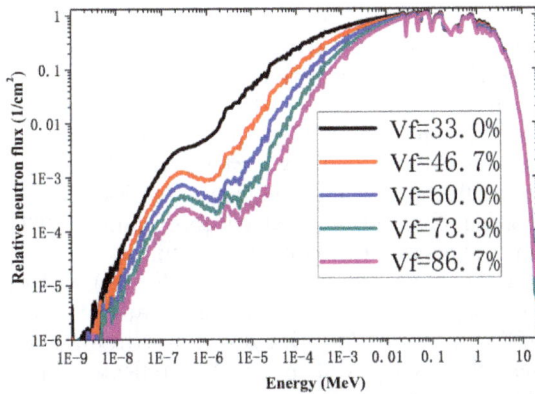

Fig. 4. Neutron spectrum dependent with Vf (case 1).

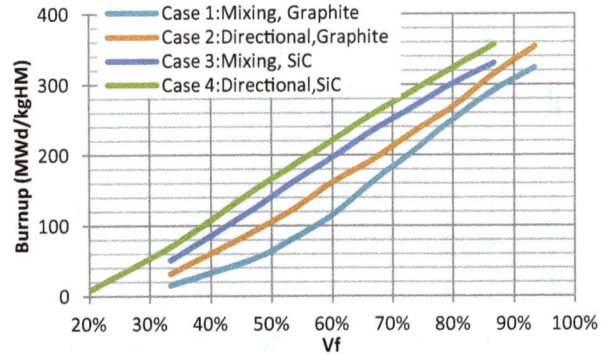

Fig. 5. Discharge burnup for thorium fuel self-sustainability in different cases as a function of Vf.

with Vf are investigated in this section. For a further analysis of neutronic performance, properties of ^{6}Li, ^{233}Pa effect and radial power distribution are also studied. Finally, preliminary thermal hydraulic is analyzed to give the boundaries of power density and Vf.

3.1 Neutron spectrum

Neutron spectrum provides a vital role for breeding or self-sustainability calculation. In the following analysis, Vf is a main parameter to adjust neutron spectrum. As shown in Figure 4, neutron spectrum varies from quasi-fast spectrum to fast spectrum with the increase of Vf. There are several dips around high energy range, corresponding to the main elastic scattering resonance of ^{7}Li and ^{19}F. In addition, there is a low peak at about 0.2 eV caused by thermal scattering of carbon from SiC and graphite reflector, but note that the peak is two or three orders of magnitude lower than the fast flux.

3.2 Discharge burnup for thorium fuel self-sustainability

In this section, the aim is to find the feasible region of thorium fuel self-sustainability and further to investigate the burnup characteristic of thorium fuel self-sustainability.

Discharge burnup for thorium fuel self-sustainability with different cases is shown in Figure 5. Discharge burnup is a function of Vf. High Vf can linearly improve the discharge burnup. For Vf lower than 20%, the discharge burnup is near to zero, which implies that it may not be feasible to breed for thorium fuel in PB-FHRs when Vf is below 20%.

The mechanism of discharge burnup for thorium fuel self-sustainability variation with V_f, can be understood in terms of the one-group cross-section ratios of thorium-uranium and UTR (Fig. 6). The one-group absorption cross-section ratio of thorium to uranium (XS(Tha)/XS (U3a)) reflects the conversion capability of thorium. The one-group fission-absorption ratio of ^{233}U (XS(U3f)/XS (U3a)) reflects ^{233}U fuel burning efficiency. As shown in Figure 6, a hard neutron spectrum provides high XS(Tha)/

XS(U3a) and high XS(U3f)/XS(U3a), which show high conversion capability and high fuel burning efficiency. In the low Vf region, UTR increases mainly due to the increase of XS(Tha)/XS(U3a) to keep reactor criticality, while in high Vf region, higher fuel burning efficiency and lower neutron absorption cross-section of fission products will allow lower fresh UTR (Fig. 6). When XS(Tha)/XS(U3a) is higher than UTR, breeding of thorium is feasible and extra ^{233}U will be produced to improve the discharge burnup of thorium fuel. The evolution of ^{233}U in case 4 with 46.7% Vf is shown in Figure 7. Concentration of ^{233}U will

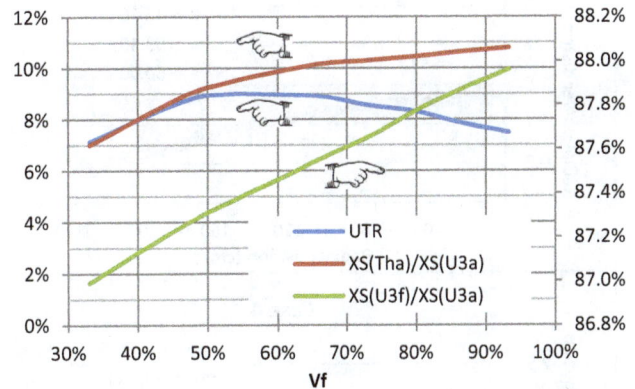

Fig. 6. One-group cross-section ratio and concentration ratio of thoirum-uranium (case 2).

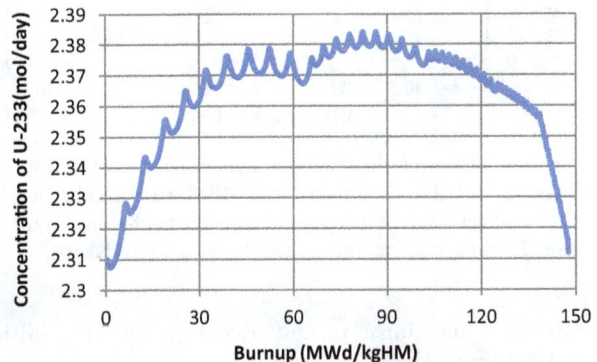

Fig. 7. Evolution of ^{233}U in case 4 with 46.7% Vf. One wave represents a single passage caused by the delay of ^{233}Pa decay.

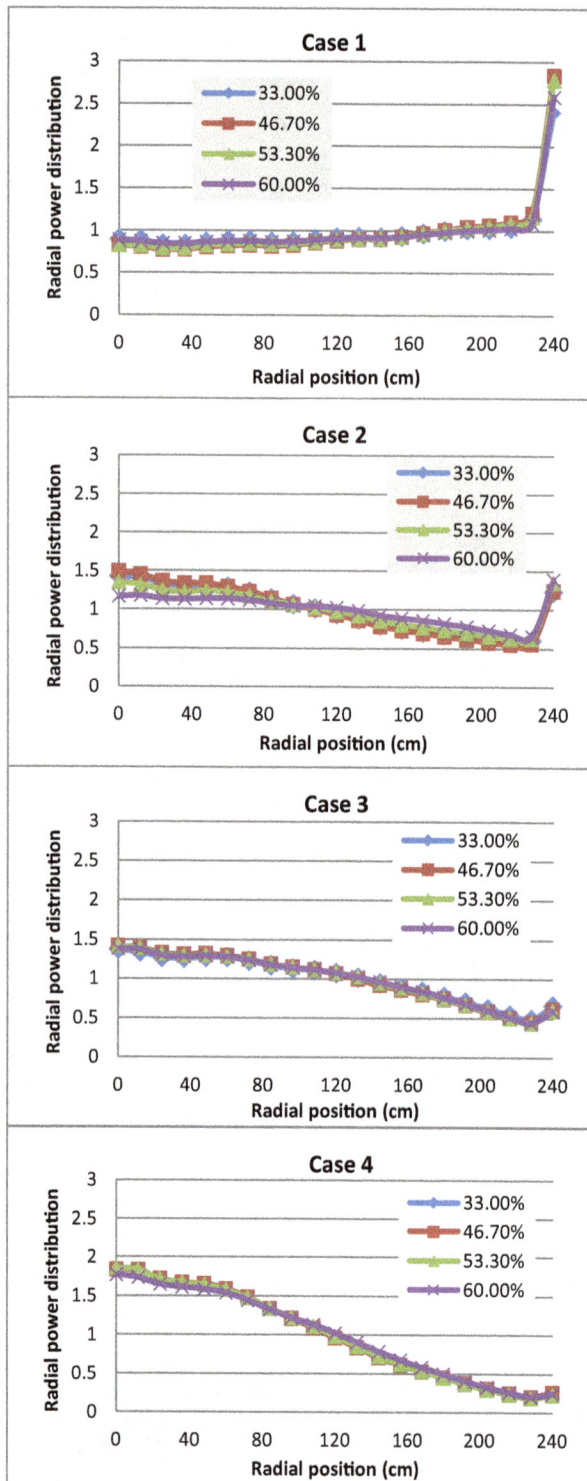

Fig. 8. Radial power distribution with different Vf and cases. Radial power distribution is tallied by TMESH card in column grid. In case 1 and case 2, power fraction in the outer channel increases because of strong slowing down effect by graphite reflector.

pattern, and a SiC reflector can provide higher discharge burnup than a graphite reflector. Reflector material has a more significant influence than flow pattern on discharge burnup in comparing case 2 and case 3. The neutron leakage rate and slowing down effect by reflector are the main contributions to differences among the four cases. In directional flow pattern, radial power fraction concentrates in the inner channel (Fig. 8) because of more ^{233}U, low fission products and consequent high flux in inner channel, which will decrease neutron leakage (Fig. 9) and lead to weak slowing down effect by reflector, and vice versa in mixing flow pattern. Additionally, power fraction in outer channel will be enhanced by graphite reflector due to the large fission cross-section caused by strong slowing down effect, which will further increase the neutron leakage (Fig. 9).

For 50% Vf dispersion fuel, a discharge burnup of 63 MWd/kgHM, 103 MWd/kgHM, 140 MWd/kgHM and 165 MWd/kgHM can be achieved in case 1 to case 4, respectively. From a view of same discharge burnup, case 4 could provide smallest Vf to reduce the manufacturing difficulty of fuel system. However, radial power peak factor case 4 is about 1.8. Case 3 is the best tradeoff between discharge burnup and radial power peak factor (about 1.4), and mixing flow pattern is the simplest refueling scheme.

3.3 Thickness of reflector

To decrease the neutron leakage, thickness of reflector is analyzed. As shown in Figure 10, the thickness of graphite reflector has an apparent positive effect on the k-eff due to the strong slowing down power, which will lead to a lower breeding capacity or discharge burnup. However, neutron leakage rate almost does not vary with the thickness of graphite reflector (Fig. 11), which may be caused by offset between the enhanced power fraction in the outer channel and the enhanced reflectivity. As analyzed above, graphite reflector seems not suitable in this reactor. From Figures 10 and 11 , 50 cm thickness seems enough for SiC reflector to prevent neutron from escaping.

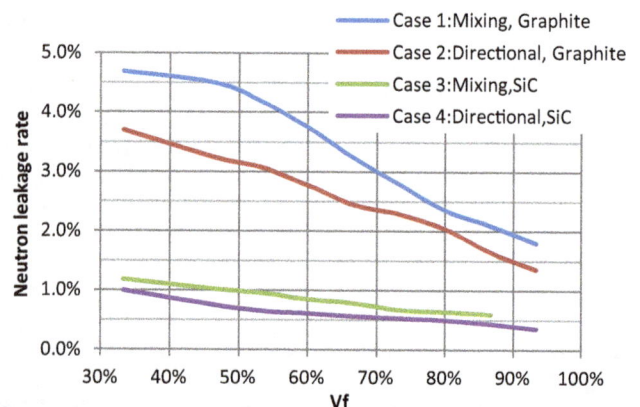

Fig. 9. Neutron leakage rate as a function of Vf with four cases. Neutron leakage rate is the escaped fraction outside the reflectors. Reflector with graphite material leads to more leakage than with SiC material.

increase at low burnup and decrease to the initial concentration at the high burnup.

As shown in Figure 5, a directional flow pattern can provide higher discharge burnup than a mixing flow

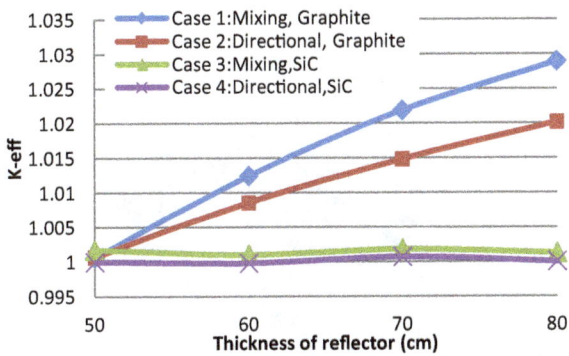

Fig. 10. k-eff with equilibrium concentrations as a function of reflector thickness and cases.

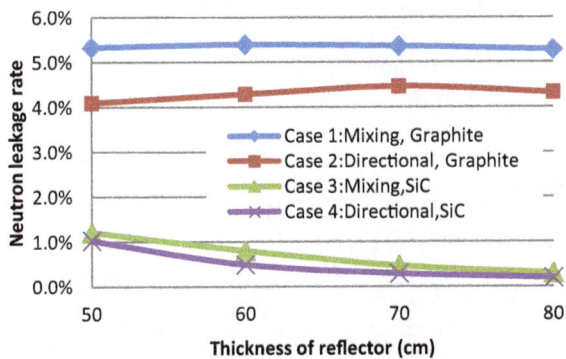

Fig. 11. Neutron leakage rate as a function of reflector thickness and cases. Neutron leakage rate is the sum of escaped fraction outside the reflector and neutron absorption rate of reflectors.

3.4 Flibe temperature reactivity coefficient

A negative Flibe TRC is necessary for PB-FHR nuclear safety. The calculated Flibe TRC is shown in Figure 12. As the increase of V_f, Flibe TRC increases. A positive Flibe TRC will happen when Vf is beyond 62%, which shows the margin of inherent safety.

Comparing different cases, case 1 and case 2 show the more negative Flibe TRC than case 3 and case 4, which could be explained by softer spectrum in case 1 and case 2.

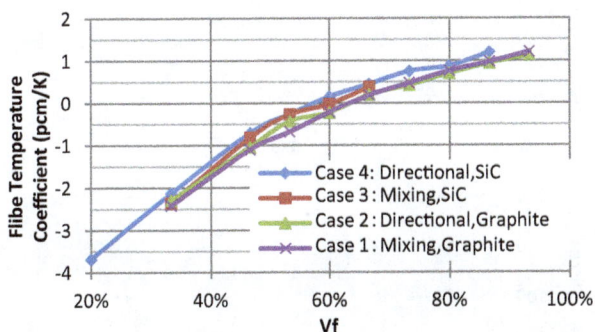

Fig. 12. Flibe salt temperature reactivity coefficient in different cases. Temperature changes from 920 K to 1050 K, and density of Flibe changes from 1.96 to 1.91 g/cm^3.

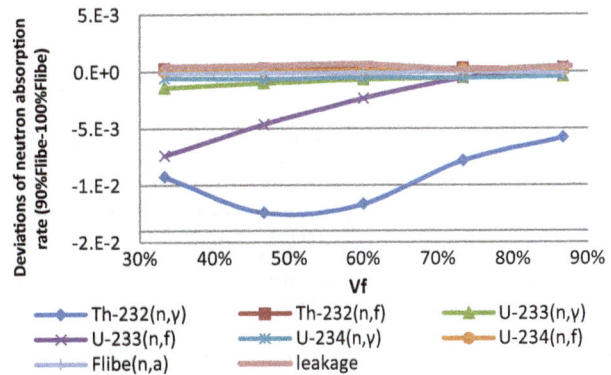

Fig. 13. Deviations of neutron absorption rate of main nuclides as a function of Vf (case 4).

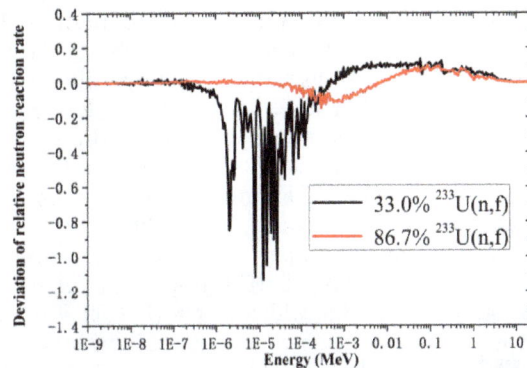

Fig. 14. Deviations of ^{233}U (n,f) reaction (90%Flibe–100%Flibe) as a function of neutron energy.

The mechanism of Flibe TRC is analyzed in models with 10% voided Flibe. The deviations of neutron absorption rate of main nuclides are shown in Figure 13. The ^{232}Th (n,γ) reaction and ^{233}U(n,f) reaction make great contributions to Flibe TRC. ^{232}Th(n,γ) makes Flibe TRC more positive, while ^{233}U(n,f) makes Flibe TRC more negative. With the increase of V_f, the deviation of ^{233}U(n,f) reaction approaches zero, as can be explained with reference to Figure 14. In 33.0% V_f, the ^{233}U(n,f) reaction in the resonance region has obvious shortfalls when slowing down by Flibe, while in 86.7% V_f, the deviation in resonance region vanishes. The same situation happens with ^{232}Th (n,γ). This indicates that some level of slowing down is required to keep a negative Flibe TRC and this could not be achieved for solid thorium fuel in a fast neutron spectrum.

Figure 13 also shows the contribution of neutron leakage, Flibe absorption rate and other reaction rates to Flibe TRC. Neutron leakage makes the Flibe TRC a little negative, while Flibe absorption rate and other reaction rates make the Flibe TRC a little positive.

3.5 Equilibrium concentration of Li-6 and production rate of H-3

The absorption rates of each nuclide in Flibe are shown in Figure 15. The ^{9}Be(n,2n) reaction rate is predominant,

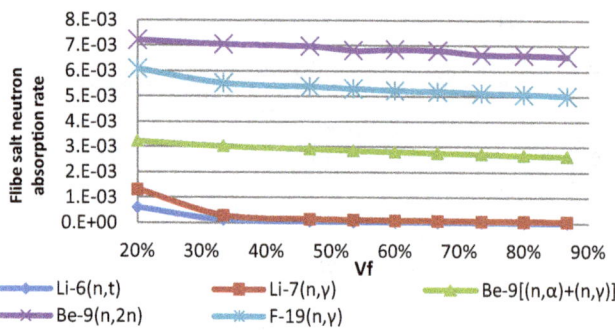

Fig. 15. Absorption rate of each nuclide in Flibe as a function of Vf (case 4).

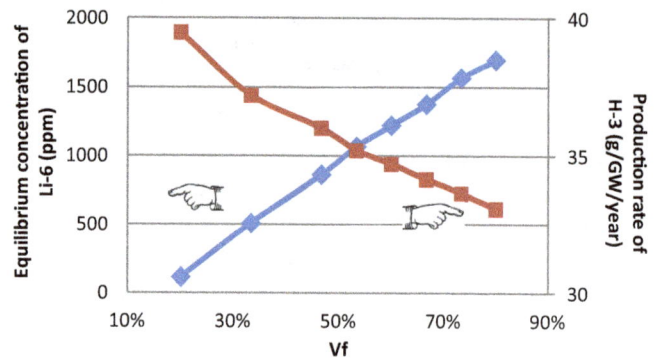

Fig. 17. Equilibrium concentration of Li-6 and production rate of H-3 as functions of Vf in case 4.

which could help reduce contribution to positive Flibe TRC. The $^{19}F(n,\gamma)$ reaction is apparent for several capture resonance peaks in fast spectrum. Notably, different from thermal spectrum, ^{6}Li and ^{7}Li show the low neutron absorption characteristics in a quasi-fast spectrum. The discharge burnup and Flibe TRC variations with concentration of ^{6}Li are shown in Figure 16. With the increase of ^{6}Li, discharge burnup decreases, while Flibe TRC does not change until beyond 3000 ppm. But it notes that discharge burnup only has a 6 MWd/kgHM drop when ^{6}Li increases from 22 ppm to 500 ppm, which shows that 99.95 at.% ^{7}Li at least is compatible for sustainability of thorium-uranium in PB-FHR. This indicates that the cost of Flibe in quasi-fast reactor can be sharply reduced by lower enrichment of ^{7}Li.

In fact, the equilibrium concentration of ^{6}Li in a quasi-fast spectrum is very much larger than in a thermal spectrum. This can be calculated by equation (9), by assuming that the concentration of ^{9}Be in the core is constant.

$$N_{Li-6} = \frac{(n,\alpha)\text{reaction rate of Be} - 9}{\text{absorption rate of Li} - 6(22\text{ ppm})}.22 \text{ ppm.} \quad (9)$$

As shown in Figure 17, the equilibrium of ^{6}Li increases as the increase of Vf due to the faster decline of one-group

absorption cross-section of ^{6}Li than that of ^{9}Be. 500 ppm of ^{6}Li can be achieved for 33% V_f, which implies that enriching the ^{7}Li to more than 99.95 at.% level for improving the discharge burnup is unnecessary.

The product rate of ^{3}H in equilibrium state can be estimated by equation (10). Number of ^{233}U fission neutron is assumed to be 2.5, fission energy of ^{233}U is assumed to be 200 MeV. The product rate of ^{3}H is equal to the (n,α) reaction rate of ^{9}Be. Figure 17 shows that the product rate of ^{3}H in equilibrium state decreases as the increase of V_f, which is in keeping with the (n,α) reaction rate of ^{9}Be shown in Figure 15. Since the (n,α) reaction rate of ^{9}Be in quasi-fast spectrum is low, the product rate of ^{3}H is only about 30–40 g/GW/year, which is not proportional to the concentration of ^{6}Li.

$$P_{\text{H}-3} = \frac{(n,\alpha)\text{reaction rate of Be} - 9}{200 \text{ MeV}}.2.5. \quad (10)$$

3.6 Effect of ^{233}Pa

In the conversion process of thorium, some of the ^{233}U will be lost by the irradiation of ^{233}Pa. This effect can be enhanced by a high neutron flux. As shown in Figure 18, discharge burnup has a 30 MWd/kgHM drop when power

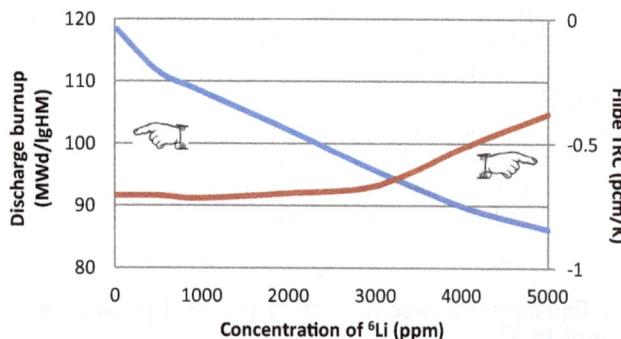

Fig. 16. Discharge burnup and Flibe temperature reactivity coefficient as functions of ^{6}Li concentration (30 MW/m^3 power density with 46.7% Vf in case 4).

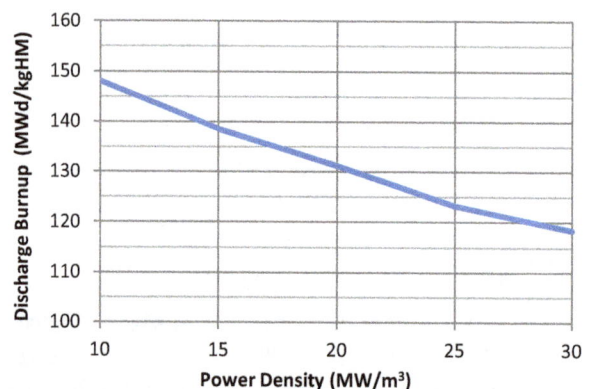

Fig. 18. Discharge burnup as a function of power density (46.7% Vf in case 4).

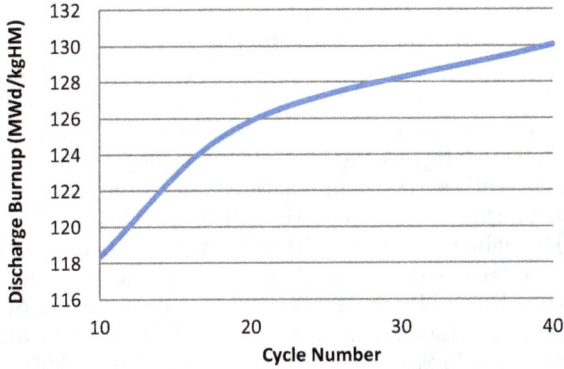

Fig. 19. Discharge burnup as a function of pebble cycle number in each channel. 30 MW/m^3 power densities with 46.7% Vf in case 4.

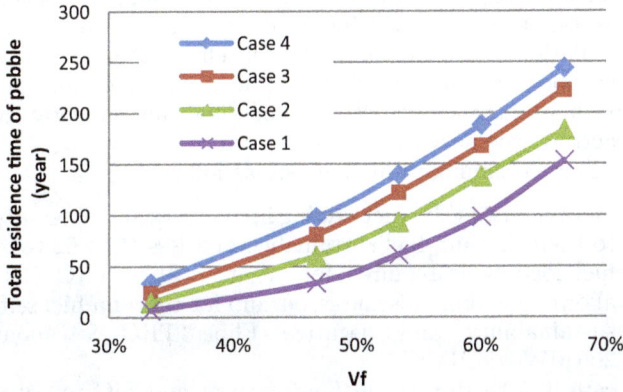

Fig. 20. Total residence time of pebble as a function of Vf and cases (10 MW/m^3 power density).

density increases from 10 to 30 MW/m^3. In PB-FHRs, ^{233}Pa has opportunity to decay away by periodically removing pebbles from the core. Figure 19 shows that the discharge burnup in high power density condition can be improved by increasing the number of times each pebble is cycled through each channel. However, this effect becomes weak when number of cycles in each channel extends beyond 20.

Because of the low power density and high fuel loading, the residence time of each pebble in this reactor is very long (Fig. 20). It is necessary to reduce the residence time by increasing the power density and decreasing the fuel loading or Vf. For 46.7% Vf in case 4, if the core power density is 40 MW/m^3, the discharge burnup may drop from 148 to about 100 MWd/kgHM for the effect of ^{233}Pa, the residence time of a pebble will be 17 years.

3.7 Thermal hydraulic analysis

In this section, Vf and power density will further be limited by thermal hydraulics considerations. For dispersion fuel system, a one-dimensional sphere geometry with equivalent

thermal conductivity is used. By reference to HTGRs [34], the limit temperature of fuel in normal conditions is assumed to be 1250 °C, and the limit temperature of fuel in accident conditions is supposed to be 1600 °C.

The maximum kernel temperature can be deduced from the maximum temperature of mixed fuel region:

$$T_k = T_f + \frac{P \cdot f \cdot r_1^2}{6k_{fuel}(1-\varepsilon) \cdot V_f} + \frac{P \cdot f \cdot r_1^3}{3k_{buffer}(1-\varepsilon) \cdot V_f(1/r_1 - 1/r_2)}$$
$$+ \frac{P \cdot f \cdot r_1^3}{3k_{SiC}(1-\varepsilon) \cdot V_f(1/r_2 - 1/r_3)}. \quad (11)$$

T_f is maximum temperature of mixed fuel region, P is the average power density, f is the total power peak factor (assumed as $1.4 \times 1.4 \approx 2$ in the following calculation), ε is porosity of pebble bed, k_{fuel} is thermal conductivity of thorium–uranium fuel, k_{buffer} is thermal conductivity of buffer, k_{SiC} is thermal conductivity of SiC cladding, r_1 is the radius of fuel kernel, r_2 is the outer radius of buffer and r_3 is the outer radius of SiC cladding.

The maximum temperature of mixed fuel region can be obtained by:

$$T_f = T_s + \frac{P \cdot f \cdot R^2}{6k(1-\varepsilon)}. \quad (12)$$

T_s is the surface temperature of pebble; R is the radius of pebble; k is equivalent thermal conductivity, in this paper, it is the volume average thermal conductivity, which will vary with Vf.

The surface temperature of pebble can be obtained by heat convection equation:

$$T_s = T_c + \frac{4P \cdot f \cdot R^2}{3k_c Nu(1-\varepsilon)}, Nu = 2 + 1.1Re^{0.6}Pr^{1/3}. \quad (13)$$

T_c is the average temperature of Flibe, k_c is the thermal conductivity of Flibe, Nu is nusselt number cited from Wakao [35], Re is Reynolds number, and Pr is Prandtl number.

$$Re = \frac{\rho U \cdot 2R}{\mu(1-\varepsilon)}, Pr = \mu C_p/k_c. \quad (14)$$

ρ is the density of Flibe, μ is dynamic viscosity, C_p is heat capacity, U is superficial velocity of Flibe. U can be calculated by:

$$U = \frac{P \cdot h}{C_p \cdot \rho \cdot (T_{outlet} - T_{inlet})}, \quad (15)$$

h is the height of core, T_{outlet} is outlet temperature of Flibe, and T_{inlet} is inlet temperature of Flibe.

The physical property parameters are listed in Table 4, the thermal conductivity of SiC is very high even after a long period of irradiation time, the thermal conductivity of Th_2CO_3/U_2CO_3 is referred from that of ThO_2, which shows a little higher thermal conductivity than UO_2.

Table 4. Constants for thermal hydraulic calculations.

Parameter	Value
ε	0.4
k_c	$1\,\mathrm{W/m^\circ C}$
k_{SiC}	$30\,\mathrm{W/m^\circ C}$
k_{fuel}	$4\,\mathrm{W/m^\circ C}$
k_{buffer}	$9\,\mathrm{W/m^\circ C}$
ρ	$1.96 \times 10^3\,\mathrm{kg/m^3}$
μ	$8.153 \times 10^{-3}\,\mathrm{Pa{\cdot}s}$
C_p	$2.38 \times 10^3\,\mathrm{J/kg/^\circ C}$
h	$5\,\mathrm{m}$
T_{inlet}	$600\,^\circ\mathrm{C}$
T_{outlet}	$700\,^\circ\mathrm{C}$
T_c	$650\,^\circ\mathrm{C}$
r_1	$700\,\mu\mathrm{m}$
r_2	$758\,\mu\mathrm{m}$
r_3	$819\,\mu\mathrm{m}$

The results are shown in Table 5. As the increase of V_f, the equivalent thermal conductivity decreases, as a result, the maximum temperature of fuel increases. However, even in 60% V_f, the maximum temperature of fuel is still below 1250 °C for 6 cm pebble under 10 MW/m^3 power density. On the other hand, the allowable power density for 6 cm pebble will not extend beyond 21 MW/m^3 if the maximum temperature of fuel is below 1250 °C. Reducing the diameter of pebble is an effective means of improving the power density, as shown in Table 5, 60 MW/m^3 power density is allowable in 40% Vf for 3 cm pebble.

As analyzed above, thermal conductivity is sensitive to the maximum temperature of the fuel. ThC may be a good candidate ceramic fuel due to the high density and high thermal conductivity.

Loss of Forced Cooling (LOFC) and Anticipated Transient Without Scram (ATWS) are the most important accidents for PB-FHRs. The decay heat removal system

with Pool Reactor Auxiliary Cooling (PRAC) heat exchangers (PHX) modules in the PB-AHTR could be applied to this work. In an LOFC accident, even under 40 MW/m^3 power density, the outlet temperature of Flibe will not rise by as much as 50 °C, and the temperature of fuel will quickly drop to the level of Flibe [36]. In an ATWS accident with a 1000 pcm reactivity insertion, the temperature of the fuel will not rise by as much as 200 °C to 1450 °C, which is still lower than 1600 °C (−5 pcm/K of fuel TRC is calculated in 46.7% V_f, case 4). In addition, the negative Flibe TRC is more effective to decrease the outlet temperature than a more negative fuel TRC, and the outlet temperature in this case will not rise by 200 °C [36].

4 Conclusions

This work investigated the sustainability of thorium fuel in PB-FHR. Dispersion fuel with SiC cladding and SiC matrix was used to increase the fuel loading. A novel equilibrium calculation method of thorium fuel self-sustainability was developed to analyze discharge burnup. The mechanism of breeding and the characteristic of Flibe salt temperature reactivity coefficient are both performed.

Some preliminary findings are as follows:

- more than 20 vol% fuel loading in fuel system is necessary to keep thorium fuel sustainable, and less than 62 vol% fuel loading is required for negative Flibe TRC. The allowed maximal discharge burnup for thorium fuel self-sustainability and negative Flibe TRC is about 200 MWd/kgHM;
- case 4 with directional flow pattern and SiC reflector displays superior burnup characteristics due to having the hardest neutron spectrum and lowest neutron leakage. While case 3 with mixing flow pattern and SiC reflector shows the best tradeoff between discharge burnup and radial power peak factor. For 50% Vf dispersion fuel, case 3 could provide 140 MWd/kgHM burnup and about 1.4 radial power peaking factor;
- the ^{232}Th(n,γ) reaction and ^{233}U(n,f) reaction are main contributions to Flibe TRC. It indicates that some level of slowing down is required to keep a negative void

Table 5. Temperature distribution in pebble.

V_f	20%	30%	40%	50%	60%
Equivalent thermal conductivity (J/cm/K)	0.24	0.21	0.18	0.15	0.12
6 cm					
$\quad T_s$-T_c (°C)	80	80	80	80	80
$\quad T_f$-T_s (°C)	208	237	277	332	415
$\quad T_k$-T_f (°C)	3.7	2.5	1.8	1.5	1.2
$\quad$Max. T_k (°C)	941	970	1009	1063	1146
$\quad$Allowable Avr. P.D. (MW/m^3)	21	19	17	15	12
3 cm					
$\quad$Max. T_k (°C)	733	740	750	764	784
$\quad$Allowable Avr. P.D. (MW/m^3)	72	67	60	53	45

reactivity coefficient, which provides a new insight for coolant in quasi-fast reactor. Flibe salt shows good neutron properties as coolant of quasi-fast reactor. The equilibrium concentration of ^{6}Li in fast spectrum is around 1000 ppm, which decreases the cost of enrichment, and the neutron absorption of ^{6}Li is still low. 99.95% ^{7}Li is compatible for sustainability of thorium-uranium in PB-FHR. In addition, the production rate of ^{3}H in quasi-fast spectrum is about 30–40 g/GW/year, usually lower than in thermal spectrum;

– effect of ^{233}Pa is significant in the high power density condition. A 30 MWd/kgHM drop in discharge burnup is obtained when power density increases from 10 to 30 MW/m^3. Increasing the number of time each thorium pebble is cycled through each channel can increase discharge burnup. The greatest challenge of this reactor is the very long irradiation time of the pebble fuel. Increasing power density can apparently decrease the irradiation time, but discharge burnup will also obviously decrease, and as a result, the reactor may not be competitive;

– thermal hydraulic calculations show good safety margin. 20 MW/m^3 is allowable for 6 cm pebble, and 60 MW/m^3 is allowable for 3 cm pebble. Vf affects the thermal conductivity, and a value lower than 50% is recommended.

In further analysis, we will focus on the high power density case, investigate how to reduce the effect of ^{233}Pa, and also perform a detail thermal hydraulic analysis.

This paper is supported by the "Strategic Priority Research Program" of the Chinese Academy of Sciences (Grant No. XDA02010200), and Science and Technology Commission of Shanghai Municipality (Grant No. 11JC1414900). Thanks for the suggestions from David W. Dean and reviewers.

Nomenclature

PB-FHR	Pebble-Bed Fluoride salt-cooled High temperature Reactor
V_f	Fuel volume dividing the volume of fuel system
UTR	Fresh ^{233}U/^{232}Th ratio
XS(Tha)/XS(U3a)	One-group absorption cross-section ratio of thorium-uranium
XS(U3f)/XS(U3a)	One-group fission cross-section of ^{233}U over one-group absorption cross-section of ^{233}U
TRC	Temperature Reactivity Coefficient

References

1. IAEA, Role of thorium to supplement fuel cycles of future nuclear energy systems, Nuclear Energy Series No. NF-T-2.4, Vienna, 2012

2. B.A. Lindley et al., Thorium breeder and burner fuel cycles in reduced-moderation LWRs compared to fast reactors, Prog. Nucl. Energy 77, 107 (2014)

3. S. Permana, N. Takaki, H. Sekimoto, Preliminary study on feasibility of large and small water cooled thorium breeder reactor in equilibrium states, Prog. Nucl. Energy 50, 320 (2008)

4. S. Permana, N. Takaki, H. Sekimoto, Breeding capability and void reactivity analysis of heavy-water-cooled thorium reactor, J. Nucl. Sci. Technol. 45, 589 (2008)

5. S. Permana, N. Takaki, H. Sekimoto, Breeding and void reactivity analysis on heavy metal closed-cycle water cooled thorium reactor, Ann. Nucl. Energy 38, 337 (2011)

6. S. Sahin et al., Investigation of CANDU reactors as a thorium burner, Energy Convers. Manag. 47, 1661 (2006)

7. A. Kumar, P.V. Tsvetkov, Optimization of U–Th fuel in heavy water moderated thermal breeder reactors using multivariate regression analysis and genetic algorithms, Ann. Nucl. Energy 85, 885 (2015)

8. Y. Yulianti, Z. Su'ud, N. Takaki, Accident analysis of heavy water cooled thorium breeder reactor, in The 5th Asian physics symposium (APS 2012), (AIP Publishing, 2015), Vol. 1656

9. F. Wols et al., Core design and fuel management studies of a thorium-breeder pebble bed high-temperature reactor, Nucl. Technol. 186, 1 (2014)

10. E.S. Bettis, R.C. Robertson, The design and performance features of a single-fluid molten-salt breeder reactor, Nucl. Appl. Technol. 8, 190 (1970)

11. A. Nuttin et al., Potential of thorium molten salt reactors detailed calculations and concept evolution with a view to large scale energy production, Prog. Nucl. Energy 46, 77 (2005)

12. J. Serp, M. Allibert, O. Benes et al., The molten salt reactor (MSR) in generation IV: overview and perspectives, Prog. Nucl. Energy 77, 308 (2014)

13. C.W. Forsberg, P.F. Peterson, R.A. Kochendarfer, Design options for the advanced high-temperature reactor, in Proceedings of ICAPP '08, Anaheim, USA (2008), Paper 8026

14. F.-P. Fardin, F. Koenig, Preliminary study of the pebble-bed advanced high temperature reactor (University of California, Berkeley, California, 2006)

15. M. Fratoni, Development and applications of methodologies for the neutronic design of the pebble bed advanced high temperature reactor (PB-AHTR) (University of California, Berkeley, California, 2008)

16. R. Hong et al., Reactor safety and mechanical design for the annular pebble-bed advanced high temperature reactor (University of California, Department of Nuclear Engineering, Berkeley, California, 2009)

17. A. Lafuente, M. Piera, Exploring new coolants for nuclear breeder reactors, Ann. Nucl. Energy 37, 835 (2010)

18. M.K. Meyer, R. Fielding, J. Gan, Fuel development for gas-cooled fast reactors, J. Nucl. Mater. 371, 281 (2007)

19. R. Stainsby et al., Gas cooled fast reactor research in Europe, Nucl. Eng. Des. 241, 3481 (2011)

20. L.L. Snead et al., Handbook of SiC properties for fuel performance modeling, J. Nucl. Mater. 371, 329 (2007)

21. L.L. Snead, Y. Katoh, S. Connery, Swelling of SiC at intermediate and high irradiation temperatures, J. Nucl. Mater. 367, 677 (2007)

22. Y. Katoh et al., Radiation effects in SiC for nuclear structural applications, Curr. Opin. Solid State Mater. Sci. 16, 143 (2012)

23. Y. Katoh et al., Stability of SiC and its composites at high neutron fluence, J. Nucl. Mater. 417, 400 (2011)

24. Y. Katoh et al., Mechanical properties of advanced SiC fiber composites irradiated at very high temperatures, J. Nucl. Mater. **417**, 416 (2011)

25. J.A. Jung et al., Feasibility study of fuel cladding performance for application in ultra-long cycle fast reactor, J. Nucl. Mater. **440**, 596 (2013)

26. K. Fukuda, K. Iwamoto, Diffusion behavior of fission product in pyrolytic silicon carbide, J. Nucl. Mater. **75**, 131 (1978)

27. D.J. Cumberland, R.J. Crawford, The packing of particles, in *Handbook of powder technology* (Elsevier, Amsterdam, 1987) Vol. 6, p. 45

28. A.T. Cisneros, E. Greenspan, P. Peterson, Use of thorium blankets in a pebble bed advanced high temperature reactor, in *Proceedings of the 2010 International Congress on Advances in Nuclear Power Plants-ICAPP'10*, (2010)

29. R. Hong et al., *Reactor safety and mechanical design for the annular pebble-bed advanced high temperature reactor* (University of California, Department of Nuclear Engineering, Berkeley, California, 2009)

30. G. Zhu, Y. Zou, M. Li et al., Development of burnup calculation code for pebble-bed high temperature reactor at equilibrium state, Atomic Energy Sci. Technol. **49**, 890 (2015)

31. E. Teuchert, U. Hansen, K.-A. Haas, VSOP-Computer code system for reactor physics and fuel cycle simulation, Kernforschungsanlage Juelich GmbH (Germany, FR), Institut fuer Reaktorentwicklung, 1980

32. H.D. Gougar, M.O. Abderrafi, W.K. Terry, *Advanced core design and fuel management for pebble-bed reactors* (Idaho National Laboratory, 2004), No. INEEL/EXT-04-02245

33. A.T. Jr., Cisneros, *Pebble bed reactors design optimization methods and their application to the Pebble Bed Fluoride Salt Cooled High Temperature Reactor (PB-FHR)* (University of California, Berkeley, California, 2013)

34. D. Hanson et al., *Development plan for advanced high temperature coated-particle fuels* (General Atomics, San Diego, CA, 2004), PC-000513, Rev. 0

35. N. Wakao, T. Funazkri, Effect of fluid dispersion coefficients on particle-to-fluid mass transfer coefficients in packed beds: correlation of Sherwood numbers, Chem. Eng. Sci. **33**, 1375 (1978)

36. A. Griveau et al., Transient thermal response of the PB-AHTR to loss of forced cooling, in *Global 2007, UC Berkeley and INL, Boise, Idaho, 9th–13th September, 2007* (2007)

10

Investigation of the relationships between mechanical properties and microstructure in a Fe-9%Cr ODS steel

Benjamin Hary[1*], Thomas Guilbert[1], Pierre Wident[1], Thierry Baudin[2], Roland Logé[3], and Yann de Carlan[1]

[1] Service de Recherches Métallurgiques Appliquées, CEA Saclay, 91191 Gif-sur-Yvette Cedex, France
[2] Institut de Chimie Moléculaire et des Matériaux d'Orsay, UMR CNRS 8182, SP2M, Université Paris-Sud, 91405 Orsay Cedex, France
[3] Laboratoire de Métallurgie Thermomécanique, École Polytechnique Fédérale de Lausanne, rue de la Maladière, 71b, CP 526, CH-2002, Neuchâtel, Switzerland

Abstract. Ferritic-martensitic Oxide Dispersion Strengthened (ODS) steels are potential materials for fuel pin cladding in Sodium Fast Reactor (SFR) and their optimisation is essential for future industrial applications. In this paper, a feasibility study concerning the generation of tensile specimens using a quenching dilatometer is presented. The ODS steel investigated contains 9%Cr and exhibits a phase transformation between ferrite and austenite around 870 °C. The purpose was to generate different microstructures and to evaluate their tensile properties. Specimens were machined from a cladding tube and underwent controlled heat treatments inside the dilatometer. The microstructures were observed using Electron Backscatter Diffraction (EBSD) and tensile tests were performed at room temperature and at 650 °C. Results show that a tempered martensitic structure is the optimum state for tensile loading at room temperature. At 650 °C, the strengthening mechanisms that are involved differ and the microstructures exhibit more similar yield strengths. It also appeared that decarburisation during heat treatment in the dilatometer induces a decrease in the mechanical properties and heterogeneities in the dual-phase microstructure. This has been addressed by proposing a treatment with a much shorter time in the austenitic domain. Thereafter, the relaxation of macroscopic residual stresses inside the tube during the heat treatment was evaluated. They appear to decrease linearly with increasing temperature and the phase transformation has a limited effect on the relaxation.

1 Introduction

Research works performed during recent years have revealed that ODS (Oxide Dispersion Strengthened) steels are promising materials for fuel pin cladding in Sodium Fast Reactors [1,2]. It appears that the bcc ferritic-martensitic lattice allows for a high resistance to irradiation swelling up to a dose of around 150 displacements per atom (dpa) and nano-oxides significantly improve creep and tensile properties at high temperature (650 °C) by blocking the dislocations motion.

ODS steels are created by powder metallurgy and mechanical alloying [3] in order to obtain a fine homogeneous dispersion of the nano-oxides within the matrix. Afterwards, the powder is compacted in a soft steel can and hot extruded. The soft steel is removed from the raw bar obtained and only the ODS steel remains. Then, the bar is cold-worked into the shape of a cladding tube by several passes of rolling. This manufacturing process tends to create a crystallographic (α fiber <110>) and a morphologic texture into the material. These passes also induce important residual stresses that can be limited or annealed by intermediate heat treatments that decrease hardness and prevent the tube from being damaged.

A martensitic ODS tube with 9%Cr has been studied. With a heating rate of 5 °C/s, this grade exhibits a phase transformation from ferrite to austenite between 870 °C (A_s) and 960 °C (A_f) that enables a total recovery of the microstructure and facilitates cold-working of the tube [4,5], since the material does not recrystallise in the ferritic state [6]. Moreover, it is possible to obtain different microstructures from ferrite to martensite by applying various cooling rates from the austenitic domain. This investigation focused on an analytic method to treat tensile specimens in order to generate different microstructures. It employed a dilatometer to precisely control thermal cycles and to measure the dimensional variations of the sample.

* e-mail: benjamin.hary@cea.fr

Table 1. Chemical composition of the ODS steel grade K30-M1.

	Cr	W	Ti	Y	Ni
wt.%	9.08	1.05	0.2	0.19	0.2
	Mn	Si	C	O	N
wt.%	0.29	0.23	0.109	0.12	0.022

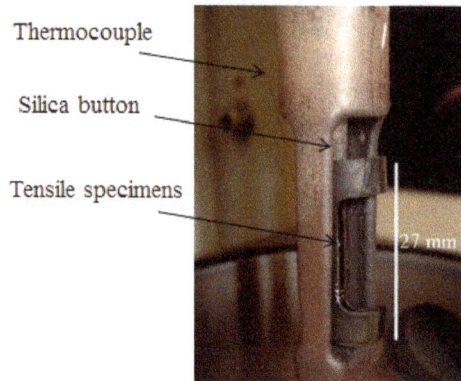

Fig. 1. Experimental device.

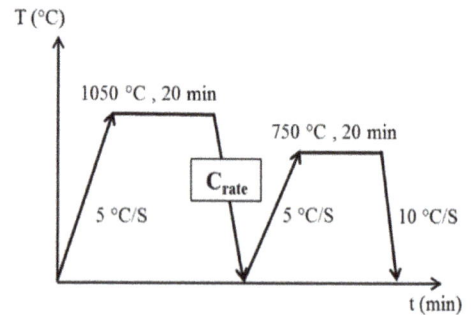

Fig. 2. Applied heat treatment for microstructure generation.

The aim was to perform different final heat treatments, assess the mechanical properties and determine the best compromise between ductility and tensile strength. Another purpose of the study was to understand the macroscopic residual stresses relaxation (1st order stresses) inside the cladding tube during the heat treatments.

2 Microstructural characterisation

2.1 Generation of microstructures for mechanical assessments

The chemical composition of the ODS steel tube investigated is presented in Table 1. The raw bar was extruded at 1100 °C and a tempering treatment was performed at 1050 °C for 30 min. Then, the soft steel was removed by chemical dissolution. In order to obtain the cladding tube, the bar was cold-rolled in the ferritic state with intermediate heat treatments in the austenitic domain. After each intermediate heat treatment, the tube was cooled at a slow rate (0.05 °C/s). At the end of the manufacturing process, the cladding tube exhibits a cold-rolled ferritic microstructure. In the following, this tube will be named K30-M1.

Tensile specimens ($27 \times 2 \times 0.5 \, \text{mm}^3$) were machined from the ferritic tube in the axial direction before undergoing a controlled heat treatment in a dilatometer under helium atmosphere. This high-speed Adamel-Lhomargy DT1000 dilatometer, retrofitted by AET Technologies, provides access to a broad range of cooling rates, from 0.1 °C/s to 100 °C/s using a cryogenic system with liquid nitrogen. Two thermocouples (Fig. 1) allow measurement of the real specimen temperature on the specimens throughout the experiment. In order to prevent welding defects on the

specimen form affecting the results of the mechanical test, thermocouples are welded onto a second specimen which is not submitted to this test. A sensor motion (silica probe) is used to measure the dimensional variation of the specimen and to identify the allotropic phase transformations during the heat treatment. The applied heat treatment presented in Figure 2 was an austenitisation plateau at 1050 °C for 20 minutes, followed by a cooling where the rate is carefully chosen. Then, a tempering treatment was performed at 750 °C for 20 minutes to allow carbide precipitation inside the martensite. Three different cooling rates (C_{rate}) were chosen from the CCT diagram [7] in order to obtain the following microstructures: tempered martensite (10 °C/s), dual-phase 50% martensite-50% ferrite (2 °C/s) and ferrite (0.1 °C/s).

Once the specimens were treated, tensile tests were performed in the longitudinal direction at room temperature and at 650 °C with a strain rate of 7×10^{-4}/s. Observation of the fracture surfaces has enabled identification of the rupture mechanisms. The same treatments were applied to cylinders cut from the tube in order to characterise each microstructure using an EBSD (Electron Backscatter Diffraction) system installed on a FEG-SEM. The samples were prepared by vibratory auto-polishing using non-crystalline colloidal silica for several hours. Analyses were made in the rolling plane, along the axial direction of the tube.

2.2 Results

According to the dilation curves presented in Figures 3a, 4a and 5a, the microstructures have been generated as expected. The expansion during heating is the same for the three samples, with an austenitisation around 870 °C (A_s). On the other hand, significant changes can be

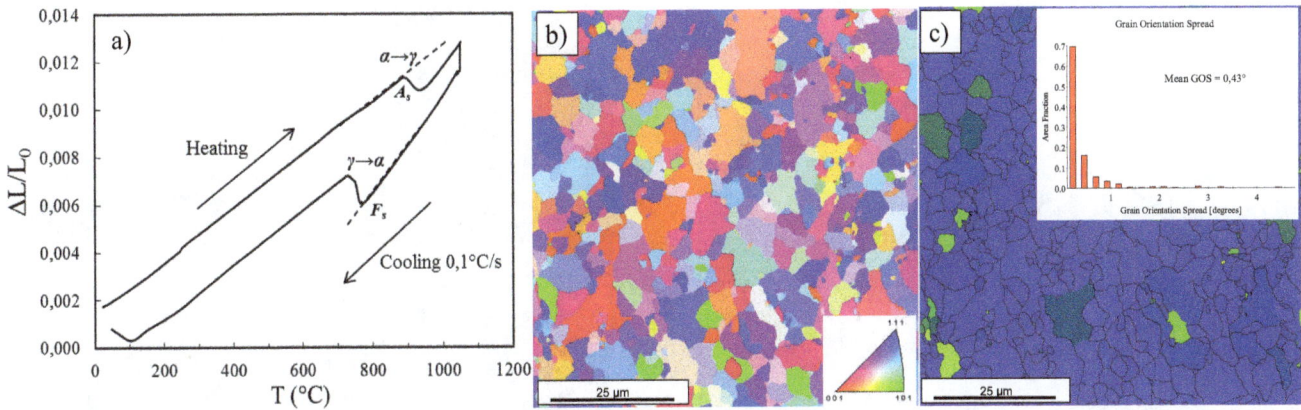

Fig. 3. (a) Dilation curve, (b) grain orientation map showing normal of crystalline planes parallel to the rolling direction ($75 \times 75~\mu\mathrm{m}^2$, scanning step: $0.1~\mu\mathrm{m}$, correctly indexed pixels: 100%), and (c) GOS map for the ferritic microstructure.

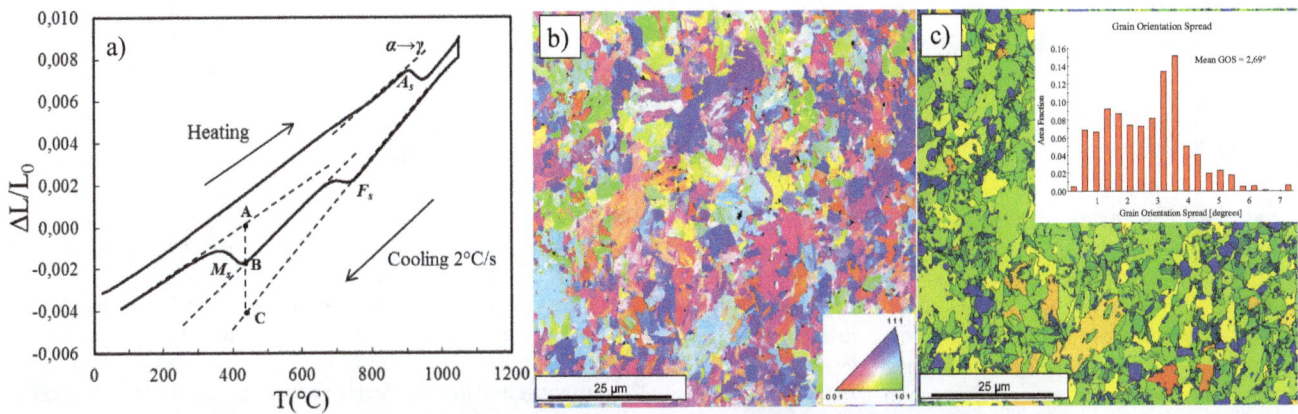

Fig. 4. (a) Dilation curve, (b) grain orientation map showing normal of crystalline planes parallel to the rolling direction ($75 \times 75~\mu\mathrm{m}^2$, scanning step: $0.1~\mu\mathrm{m}$, correctly indexed pixels: 99.4%) and (c) GOS map for the dual-phase microstructure.

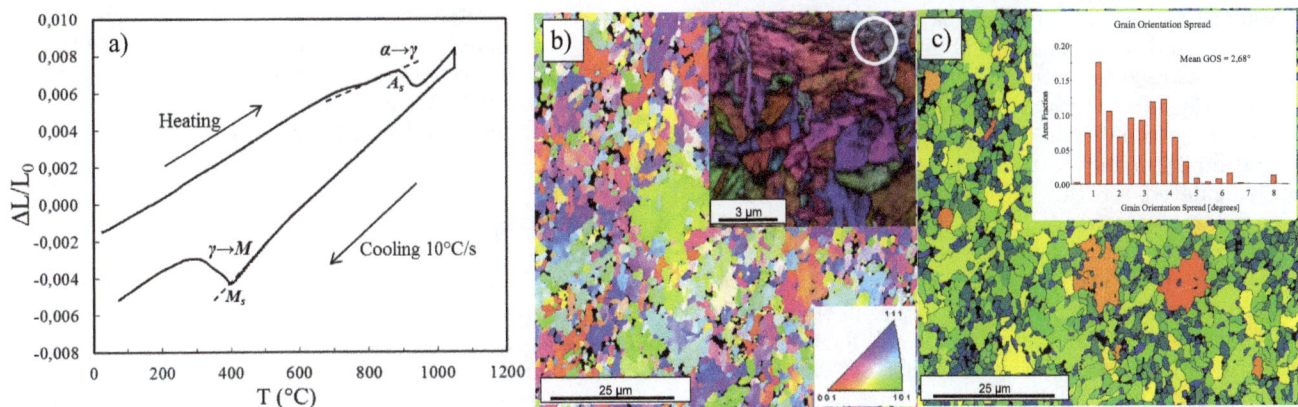

Fig. 5. (a) Dilation curve, (b) grain orientation map showing normal of crystalline planes parallel to the rolling direction ($75 \times 75~\mu\mathrm{m}^2$, scanning step: $0.1~\mu\mathrm{m}$, correctly indexed pixels: 94%), and (c) GOS map for the martensitic microstructure.

observed during cooling. In Figure 3a, only the ferritic transformation (F_s) around $780~^\circ\mathrm{C}$ is apparent. Figure 4a shows that both ferritic and martensitic transformations have occurred, and their proportion can be graphically estimated by comparing the AB and BC segments. The uncertainty in the fraction phases is about 10% using this method. Here, the two segments are equivalent and thus the fractions of phases: about 50% ferrite and 50% martensite.

Then, Figure 5a shows that only martensite is created from temperature M_s. EBSD data were treated with the OIM Analysis software, developed by the EDAX society. The cleanup procedure used to analyse the data was a Grain Dilation (one iteration, minimum grain size = 5, grain tolerance angle = 5) followed by a Grain CI Standardisation (same parameters). Then, only the points with a confidence index higher than 0.1 were taken into account.

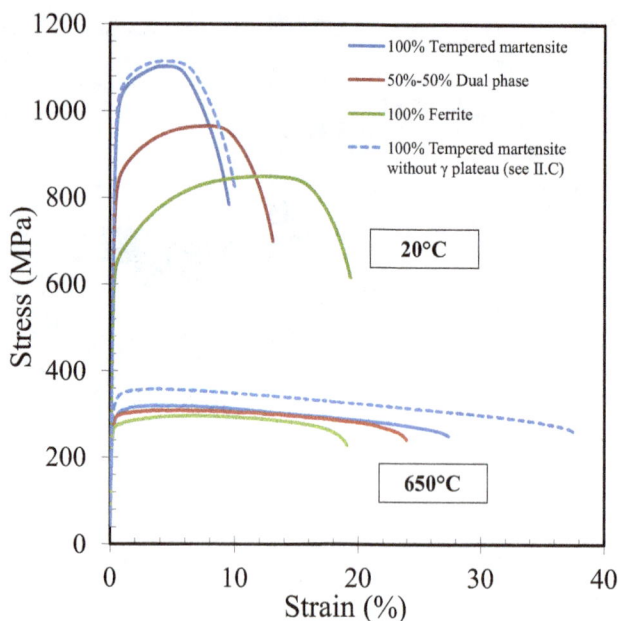

Fig. 6. Tensile properties of the microstructures created using the dilatometer.

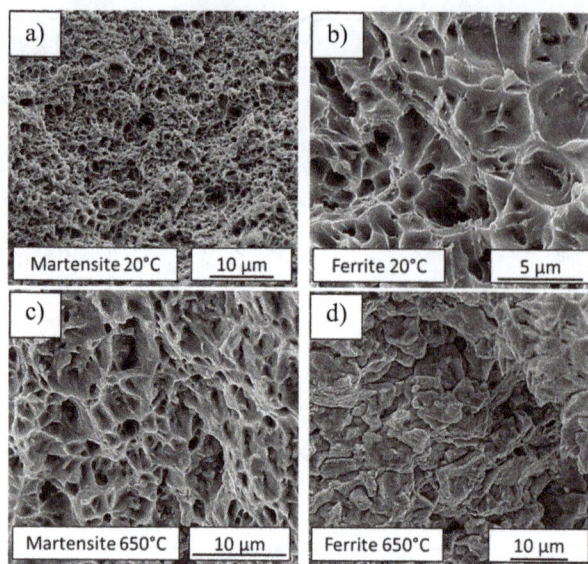

Fig. 7. Fracture surfaces of tensile specimens.

All of the grain orientation maps show neither crystallographic texture nor morphologic texture, which confirmed the reset effect of the phase transformation during the heat treatment. Figure 3b shows a recrystallised ferritic structure composed of equiaxed grains with a mean size of 5.5 μm whereas the martensitic structure (Fig. 5b) is composed of smaller grains with a mean size of 3.6 μm. The scan performed at a smaller scale with an analysis step of 10 nm enables observation of substructures in the martensite (see the white circle) which could be identified as laths, blocks or packets. Figure 4b shows the microstructure of the dual-phase sample. The fine grains make the identification of ferrite and martensite difficult, and an alternative method was used to distinguish the two phases: the Grain Orientation Spread (GOS). This method uses the EBSD dataset to estimate the intragranular misorientation of each grain in the microstructure [8]. The GOS distribution was calculated for the three microstructures. The misorientation inside ferritic grains is very low (mean GOS 0.4°) compared to the one in martensitic grains (mean GOS 2.7°). This can be attributed to the fact that just after the cooling, the dislocation density is higher in the martensite than in the ferrite. Thus, martensite and ferrite grains can be distinguished on the basis of their GOS value.

Considering this, the dual-phase microstructure seems to contain much more martensite than ferrite, which is a surprising result according to the dilation curve. This is discussed in the following sections.

From Figure 6, tensile tests performed at room temperature show a significant strain hardening. The yield strength increases with increasing martensitic content, and the maximum uniform elongation decreases.

The fracture surfaces on the three microstructures showed numerous dimples (Figs. 7a and 7b), characteristic of a ductile behaviour. One can point out a strong relationship between the microstructure and the tensile properties at this temperature. The results obtained at high temperature are quite different: On the one hand, the dependence of the yield strength on the microstructure is highly reduced. In addition, the strain hardening is much less significant at this temperature for the three microstructures. On the other hand, the fracture strain of the annealed martensite increases significantly from 10% to almost 30%, whereas it remains the same as at room temperature for the ferrite, around 20%. The maximum uniform elongation remains higher for the ferrite but is divided by two (12% at 20 °C and 6.7% at 650 °C), whereas it remains almost the same for martensite, around 4.5%. Analysis of fracture surfaces shows dimpled features in the martensite (Fig. 7c) and in the dual-phase, but some intergranular decohesion areas in the ferrite (Fig. 7d). This may be responsible for the less ductile behaviour of ferrite at 650 °C. In the literature [9], intergranular decohesion mechanisms have already been observed in ferritic ODS (14%Cr) steels above 600 °C caused by cavities lining up along the grain boundaries.

2.3 Discussion

Based on these mechanical tests, the martensitic structure seems to be the optimum state to withstand the tensile loading at room temperature. In fact, it shows the highest strength and a ductile behaviour. Several contributions can be identified to explain this difference. According to the Hall-Petch effect, the presence of the finer grains in martensite induces a higher yield strength as compared to ferrite. In addition, the higher dislocation density created by the displacive transformation is also known to reinforce the material. Finally, the precipitation of carbides in 9%Cr ODS steel can vary significantly between the microstructures, as observed by Klimiankou et al. [10]. In ferrite, they tend to nucleate at the grain boundaries and are essentially coarse $M_{23}C_6$ (M = Fe, Cr, W) or TiC carbides. On the other hand, for a tempered martensitic structure, the carbides are likely to nucleate more homogeneously in the microstructure.

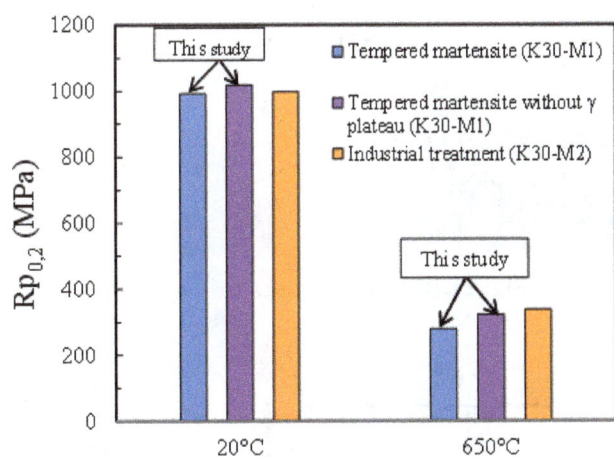

Fig. 8. Yield strength comparison at 20 °C and 650 °C of different 9%Cr ODS steel cladding tubes investigated at CEA.

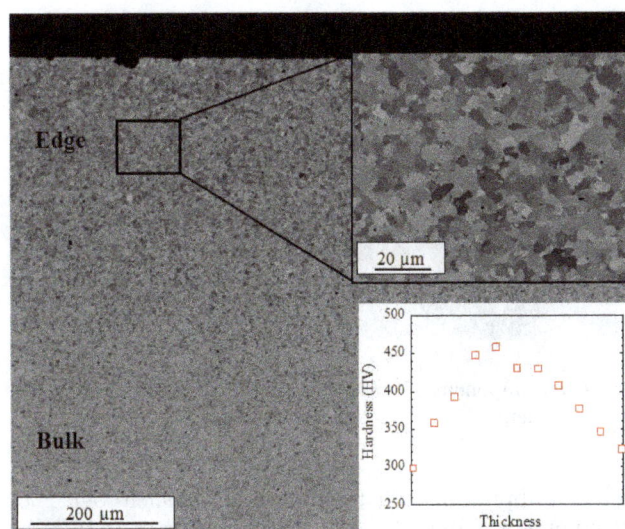

Fig. 9. SEM image of the microstructure of the dual-phase sample, from the edge of the tube to the core and hardness distribution across the thickness of the tube.

This can be attributed to a smaller grain size and a more significant number of interfaces. Considering an Orowan strengthening mechanism, the lower the distance between particles, the higher the strengthening effect will be. These hardening mechanisms could explain the higher mechanical resistance of martensite at room temperature.

At 650 °C, the relationships between the mechanical properties and the microstrutures are more difficult to explain. The yield strength of martensite is almost the same as that of ferrite. This weak variation of the yield strength with tensile loading at high temperature between different 9%Cr ODS steel microstructures has already been observed [11]. It suggests that the strengthening mechanisms that make martensite much more resistant than ferrite at room temperature are not the same at 650 °C. In different studies [12–14], the presence of residual ferrite was found. This phase did not undergo the austenitisation and TEM analysis showed that it contains a higher density and a finer diameter of nano-particles than martensite. It leads to a more important pinning of the dislocations considering an Orowan mechanism, which would become predominant at high temperature, and so a hardening of the ferrite. However, there is no presence of residual ferrite in the present work. The samples seemed to undergo a full austenitisation according to the recrystallised ferritic structure in Figures 3b and 3c. This may be due to a lower content of alphagen alloying elements in K30-M1, increasing the driving force for austenitisation. Consequently, the most plausible explanation for the similar yield strength at 650 °C would be a very low contribution of dislocations and Hall-Petch effect in the strengthening mechanisms.

In order to evaluate the efficiency of controlled treatments in the dilatometer, the tensile properties of the martensitic sample were compared to those of another ODS 9%Cr cladding tube (named K30-M2) studied at CEA [15,16], created from the same powder, and presenting the same chemical composition. The cladding tube of this grade was treated in a classical (industrial) furnace at 1050 °C for 30 min and cooled at 70 °C/min. Then, a softening treatment at 750 °C for one hour was performed. Tensile specimens were machined afterwards. This tube will be called "industrial grade" in the following. The cooling rate is a parameter of the furnace and was not measured directly on the tube during the treatment. The atmosphere in the furnace is a primary vacuum. The microstructure of the industrial grade was observed and essentially showed laths of martensite. One should note that the industrial grade has undergone a lower cooling rate under a less controlled atmosphere than the K30-M1 martensitic specimen heat treated in the dilatometer. Despite these considerations, Figure 8 shows that the yield strength of tempered martensite K30-M1 (blue) at 650 °C is about 60 MPa lower than that of the industrial grade (orange). The experimental uncertainty on the yield strength was considered to be 10 MPa.

To explain these results, a decarburisation inside the dilatometer during the austenitisation seems to be the most probable hypothesis. In fact, observations on the dual-phase cylindrical sample using SEM with the back-scattering electron detector in Figure 9 showed only large grains on the edge of the sample. It is known that low carbon content promotes growth of ferrite [13] and thus increases the quench critical rates, which determine the formation domain of the different microstructures. To support this hypothesis, micro-hardness measurements (load 100 g) were performed across the thickness of the tube. The maximum hardness (450 HV) is located at the half-thickness, whereas the minimum (300 HV) is located at the edges. Thus, one can suggest that during cooling, ferrite nucleated at the edge where the carbon content was low and martensite appeared in the bulk of the sample.

This is in agreement with the GOS map in Figure 4c where most of the grains showed a high stored energy and are probably grains of martensite (EBSD analysis was performed in the bulk). In order to have a more accurate idea of the decarburisation, carbon content after the heat treatment could be measured using EPMA (Electron MicroProbe Analysis) or a melting method (LECO) and be compared with the initial content in the tube (0.109%).

The decarburisation thickness can be estimated by calculating the diffusion length of carbon into the material.

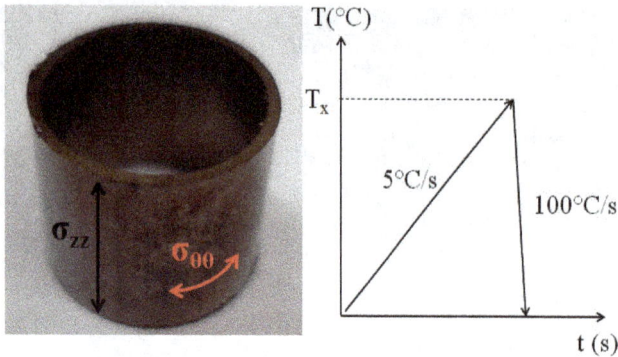

Fig. 10. Component of stress tensor inside the tube and applied heat treatment.

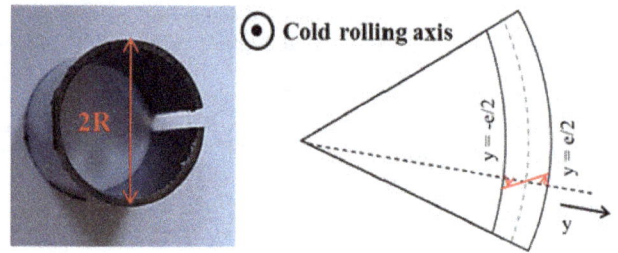

Fig. 11. Orthoradial stress measurement.

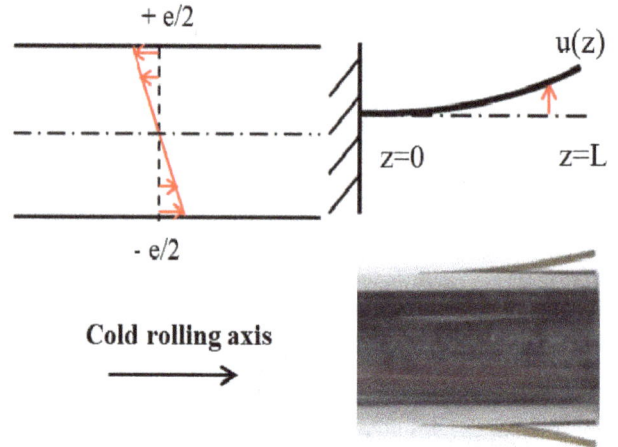

Fig. 12. Longitudinal stress measurement.

As a first approximation, the diffusion coefficient of carbon within pure iron is used. According to Bakker et al. [17], it is estimated as $3.8 \times 10^{-11}\,\mathrm{m^2/s}$ at $1050\,°C$. Considering this approximation and a diffusion time of 20 minutes, one obtains a diffusion length around $300\,\mu m$. Knowing that the tensile specimens are only $500\,\mu m$ thick, a significant amount of carbon is likely to escape from the samples. To prevent this decarburisation, a new heat treatment without austenitisation plateau was performed into the dilatometer to get a tempered martensitic microstructure. The increased carbon content in this sample as compared to that with the initial treatment has been confirmed by looking at the two martensitic start temperatures (M_s). On the dilation curves, it can be seen that the austenitization plateau of the initial treatment induces a shift of M_s around $30\,°C$ towards higher temperatures to $400\,°C$ (Fig. 4a). The Andrews relation [18] gives:

$$M_s(°C) = 539 - 423\%C - 30.4\%Mn - 17.7\%Ni -$$
$$12.1\%Cr - 11\%Si - 7\%Mo.$$

Using this formula and the chemical composition of the material, one finds a carbon loss of approximately 70% due to the austenitisation for 20 minutes at $1050\,°C$. Therefore, it can be noted in Figure 8 that the new treatment induces stronger mechanical properties for the tempered martensite (purple) than the initial one. The increase of the yield strength is $27\,\mathrm{MPa}$ at room temperature and $44\,\mathrm{MPa}$ at $650\,°C$. This improvement of the mechanical properties between these two tempered martensitic samples shows that the precipitation of carbides has a significant reinforcement role within the material, particularly at high temperature. This is in agreement with a predominant role of the Orowan mechanism at $650\,°C$, as previously mentioned.

3 Macroscopic residual stresses relaxation

3.1 Experimental procedure

The dilatometer was used to evaluate the efficiency of simple heat treatments on the relaxation of first order residual stresses after cold-rolling. Both orthoradial ($\sigma_{\theta\theta}$) and longitudinal (σ_{zz}) macroscopic residual stress have

been sources of interest. The measurements were made with the calculation method proposed by Béchade et al. [19]. An elastic behaviour model was used. In the framework of this study, the hardening during cold-rolling was assumed to be isotropic (no kinematical hardening).

Moreover, the following hypotheses were considered: a transversal isotropic stress state and a linear gradient of the stresses in the thickness of the tube, as presented in Figures 11 and 12. To perform this experiment, 9 mm long cylinders were cut from the as-rolled tube and heated at different temperatures (T_x on Fig. 10) between $400\,°C$ and $950\,°C$. Then a very rapid quench ($100\,°C/s$) was applied to freeze the microstructure and thus the stress state. Residuals stresses were measured at room temperature. Samples were cut with an aluminum oxide grindstone. Deformations were measured with a micrometer (uncertainty of $10\,\mu m$ induces an uncertainty of $10\,\mathrm{MPa}$ on the stress).

3.1.1 Orthoradial stresses

The orthoradial residual stresses were estimated by cutting the cylinder along the longitudinal direction. The measurement of the opening can give access to the maximal residual stress using the following formula [20]:

$$\sigma_{\theta\theta}^{\max} = E \cdot \frac{e}{2} \cdot \left(\frac{1}{R_0} - \frac{1}{R} \right), \tag{1}$$

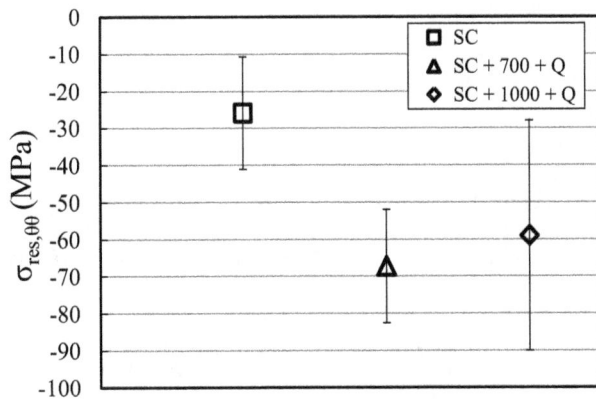

Fig. 13. Macroscopic residual stress evolution during the heat treatment.

where E is the Young's Modulus of the steel (225 GPa), e is the thickness of the tube, R_0 is the radius before cutting and R is the radius after cutting.

3.1.2 Longitudinal stresses

Two stripes were diametrically cut along the cylinder. Measurement of the spire enables calculation of the maximal longitudinal residual stress [4,10]:

$$\sigma_{zz}^{max} = -\frac{u(L) \cdot E}{L^2} e, \qquad (2)$$

where $u(L)$ is the spire, L is the length of the stripes and e is the thickness of the tube.

3.2 Results and discussion

To discuss the results on a sound basis, the contribution of quenching to the orthoradial residual stresses was also studied (for experimental reasons, it was not possible to do this for the longitudinal stresses). Stresses were measured after three different heat treatments:

– Austenitisation + slow cooling at 0.1 °C/s (SC);
– Austenitisation + slow cooling + heating to 700 °C + quench at 100 °C/s (SC + 700 + Q);
– Austenitisation + slow cooling + heating to 1000 °C + quench at 100 °C/s (SC + 1000 + Q).

According to the literature [20], the first treatment should give a zero stress state, while the second and third treatments should give contributions of quenching from the ferritic domain and from the austenitic domain, respectively. From Figure 13, one can conclude that orthoradial stresses after slow cooling are negligible taking into account the uncertainty. The slight difference from zero may have been introduced by the cutting method. On the other hand, compressive stresses of about 60 MPa are measured after quenching from 700 °C and 1000 °C. There does not appear to be a difference between quenching from 700 °C (ferritic domain) or from 1000 °C (austenitic domain), so the effect of martensitic transformation is

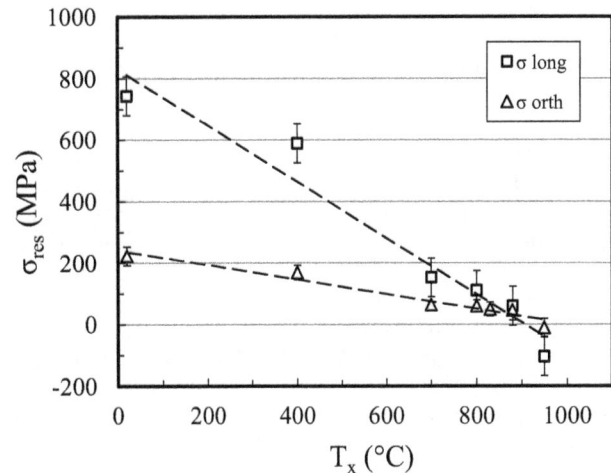

Fig. 14. Contribution of quenching on the orthoradial residual stresses.

negligible, at least for the first order stresses. These quenching stresses should be taken into account in the measures.

Figure 14 shows the macroscopic residual stresses measured at room temperature inside the tube after a heating at temperature T_x. At room temperature, longitudinal stresses are much more significant than the orthoradial stresses and reach 750 MPa. With increasing temperature, both longitudinal and orthoradial stresses decrease linearly when the material is in the ferritic state. The macroscopic residual stresses are almost removed at the austenitisation start temperature (A_s), around 880 °C. The slight compressive value for the treatment at 950 °C can be attributed to quenching: once residual stresses due to cold-rolling are relaxed, only the contribution of quenching remains.

One can conclude that the phase transformation is not responsible for the relaxation of the first order residual stresses, which is an unexpected result. Indeed, one could have expected a plateau from room temperature to A_s and then a sharp decrease of the residual stresses due to the phase transformation. To understand this phenomenon, further experiments using X-ray diffraction would be necessary to obtain the residual stresses tensor and identify the relaxation mechanisms. In fact, different hypotheses could be considered, such as small intragranular dislocation motions or rearrangements at grains boundaries. It would also be interesting to perform this analytic experiment on a ferritic ODS 14%Cr steel, that does not present a phase transformation.

4 Conclusion

A quenched dilatometer was used to generate microstructures from ferrite to martensite in a 9%Cr ODS steel cladding tube. Microstructures were observed and tensile tests were carried out at room temperature and at 650 °C. The keys findings are as follows:

– EBSD data analyses have enabled us to distinguish martensite from ferrite according to the intragranular

misorientation. Tensile tests have shown that the tempered martensitic microstructure is optimal for tensile loading at room temperature. At 650 °C, the mechanisms governing the mechanical resistance are different. The yield strength of ferrite becomes almost equivalent to that of martensite, but its behaviour is less ductile. This could be due to a similar contribution of the Hall-Petch effect and dislocations in the microstructures;
– an experimental artefact, decarburisation inside the dilatometer, led to weaker mechanical properties as compared to the industrial grade and heterogeneous microstructure in the dual-phase sample, with ferrite in the edges and martensite in the bulk. This decarburisation can be corrected by removing the austenitisation plateau from the heat treatment. It shows encouraging results for the tempered martensite, since one can achieve higher yield strength.

The macroscopic residual stress relaxation inside the tube during the heat treatment was measured. The investigation shows that it decreases linearly in the ferritic state with increasing temperature, and reaches a zero stress state at 950 °C. Thus, the relaxation mechanisms are not induced by the phase transformation. X-ray diffraction would be interesting to more completely understand this phenomenon.

The authors would like to thank Jean-Luc Flament for the realisation of the mechanical tests, Annick Bougault for her help in analysing the fracture surfaces and Patrick Bonnaillie for the SEM images.

References

1. Y. de Carlan, J.-L. Béchade, P. Dubuisson et al., CEA development of new ferritic ODS alloys for nuclear application, J. Nucl. Mater. **386-388**, 430 (2009)

2. P. Dubuisson, Y. de Carlan, V. Garat, M. Blat, ODS Ferritic/martensitic alloys for Sodium Fast Reactor fuel pin cladding, J. Nucl. Mater. **428**, 6 (2012)

3. J.S. Benjamin, Dispersion strengthened superalloys by mechanical alloying, Metall. Trans. **1**, 2943 (1970)

4. L. Toualbi, C. Cayron, P. Olier, R. Loge, Y. de Carlan, Relationships between mechanical behavior and microstructural evolutions in Fe 9Cr–ODS during the fabrication route of SFR cladding tubes, J. Nucl. Mater. **42**, 410 (2013)

5. L. Toualbi, C. Cayron, P. Olier et al., Assessment of a new fabrication route for Fe–9Cr–1W ODS cladding tubes, J. Nucl. Mater. **428**, 47 (2012)

6. H.R.Z. Sandim, R.A. Renzetti, A.F. Padilha, D. Raabe, M. Klimenkov, R. Lindau, A. Moslang, Annealing behavior of ferritic-martensitic 9%Cr-ODS-Eurofer steel, Mater. Sci. Eng. A **527**, 3602 (2010)

7. P. Moayeart, Investigation of the martensitic transformation using dilatometry, CEA Report, DEN/DANS/DMN/SRMA/LA2M, 2013

8. A. Ayad, N. Allain-Bonasso, N. Rouag, F. Wagner, Grain Orientation Spread values in IF steels after plastic deformation and recrystallization, Mater. Sci. Forum **702-703**, 269 (2012)

9. M. Praud, F. Mompiou, J. Malaplate, D. Caillard, J. Garnier, A. Steckmeyer, B. Fournier, Study of the deformation mechanism in Fe-14% Cr ODS alloys, J. Nucl. Mater. **428**, 90 (2012)

10. M. Klimiankou, R. Lindau, A. Möslang, Direct correlation between morphology of $(Fe, Cr)_{23}C_6$ precipitates and impact behavior on ODS steels, J. Nucl. Mater. **367-370**, 173 (2007)

11. S. Noh, B.-K. Choi, C.-H. Han, S.H. Kang, J. Jang, Y.-H. Jeong, T.K. Kim, Effects of heat treatments on microstructures and mechanical properties of dual phase ODS steels for high temperature strength, Nucl. Eng. Technol. **45**, 821 (2013)

12. S. Ukai, S. Ohtsuka, T. Kaito, H. Sakasegawa, N. Chikata, S. Hayashi, S. Ohnuki, High-temperature strength characterization of advanced 9Cr-ODS ferritic steels, Mat. Sci. Eng. A Struct. **510-511**, 115 (2009)

13. R. Miyata, S. Ukai, X. Wu, N. Oono, S. Hayashi, S. Ohtsuka, T. Kaito, Strength correlation with residual ferrite fraction in 9Cr-ODS ferritic steel, J. Nucl. Mater. **442**, S138 (2013)

14. M. Yamamoto, S. Ukai, S. Hayashi, T. Kaito, S. Ohtsuka, Formation of residual ferrite in 9Cr-ODS ferritic steels, Mat. Sci. Eng. A **527**, 4418 (2010)

15. S. Vincent, J. Ribis, Microstructural and mechanical characterization of CEA ODS steels, Technical report, DEN/DANS/DMN/SRMA/LC2M&LA2M/NT/2013/3393/A, 2013

16. C. Cayron, A. Montani, D. Venet, N. Herve, Y. de Carlan, Microstructural characterization of ODS steels after manufacturing and after creep loading, Technical report DEHT/DL/2013/122, 2013

17. H. Bakker, H.P. Bonzel, C.M. Bruff, *"LANDOLT-BÖRNSTEIN", Group III: Crystal and solid State Physics, Volume 26, Diffusion in Solid Metals and Alloys* (Springer-Verlag, 1990), p. 481

18. C.Y. Kung, J.J. Rayment, An examination of the validity of existing empirical formulae for the calculation of M_s temperature, Met. Trans. A **13A**, 328 (1982)

19. J.L. Béchade, L. Toualbi, S. Bosonnet, Macroscopic and microscopic determination of residual stresses in thin Oxides Dispersion Strengthened steel tubes, Mater. Sci. Forum **768-769**, 296 (2014)

20. L. Toualbi, Improvement of the manufacturing route of ODS steels cladding tubes, translated from "Optimisation de la gamme de fabrication de tubes en aciers renforcés par dispersion nanométriques d'oxydes (ODS)", PhD thesis, CEMEF-Mines ParisTech, 2012

11

Comparison of SERPENT and SCALE methodology for LWRs transport calculations and additionally uncertainty analysis for cross-section perturbation with SAMPLER module

Antonella Labarile[*], Nicolas Olmo, Rafael Miró, Teresa Barrachina, and Gumersindo Verdú

Institute for Industrial, Radiophysical and Environmental Safety (ISIRYM), Universitat Politècnica de València, Camí de Vera s/n, 46022, Valencia, Spain

Abstract. In nuclear safety research, the quality of the results of simulation codes is widely determined by the reactor design and safe operation, and the description of neutron transport in the reactor core is a feature of particular importance. Moreover, for the long effort that is made, there remain uncertainties in simulation results due to the neutronic data and input specification that need a huge effort to be eliminated. A realistic estimation of these uncertainties is required for finding out the reliability of the results. This explains the increasing demand in recent years for calculations in the nuclear fields with best-estimate codes that proved confidence bounds of simulation results. All this has lead to the Benchmark for Uncertainty Analysis in Modelling (UAM) for Design, Operation and Safety Analysis of LWRs of the NEA. The UAM-Benchmark coupling multi-physics and multi-scale analysis using as a basis complete sets of input specifications of boiling water reactors (BWR) and pressurized water reactors (PWR). In this study, the results of the transport calculations carried out using the SCALE-6.2 program (TRITON/NEWT and TRITON/KENO modules) as well as Monte Carlo SERPENT code, are presented. Additionally, they have been made uncertainties calculation for a PWR 15×15 and a BWR 7×7 fuel elements, in two different configurations (with and without control rod), and two different states, Hot Full Power (HFP) and Hot Zero Power (HZP), using the TSUNAMI module, which uses the Generalized Perturbation Theory (GPT), and SAMPLER, which uses stochastic sampling techniques for cross-sections perturbations. The results obtained and validated are compared with references results and similar studies presented in the exercise I-2 (Lattice Physics) of UAM-Benchmark.

1 Introduction

This work takes part in the framework of the Organization for Economic Cooperation and Development/Nuclear Energy Agency (OECD/NEA) Benchmark, for Uncertainty Analysis in Best-Estimate Modelling (UAM), for Design, Operation and Safety Analysis of LWRs. The groundwork was launched in 2005 with the objective to prepare a benchmark work program with steps (exercises) that would be needed to define the uncertainty and modelling task for the development of uncertainty analysis methodologies, for multi-physics and multi-scale simulation. The final goal will create a roadmap along with schedule and organization, for the development and validation of method and codes required for uncertainty and safety analysis in LWR design [1].

Reference systems and scenarios for coupled code analysis are defined to study the uncertainty effects for all stages of the system calculations. Measured data from plant operation and experimental reference data are available for the chosen scenarios. The full chain of uncertainty propagation from basic data, engineering uncertainties, across different scales (multi-scale), and physics phenomena (multi-physics) is tested on some benchmark exercises for which experimental data are available and for which the power plant details have been released. The general frame of the OECD UAM LWR Benchmark consists of three phases with different exercises for each phase: Phase 1 (neutronics phase), Phase 2 (core phase) and Phase 3 (system phase). The focus of Phase 1 is on propagating uncertainties in stand-alone neutronics calculations and consists of the following three exercises:

- Exercise I-1: "Cell Physic" focused on the derivation of the multigroup microscopic cross-section libraries and associated uncertainties;

* e-mail: alabarile@iqn.upv.es

– Exercise I-2: "Lattice Physics" focused on the derivation of the few-group macroscopic cross-section libraries and associated uncertainties;
– Exercise I-3: "Core Physics" focused on the core steady state stand-alone neutronics calculations and associated uncertainties.

The present paper deals with Cell Physic and Lattice Physics exercises, determining uncertainties associated with basic nuclear data, method and modelling approximation used in lattice physics codes.

This document is structured as follows: the Introduction in Section 1. Section 2 is focused on the description of the codes. The description of the models is shown in Section 3. The results for different codes are presented in Section 4. Finally, the conclusions are shown in Section 5.

2 Codes description

2.1 Transport calculation

In this work, two- and three-dimensional lattice codes (deterministic and stochastic) were selected to perform transport calculations: SCALE-6.2 with TRITON/NEWT and TRITON/KENO modules and SERPENT-2.1.22 code.

The SCALE code system [2] is a collection of computational modules whose execution can be linked by various "sequences" to solve a wide variety of applications.

TRITON (Transport Rigor Implemented with Time-dependent Operation for Neutronics depletion) is a multipurpose SCALE control module for transport and depletion calculations for reactor physics applications. TRITON is used to provide automated, problem-dependent cross-sections processing followed by multigroup neutron transport calculation for one-, two- or three-dimensional configuration [3].

NEWT (New ESC-based Weighting Transport code) is a two-dimensional (2D) discrete ordinates transport code developed at Oak Ridge National Laboratory. It is based on the Extended Step Characteristic (ESC) approach for spatial discretization in an arbitrary mesh structure. This discretization scheme makes NEWT an extremely powerful and versatile tool for deterministic calculation in real-world non-orthogonal problem domains. The NEWT computer code has been developed to run on SCALE. Thus, NEWT uses AMPX-formatted cross-sections processed by other SCALE modules [4].

The implemented methodology of these coupled modules of SCALE allows carrying out transport calculation with the computation of energy collapsed and homogenized macroscopic cross-sections. In TRITON, NEWT is used to calculate weighted burnup-dependent cross-sections that are employed to provide localized fluxes used for multiple depletion regions. Additionally, TRITON uses a two-pass cross-section update approach to perform fuel assembly burnup calculations and generates a database of cross-sections and other burnup-dependent physics data that can be used for full-core analysis [5].

KENO is a functional module in the SCALE system and a Monte Carlo criticality program used to calculate the k_{eff} of three-dimensional (3D) system [6]. It uses the SCALE Generalized Geometry Package (SGGP), which offers a powerful geometric representation. KENO was one of the oldest criticality safety analysis tools in SCALE. The primary purpose of its employment in this work is to determine k_{eff} calculations and compare KENO results with TRITON/NEWT and SERPENT-2 calculation.

SERPENT is a three-dimensional continuous-energy code, based on the Monte Carlo method, for reactor physics burnup calculation [7,8]. The SERPENT project started in 2004 at the VTT Technical Research Centre of Finland. The first version of the code was available to universities and research institutes from 2008 and currently it is still under development. The suggested applications of SERPENT include, among other applications, the spatial homogenization and constant group generation for deterministic reactor calculations and the validation of deterministic lattice transport codes.

2.2 Sensitivity and uncertainty calculation

Sensitivity analysis and propagation of uncertainties of cross-sections have been carried out using TSUNAMI and SAMPLER modules.

TSUNAMI-2D (Tools for Sensitivity and Uncertainty Analysis Methodology Implementation in Two Dimension) is a SCALE sequence for calculating sensitivity coefficients and response uncertainties to nuclear systems analyses for criticality safety applications. TSUNAMI uses the Generalized Perturbation Theory (GTP) that performs similarity analysis and consolidates experimental and computational results through data adjustment. The uncertainties, resulting from uncertainties in the basic data, are estimated using energy-dependent cross-section-covariance matrices [9]. The SAMS module is used to determine the sensitivities of calculated value of k_{eff} and other system responses to the nuclear data used in the calculations as a function of a nuclide, reaction type, and energy. This sensitivity of k_{eff} to the number density is equivalent to the sensitivity of k_{eff} to the total cross-section, integrated over energy. Because the total cross-section sensitivity coefficient tests much of the data used to compute all other sensitivity coefficients, it is considered an adequate test for verification. For each sensitivity coefficient examined by direct perturbation, the k_{eff} of the system is computed first with the nominal values of the input quantities, and then with a selected nominal input value increased by a certain percentage, and then with the nominal value decreased by the same percentage. The direct perturbation sensitivity coefficient of k_{eff} to some input value α is computed as:

$$S_{k,\alpha} = \frac{\alpha}{k} \times \frac{dk}{d\alpha} = \frac{\alpha}{k} \times \frac{k_{\alpha^+} - k_{\alpha^-}}{\alpha^+ - \alpha^-}, \qquad (1)$$

where α^+ and α^- represent the increased and decreased values, respectively, of the input quantity α, and k_{α^+} and k_{α^-} represent the corresponding values of k_{eff}.

Statistical uncertainties in the computed values of k_{eff} are propagated to direct perturbation sensitivity coefficients by standard error propagation techniques as:

$$\sigma_S = \sqrt{\left(\frac{\left(\sigma_{k^+}^2 + \sigma_{k^-}^2\right)}{(k^+ - k^-)} + \frac{\sigma_k^2}{k^2}\right) \times \left(\frac{k^+ - k^-}{k}\right)^2} \\ \times \frac{\alpha}{\alpha^+ - \alpha^-}. \qquad (2)$$

It is important in sensitivity calculations to ensure that the k_{eff} value of the forward and adjoint solutions closely agree, and typically, the transport calculation of concern is the adjoint calculation. More details of the GPT methodology are provided in the SAMS manual [10].

SAMPLER is a module for statistical uncertainty analysis of any SCALE sequences. The SAMPLER methodology samples probability density functions (pdf) defined by information in the SCALE multigroup covariance library by XUSA program and produces a random sample for the input computational data vector (CDV) that contains all nuclear cross-sections used in a transport calculation. After making random perturbations in input data, SAMPLER responses uncertainties are computed by statistical analysis of output responses distribution [11].

The perturbed data vector can be used in any SCALE functional module to perform a single forward solution that computes all the desired perturbed responses. The process is repeated for the specified number of random input samples, and the resulting distribution of output responses from SCALE can be analysed with standard statistical analysis tools to obtain the standard deviations and correlation coefficients for all responses. The typical approach is to assume that the generic multigroup (MG) data pdf is a multivariate normal distribution, which is completely defined by the expected values and covariance matrices for the data. An XSUSA statistical sample consists of a full set of perturbed, infinitely dilute MG data for all groups, reactions, and materials. The SCALE generic multigroup covariance data are given as relative values of the infinitely dilute cross-sections, so a random perturbation sample for cross-sections $\sigma_{x,g}(\infty)$ corresponds to $\Delta\sigma_{x,g}(\infty)/\sigma_{x,g}(\infty)$. XSUSA converts these values to a set of multiplicative perturbation factors $Q_{x,g}$ that are applied to the reference data to obtain the altered values:

$$\sigma'_{x,g} = Q_{x,g}\sigma_{x,g}, \qquad (3)$$

where

$$Q_{x,g} = 1 + \frac{\Delta\sigma_{x,g}(\infty)}{\sigma_{x,g}(\infty)}. \qquad (4)$$

Subsequently, the multiplicative perturbation factors for all data are pre-processed and stored in a data file for subsequent SCALE calculations [12].

To obtain the uncertainty and correlation coefficient, all parameters are randomly perturbed for each calculation, and the uncertainties and correlations are determined.

Mathematically, the uncertainty in an individual output parameter k is determined as:

$$\Delta k^{\exp}(i) = \hat{\mu}_i = \sqrt{\frac{1}{n-1}\sum_{a=1}^{n}\left(\left(k_{calc}^{MC}(i)\right)_a - \overline{k_{calc}^{MC}(i)}\right)^2}, \quad (5)$$

where $\Delta k^{\exp}(i)$ is the uncertainty in system i due to uncertainties in the input parameters.

$\left(k_{calc}^{MC}(i)\right)_a$ is the ath Monte Carlo (MC) sample of system i, where all uncertain input parameters have been randomly varied within the specified distribution.

The covariance between two systems, i and j, is determined as shown in equation (6).

$$\hat{\Sigma}_{ij} = \sqrt{\frac{1}{n-1}\sum_{a=1}^{n}\left(\left(k_{calc}^{MC}(i)\right)_a - \overline{k_{calc}^{MC}(i)}\right)\left(\left(k_{calc}^{MC}(j)\right)_a - \overline{k_{calc}^{MC}(j)}\right)}. \qquad (6)$$

The correlation coefficient between systems i and j can be determined from equations (5) and (6) as:

$$c_{ij} = \frac{\hat{\Sigma}_{ij}}{\hat{\mu}_i\hat{\mu}_j}. \qquad (7)$$

3 Model description

Two main LWR types have been selected for this study, based on previous benchmark experience and available data:

– Pressurized Water Reactor (PWR) - Three Mile Island 1 (TMI-1);
– Boiling Water Reactor (BWR) - Peach Bottom 2 (PB-2).

Both models have been analyzed at Hot Full Power (HFP) and Hot Zero Power (HZP) conditions.

Additionally, the two models have been designed with and without control rod.

The different fuel pin cell geometry and reference configuration are schematized in Figures 1 and 2.

SCALE calculations use the Extended Step Characteristic (ESC) approach. The entire problem domain is mapped regarding a set of finite cells. Cells sharing a given side share the value of the angular flux on that side. Once the angular flux has been determined for all sides of the cell for the given direction, it is possible to use a neutron balance to compute the average value of the angular flux within the cell. The process is then repeated for all direction. Numerical quadrature can then be used to determine the average scalar flux in each cell in the problem domain and can be used to determine fission and scattering reaction rates and to update the value of average cell source. In this way, the spatial discretization in SCALE allows to obtain satisfactory results.

The propagation of the cross-sections uncertainties across lattice physics is the main purpose of this exercise. To achieve that, in UAM-Benchmark instructions there are defined two assembly design for the models studied (a PWR

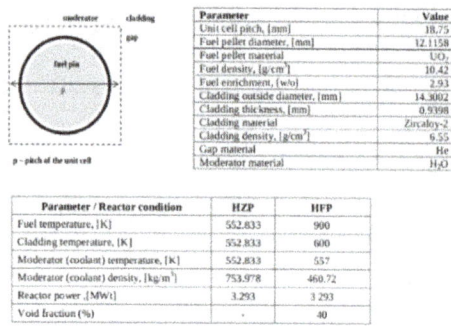

Parameter	Value
Unit cell pitch, [mm]	18.75
Fuel pellet diameter, [mm]	12.1158
Fuel pellet material	UO₂
Fuel density, [g/cm³]	10.42
Fuel enrichment, (w/o)	2.93
Cladding outside diameter, [mm]	14.3002
Cladding thickness, [mm]	0.9398
Cladding material	Zircaloy-2
Cladding density, [g/cm³]	6.55
Gap material	He
Moderator material	H₂O

Parameter / Reactor condition	HZP	HFP
Fuel temperature, [K]	552.833	900
Cladding temperature, [K]	552.833	600
Moderator (coolant) temperature, [K]	552.833	557
Moderator (coolant) density, [kg/m³]	753.978	460.72
Reactor power ,[MWt]	3.293	3.293
Void fraction (%)	-	40

Fig. 1. The configuration of PB-2 BWR unit cell.

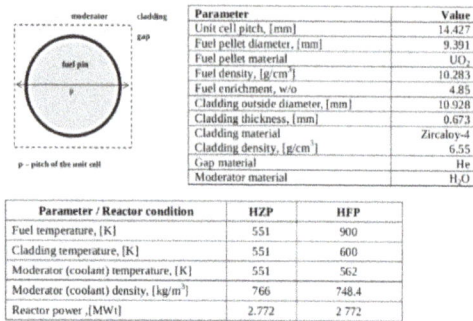

Parameter	Value
Unit cell pitch, [mm]	14.427
Fuel pellet diameter, [mm]	9.391
Fuel pellet material	UO₂
Fuel density, [g/cm³]	10.283
Fuel enrichment, w/o	4.85
Cladding outside diameter, [mm]	10.928
Cladding thickness, [mm]	0.673
Cladding material	Zircaloy-4
Cladding density, [g/cm³]	6.55
Gap material	He
Moderator material	H₂O

Parameter / Reactor condition	HZP	HFP
Fuel temperature, [K]	551	900
Cladding temperature, [K]	551	600
Moderator (coolant) temperature, [K]	551	562
Moderator (coolant) density, [kg/m³]	766	748.4
Reactor power ,[MWt]	2.772	2.772

Fig. 2. The configuration of TMI-1 PWR unit cell.

and a BWR) [13]. These assemblies are shown in Figures 3 and 4 while Table 1 shows both fuel assemblies data.

As a result of implementing both fuel assemblies in TRITON/NEWT code for transport calculation, the layout outputs for both types of assemblies have been obtained, as shown in Figures 5 and 6.

There are different methodologies used in our calculation of this work. These methods cover the deterministic approach (TRITON/NEWT) and Monte Carlo methodology (TRITON/KENO and SERPENT), as well as the Generalized Perturbation Theory (TSUNAMI) to stochastic sampling techniques (SAMPLER).

4 Results

In this section, the results of transport calculation with TRITON/NEWT, SERPENT-2 and TRITON/KENO are

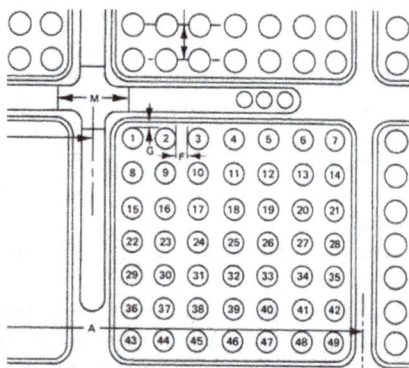

Fig. 3. PB-2 BWR assembly design.

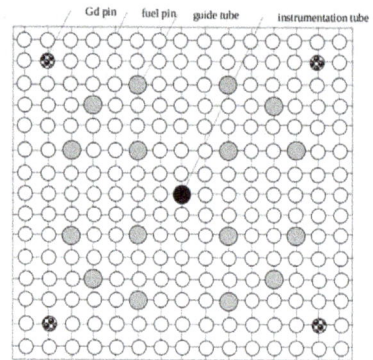

Fig. 4. TMI-1 PWR assembly design.

Table 1. Fuel assembly data for the test cases of Exercises I-2.

Parameter	BWR	PWR
FA geometry	7 × 7	15 × 15
FA pitch (mm)	152.4	218.11
Fuel rods per assembly	49	208
Number of guide tubes per FA	–	16
Number of instrumentation tubes per FA	–	1
Number of GD pins per FA	4	4
Guide tube outside diameter (mm)	–	13.462
Guide tube inside diameter (mm)	–	12.649
Instrumentation tube outside diameter (mm)	–	12.522
Instrumentation tube inside diameter (mm)	–	11.201

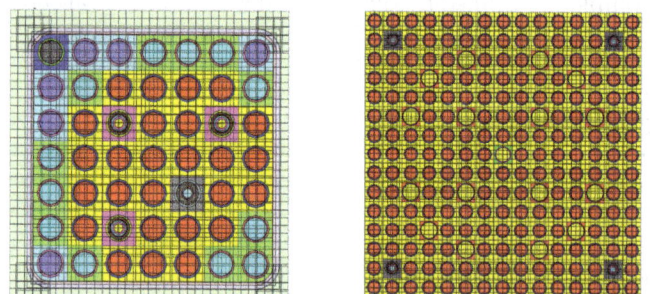

Fig. 5. BWR and PWR without control rod 2D assembly.

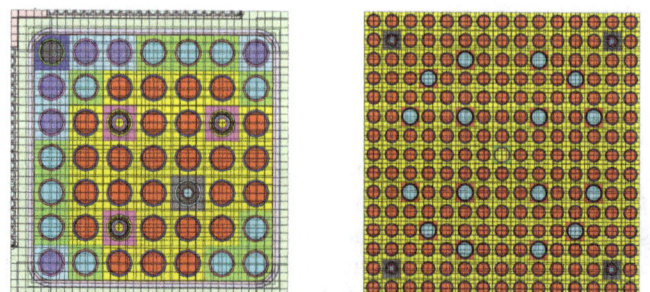

Fig. 6. BWR and PWR with control rod 2D assembly.

Table 2. Parameter list to compare - Exercise I-2.

Output	Description	Units
k_assembly	Eigenvalue/multiplication factor for two-group assembly	–
fuel_maca_1/2	Macroscopic absorption cross-section for both groups	1/cm
fuel_macf_1/2	Macroscopic fission cross-section for both groups	1/cm
diff_1/2	Diffusion coefficient for both groups	cm
flux_1/2	Neutron flux for both groups	$1/cm^2s$

presented, as well as the sensitivity analysis and propagation of uncertainties results that were performed using TSUNAMI and SAMPLER modules. All calculations were carried out for four assemblies models (like shown in Figs. 5 and 6) and two different states (HFP and HZP), with a total of eight configurations for this test case in UAM-Benchmark, Exercise I-2.

4.1 TRITON/NEWT and SERPENT-2 results

Results for the k_{eff} and cross-section values are summarized in this section. The aim is to compare TRITON/NEWT and SERPENT-2 results with the average of Benchmark participation values. In the TRITON/NEWT modules, the 238-group nuclear data library collapse was used. Otherwise, the computation with the SERPENT-2.1.22 code was carried out with two libraries, JEFF-3.1 and ENDF/B-VII, to compare results with SCALE (ENDF/B-VII.1 library).

The output values compared in this paper are listed in Table 2. Comparison of TRITON/NEWT and SERPENT-2 are carried out for the multiplication factor (k_{eff}), the Absorption cross-section, the Fission cross-section, the Diffusion cross-section and the Flux. All cross-section values are presented for both groups, Fast and Thermal.

In Tables 3 and 4, the first column shows the output values compared in this paper, the second one are the reference values found in UAM-Benchmark results. The third and fourth columns represent TRITON/NEWT calculations and its error with UAM-Benchmark references. Subsequently, the results of SERPENT-2 calculations and their comparison with TRITON/NEWT and UAM-Benchmark values are presented.

The reference values adopted in these tables have been calculated as an average of all submitted results of all benchmark participants, referring to the last submission of 2013.

It should be considered that each participant uses their own code, and this can introduce errors due to the different methodology of each code, but the objective of the benchmark program takes into account these discrepancies.

Therefore, in this work, we calculated the average results of all participants and it is intended to calculate the error between our simulation against SCALE and SERPENT codes.

It is evident from Table 3 that there is good agreement between TRITON/NEWT and reference values, especially

for the unrodded configurations. There is a slight recurring discrepancy in the flux value (in both groups, one and two), but it is assumed to be because of different normalization methods of implemented codes in this benchmark exercise.

In Table 3, comparing the SERPENT-2.1.22 results with TRITON/NEWT and UAM-Benchmark, it can be seen a good agreement in both models (BWR and PWR) as has been observed for SCALE results. The slight differences, especially in flux and diffusion coefficient results, can be due to the different methodologies (Monte Carlo) implemented in SERPENT code. In fact, in order to obtain most accurate results, the option B1 was adopted in SERPENT calculation while this option is not available in the SCALE beta version used in this work.

Additionally, we have compared SERPENT calculations using both JEFF-3.1 and ENDF/B-VII libraries, and the JEFF-3.1 results are more close to TRITON/NEWT values despite TRITON/NEWT uses the ENDF/B-VII library.

Even though the slight differences comparing SERPENT with validated and reference results, it could be a good transport calculation code, ongoing testing and validation.

With the aim to show a clearest exposition of the results, the standard deviations in SERPENT, which are lower than 30 pcm for all cases presented, are avoided in Tables 3 and 4.

Table 4 shows good agreement between TRITON/NEWT and reference values like it was presented in Table 3. Comparing SERPENT-2 results with TRITON/NEWT and UAM-Benchmark, the results are similar, as shown in PWR configuration in Table 3.

Even though a different approach is used in the SERPENT code, there is an acceptable agreement between SERPENT-2 and TRITON/NEWT and Benchmark values. In this case, JEFF-3.1 library results are a little bit better comparing with ENDF/B-VII library.

As a conclusion, it is important to give relevance to the short computational times in transport calculation for SERPENT code that was found faster than TRITON/NEWT code. In fact, in SERPENT calculation they have simulated 50,000 particles and 350 cycles in which the first 50 cycles were discarded because of its low statistical weight.

For all exercises, it was used a cluster composed of four blocks with 18 servers equipped with two processors Intel Xeon E5-4620 8c/16T and with a RAM of 64 GB DDR3, and 2 × interfaces 1 0 GbE.

Table 3. Comparison of cross-section values of PWR Benchmark configuration exercises 1-2.

Output	Benchmark	SCALE (TRITON/NEWT)	Error (%)	SERPENT-2 (JEFF)	Error (%)	SERPENT-2 vs SCALE (%)	SERPENT-2 (ENDFB)	Error (%)	SERPENT-2 vs SCALE (%)
PWR_HFP_unrodded									
k_assembly	1.398E+00	1.394E+00	0.30	1.387E+00	0.78	0.48	1.387E+00	0.82	0.52
fuel_maca_1	1.040E-02	1.077E-02	3.58	1.016E-02	2.32	6.04	1.016E-02	2.30	6.02
fuel_maca_2	1.090E-01	1.098E-01	0.78	1.109E-01	1.76	0.97	1.111E-01	1.96	1.16
fuel_macf_1	3.530E-03	3.613E-03	2.35	3.459E-03	2.00	4.44	3.459E-03	2.01	4.45
fuel_macf_2	7.720E-02	7.848E-02	1.66	7.910E-02	2.46	0.78	7.926E-02	2.66	0.98
diff_1	1.423E+00	1.430E+00	0.47	1.352E+00	4.97	5.71	1.353E+00	4.94	5.68
diff_2	3.640E-01	3.623E-01	0.46	4.080E-01	12.10	11.20	4.119E-01	13.17	12.04
flux_1	8.648E-01	8.658E-01	0.12	8.808E-01	1.86	1.70	8.811E-01	1.88	1.73
flux_2	1.352E-01	1.342E-01	0.77	1.192E-01	11.85	12.57	1.189E-01	12.04	12.81
PWR_HZP_unrodded									
k_assembly	1.413E+00	1.411E+00	0.17	1.404E+00	0.63	0.46	1.404E+00	0.61	0.45
fuel_maca_1	1.050E-02	1.060E-02	0.95	9.976E-03	4.99	6.26	9.978E-03	4.97	6.23
fuel_maca_2	1.110E-01	1.111E-01	0.09	1.127E-01	1.54	1.42	1.131E-01	1.87	1.74
fuel_macf_1	3.550E-03	3.624E-03	2.08	3.466E-03	2.36	4.55	3.465E-03	2.39	4.59
fuel_macf_2	7.870E-02	7.933E-02	0.80	8.041E-02	2.17	1.34	8.067E-02	2.51	1.67
diff_1	1.371E+00	1.416E+00	3.29	1.335E+00	2.59	6.04	1.336E+00	2.58	6.03
diff_2	3.480E-01	3.542E-01	1.77	4.025E-01	15.66	12.01	4.063E-01	16.74	12.83
flux_1	8.572E-01	8.629E-01	0.67	8.792E-01	2.56	1.85	8.795E-01	2.60	1.89
flux_2	1.428E-01	1.371E-01	4.01	1.209E-01	15.36	13.41	1.205E-01	15.62	13.76
PWR_HFP_rodded									
k_assembly	1.065E+00	1.025E+00	3.77	1.011E+00	5.04	1.34	1.012E+00	4.98	1.27
fuel_maca_1	1.290E-02	1.356E-02	5.10	1.360E-02	5.41	0.29	1.360E-02	5.45	0.33
fuel_maca_2	1.350E-01	1.403E-01	3.89	1.421E-01	5.25	1.30	1.424E-01	5.45	1.48
fuel_macf_1	3.420E-03	3.478E-03	1.71	3.427E-03	0.19	1.51	3.427E-03	0.19	1.51
fuel_macf_2	7.890E-02	8.049E-02	2.02	8.121E-02	2.93	0.89	8.138E-02	3.14	1.09
diff_1	1.430E+00	1.393E+00	2.58	1.373E+00	3.98	1.45	1.373E+00	3.98	1.45
diff_2	3.610E-01	3.659E-01	1.37	4.207E-01	16.54	13.02	4.242E-01	17.51	13.73
flux_1	9.017E-01	9.110E-01	1.03	9.147E-01	1.44	0.41	9.148E-01	1.45	0.42
flux_2	9.827E-02	8.901E-02	9.42	8.529E-02	13.21	4.37	8.518E-02	13.32	4.50
PWR_HZP_rodded									
k_assembly	1.095E+00	1.040E+00	4.99	1.037E+00	5.27	0.30	1.037E+00	5.26	0.29
fuel_maca_1	1.280E-02	1.342E-02	4.86	1.313E-02	2.61	2.19	1.314E-02	2.63	2.18
fuel_maca_2	1.360E-01	1.415E-01	4.05	1.453E-01	6.82	2.60	1.458E-01	7.17	2.91

Table 3. (continued).

Output	Benchmark	SCALE (TRITON/NEWT)	Error (%)	SERPENT-2 (JEFF)	Error (%)	SERPENT-2 vs SCALE (%)	SERPENT-2 (ENDFB)	Error (%)	SERPENT-2 vs SCALE (%)
fuel_macf_1	3.430E-03	3.493E-03	1.84	3.442E-03	0.35	1.48	3.444E-03	0.39	1.44
fuel_macf_2	7.970E-02	8.139E-02	2.13	8.284E-02	3.94	1.74	8.307E-02	4.22	2.01
diff_1	1.406E+00	1.38 1E+00	1.77	1.356E+00	3.55	1.84	1.356E+00	3.55	1.84
diff_2	3.550E-01	3.583E-01	0.94	4.141E-01	16.66	13.48	4.176E-01	17.63	14.19
flux_1	9.0l8E-01	9.084E-01	0.73	9.128E-01	1.22	0.48	9.130E-01	1.24	0.50
flux_2	9.8l8E-02	9.156E-02	6.74	8.724E-02	11.15	4.96	8.700E-02	11.38	5.24

Table 4. Comparison of cross-section values of BWR Benchmark configuration exercises I-2.

Output	Benchmark	SCALE (TRITON/NEWT)	Error (%)	SERPENT-2 (JEFF)	Error (%)	SERPENT-2 vs SCALE (%)	SERPENT-2 (ENDFB)	Error (%)	SERPENT-2 vs SCALE (%)
BWR_HFP_unrodded									
k_assembly	1.076E+00	1.080E+00	0.38	1.081E+00	0.44	0.06	1.081E+00	0.46	0.08
fuel_maca_1	6.880E-03	6.934E-03	0.79	6.848E-03	0.47	1.26	6.847E-03	0.48	1.27
fuel_maca_2	5.250E-02	5.196E-02	1.02	5.285E-02	0.66	1.67	5.304E-02	1.03	2.03
fuel_macf_1	1.780E-03	1.817E-03	2.07	1.818E-03	2.13	0.06	1.818E-03	2.15	0.08
fuel_macf_2	2.610E-02	2.708E-02	3.75	2.761E-02	5.77	1.91	2.770E-02	6.15	2.26
diff_1	1.768E+00	1.645E+00	6.98	1.611E+00	8.88	2.08	1.611E+00	8.86	2.06
diff_2	4.260E-01	4.117E-01	3.35	4.567E-01	7.22	9.86	4.626E-01	8.58	10.99
flux_1	8.015E-01	7.765E-01	3.12	7.905E-01	1.37	1.77	7.911E-01	1.30	1.85
flux_2	1.985E-01	2.235E-01	12.59	2.095E-01	5.53	6.69	2.089E-01	5.24	6.99
BWR_HZP_unrodded									
k_assembly	1.108E+00	1.106E+00	0.20	1.107E+00	0.07	0.13	1.108E+00	0.04	0.16
fuel_maca_1	7.150E-03	7.201E-03	0.71	7.098E-03	0.72	1.45	7.098E-03	0.73	1.45
fuel_maca_2	5.460E-02	5.542E-02	1.50	5.621E-02	2.95	1.41	5.646E-02	3.41	1.85
fuel_macf_1	1.900E-03	1.909E-03	0.47	1.912E-03	0.64	0.17	1.912E-03	0.63	0.16

Table 4. (continued).

Output	Benchmark	SCALE (TRITON/NEWT)	Error (%)	SERPENT-2 (JEFF)	Error (%)	SERPENT-2 vs SCALE (%)	SERPENT-2 (ENDFB)	Error (%)	SERPENT-2 vs SCALE (%)
fuel_macf_2	2.910E-02	2.852E-02	1.99	2.894E-02	0.55	1.45	2.907E-02	0.10	1.90
diff_1	1.496E+00	1.455E+00	2.74	1.387E+00	7.27	4.89	1.388E+00	7.25	4.86
diff_2	3.320E-01	3.367E-01	1.40	3.646E-01	9.81	7.66	3.696E-01	11.31	8.91
flux_1	7.330E-01	7.284E-01	0.62	7.411E-01	1.11	1.71	7.420E-01	1.22	1.83
flux_2	2.670E-01	2.716E-01	1.71	2.589E-01	3.03	4.89	2.581E-01	3.36	5.24
BWR_HFP_rodded									
k_assembly	7.870E-01	7.689E-01	2.31	7.957E-01	1.11	3.37	7.950E-01	1.02	3.29
fuel_maca_1	9.530E-03	9.121E-03	4.29	9.697E-03	1.75	5.93	9.698E-03	1.76	5.95
fuel_maca_2	7.100E-02	7.334E-02	3.30	7.210E-02	1.55	1.72	7.232E-02	1.86	1.41
fuel_macf_1	1.830E-03	1.750E-03	4.38	1.816E-03	0.74	3.67	1.817E-03	0.72	3.69
fuel_macf_2	3.070E-02	3.097E-02	0.87	3.084E-02	0.46	0.41	3.092E-02	0.70	0.16
diff_1	1.713E+00	1.698E+00	0.87	1.603E+00	6.44	5.96	1.603E+00	6.43	5.94
diff_2	4.680E-01	5.195E-01	11.01	4.892E-01	4.53	6.19	4.945E-01	5.65	5.07
flux_1	8.696E-01	8.827E-01	1.51	8.545E-01	1.73	3.30	8.550E-01	1.67	3.24
flux_2	1.304E-01	1.173E-01	10.10	1.455E-01	11.55	19.40	1.450E-01	11.18	19.13
BWR_HZP_rodded									
k_assembly	8.620E-01	8.574E-01	0.53	8.667E-01	0.54	1.06	8.662E-01	0.49	1.02
fuel_maca_1	9.790E-03	9.503E-03	2.93	9.901E-03	1.14	4.03	9.898E-03	1.10	3.99
fuel_maca_2	7.410E-02	7.622E-02	2.87	7.590E-02	2.43	0.43	7.619E-02	2.83	0.04
fuel_macf_1	1.970E-03	1.905E-03	3.31	1.932E-03	1.92	1.41	1.932E-03	1.91	1.43
fuel_macf_2	3.180E-02	3.249E-02	2.16	3.313E-02	4.20	1.95	3.323E-02	4.49	2.23
diff_1	1.466E+00	1.429E+00	2.50	1.387E+00	5.37	3.03	1.388E+00	5.34	2.99
diff_2	3.390E-01	3.467E-01	2.27	3.803E-01	12.19	8.84	3.848E-01	13.51	9.90
flux_1	8.029E-01	8.091E-01	0.77	8.069E-01	0.50	0.27	8.074E-01	0.56	0.21
flux_2	1.971E-01	1.909E-01	3.14	1.931E-01	2.02	1.14	1.926E-01	2.28	0.88

Table 5. Comparison of KENO and NEWT k_{eff} assembly results.

	TRITON/KENO	TRITON/NEWT	Error (%)
PWR_HFP_unrodded	1.40031 ± 0.00036	1.394E+00	0.45
PWR_HZP_unrodded	1.41584 ± 0.00044	1.411E+00	0.34
PWR_HFP_rodded	1.02114 ± 0.00057	1.025E+00	0.38
PWR_HZP_rodded	1.03528 ± 0.00047	1.040E+00	0.45

BWR_HFP_unrodded Uncertainty Information

The relative standard deviation of K_{eff} (%$\Delta k/k$) due to cross-section covariance data is: 0.5358 % $\Delta k/k$

contributions to uncertainty in K_{eff} (%$\Delta k/k$) by individual energy covariance matrices:

Covariance Matrix		Contributions to Uncertainty in K_{eff} (%$\Delta k/k$)
Nuclide-Reaction	Nuclide-Reaction	Due to this Matrix
$^{238}U_{n,gamma}$	$^{238}U_{n,gamma}$	3.0317E-01
$^{235}U_{nubar}$	$^{235}U_{nubar}$	1.7051E-01
$^{238}U_{n,n'}$	$^{238}U_{n,n'}$	1.6967E-01
$^{235}U_{n,gamma}$	$^{235}U_{n,gamma}$	1.4249E-01

Forward Calculation: K_{eff} = 1.08039129

Adjoint Calculation: K_{eff} = 1.08035940

BWR_HZP_unrodded Uncertainty Information

The relative standard deviation of K_{eff} (%$\Delta k/k$) due to cross-section covariance data is: 0.5061 % $\Delta k/k$

contributions to uncertainty in K_{eff} (%$\Delta k/k$) by individual energy covariance matrices:

Covariance Matrix		Contributions to Uncertainty in K_{eff} (%$\Delta k/k$)
Nuclide-Reaction	Nuclide-Reaction	Due to this Matrix
$^{238}U_{n,gamma}$	$^{238}U_{n,gamma}$	2.7926E-01
$^{235}U_{nubar}$	$^{235}U_{nubar}$	2.7820E-01
$^{235}U_{n,gamma}$	$^{235}U_{n,gamma}$	1.4214E-01
$^{235}U_{fission}$	$^{235}U_{fission}$	1.4189E-01

Forward Calculation: K_{eff} = 1.10596891

Adjoint Calculation: K_{eff} = 1.10593861

BWR_HFP_rodded Uncertainty Information

The relative standard deviation of K_{eff} (%$\Delta k/k$) due to cross-section covariance data is: 0.6012 % $\Delta k/k$

contributions to uncertainty in K_{eff} (%$\Delta k/k$) by individual energy covariance matrices:

Covariance Matrix		Contributions to Uncertainty in K_{eff} (%$\Delta k/k$)
Nuclide-Reaction	Nuclide-Reaction	Due to this Matrix
$^{238}U_{n,n'}$	$^{238}U_{n,n'}$	3.0300E-01
$^{238}U_{n,gamma}$	$^{238}U_{n,gamma}$	2.9372E-01
$^{235}U_{nubar}$	$^{235}U_{nubar}$	2.4778E-01
$^{235}U_{chi}$	$^{235}U_{chi}$	2.0790E-01

Forward Calculation: K_{eff} = 0.76879110

Adjoint Calculation: K_{eff} = 0.76878164

BWR_HZP_rodded Uncertainty Information

The relative standard deviation of K_{eff} (%$\Delta k/k$) due to cross-section covariance data is: 0.4936 % $\Delta k/k$

contributions to uncertainty in K_{eff} (%$\Delta k/k$) by individual energy covariance matrices:

Covariance Matrix		Contributions to Uncertainty in K_{eff} (%$\Delta k/k$)
Nuclide-Reaction	Nuclide-Reaction	Due to this Matrix
$^{235}U_{nubar}$	$^{235}U_{nubar}$	2.6840E-01
$^{238}U_{n,gamma}$	$^{238}U_{n,gamma}$	2.5420E-01
$^{235}U_{fission}$	$^{235}U_{fission}$	1.4959E-01
$^{238}U_{n,n'}$	$^{238}U_{n,n'}$	1.3228E-01

Forward Calculation: K_{eff} = 0.85751511

Adjoint Calculation: K_{eff} = 0.85750102

Fig. 7. Most important contributor to uncertainty in k_{eff} (%$\Delta k/k$) in BWRs.

A comparison of the total computational time between SERPENT and TRITON/NEWT calculations, for PWR_HFP_unrodded configuration, returns:

- SERPENT total computational time (seconds): 935;
- TRITON/NEWT total computational time (seconds): 1358.

4.2 TRITON/KENO and SERPENT-2 results

From TRITON/KENO calculation, the best estimate system of k_{eff} in 3D Monte Carlo methodology is obtained, and it is possible to compare these results with deterministic values as shown in Table 5. In this table, the first column shows TRITON/KENO results, the k_{eff} values obtained in TRITON/NEWT figured in the second column, and its error with TRITON/KENO was finally reported.

Analysing Table 5 is possible to corroborate a good agreement through both SCALE models since the largest error calculated has been 0.45%.

4.3 TSUNAMI-2D results

TSUNAMI module is very useful for sensitivity analysis and propagation of uncertainties of cross-sections. TSUNAMI employs the Generalized Perturbation Theory and determines the sensitivities coefficient for each nuclide. Furthermore, the sensitivity coefficients for total reaction for each nuclide and mixture can be calculated with this module.

A list of four major contributors to the uncertainty in k_{eff} by individual energy covariance matrices is presented in Figure 7 for BWR calculations and in Figure 8 for PWR calculations.

In both figures, it is possible to find out that the list of major contributors does not vary greatly from case to case, and for all cases uranium seems to be responsible for the uncertainty of k_{eff}. In particular, $^{238}U_{n,\gamma}$, $^{235}U_{nubar}$, $^{238}U_{n,n'}$, are present at the top of contributors list.

Finally, looking at these results is possible to wise up the relative standard deviation due to cross-section covariance data. The relative standard deviation has been close to 0.50% in all cases, for both BWRs and PWRs calculations.

Likewise, it is interesting to analyse the sensitivity profiles of these major contributors, as shown in Figures 9 and 10. The

PWR_HFP_unrodded Uncertainty Information

The relative standard deviation of K $_{eff}$ (% Δk/k) due to cross-section covariance data is: 0.4713 % Δk/k

contributions to uncertainty in K $_{eff}$ (%Δk/k) by individual energy covariance matrices:

Covariance Matrix		Contributions to Uncertainty in K$_{eff}$ (%Δk/k)
Nuclide-Reaction	Nuclide-Reaction	Due to this Matrix
$^{235}U_{nubar}$	$^{235}U_{nubar}$	2.6798E-01
$^{238}U_{n,gamma}$	$^{238}U_{n,gamma}$	2.5618E-01
$^{235}U_{n,gamma}$	$^{235}U_{n,gamma}$	2.0147E-01
$^{235}U_{fission}$	$^{235}U_{n,gamma}$	1.0809E-01

Forward Calculation: K$_{eff}$ = 1.39405262

Adjoint Calculation: K$_{eff}$ = 1.39403963

PWR_HZP_unrodded Uncertainty Information

The relative standard deviation of K $_{eff}$ (% Δk/k) due to cross-section covariance data is: 0.4651 % Δk/k

contributions to uncertainty in K $_{eff}$ (%Δk/k) by individual energy covariance matrices:

Covariance Matrix		Contributions to Uncertainty in K$_{eff}$ (%Δk/k)
Nuclide-Reaction	Nuclide-Reaction	Due to this Matrix
$^{235}U_{nubar}$	$^{235}U_{nubar}$	2.6906E-01
$^{238}U_{n,gamma}$	$^{238}U_{n,gamma}$	2.4790E-01
$^{235}U_{n,gamma}$	$^{235}U_{n,gamma}$	2.0090E-01
$^{235}U_{fission}$	$^{235}U_{n,gamma}$	1.0834E-01

Forward Calculation: K$_{eff}$ = 1.41055640

Adjoint Calculation: K$_{eff}$ = 1.41054241

PWR_HFP_rodded Uncertainty Information

The relative standard deviation of K $_{eff}$ (% Δk/k) due to cross-section covariance data is: 0.5552 % Δk/k

contributions to uncertainty in K $_{eff}$ (%Δk/k) by individual energy covariance matrices:

Covariance Matrix		Contributions to Uncertainty in K$_{eff}$ (%Δk/k)
Nuclide-Reaction	Nuclide-Reaction	Due to this Matrix
$^{238}U_{n,n'}$	$^{238}U_{n,n'}$	6.0318E-01
$^{238}U_{n,gamma}$	$^{238}U_{n,gamma}$	3.5015E-01
$^{235}U_{n,gamma}$	$^{235}U_{n,gamma}$	3.0614E-01
$^{235}U_{chi}$	$^{235}U_{chi}$	2.3938E-01

Forward Calculation: K$_{eff}$ = 1.06274374

Adjoint Calculation: K$_{eff}$ = 1.06274353

PWR_HZP_rodded Uncertainty Information

The relative standard deviation of K $_{eff}$ (% Δk/k) due to cross-section covariance data is: 0.5273 % Δk/k

contributions to uncertainty in K $_{eff}$ (%Δk/k) by individual energy covariance matrices:

Covariance Matrix		Contributions to Uncertainty in K$_{eff}$ (%Δk/k)
Nuclide-Reaction	Nuclide-Reaction	Due to this Matrix
$^{238}U_{n,n'}$	$^{238}U_{n,n'}$	5.7584E-01
$^{238}U_{n,gamma}$	$^{238}U_{n,gamma}$	3.3700E-01
$^{235}U_{n,gamma}$	$^{235}U_{n,gamma}$	3.0398E-01
$^{235}U_{chi}$	$^{235}U_{chi}$	2.3223E-01

Forward Calculation: K$_{eff}$ = 1.08104281

Adjoint Calculation: K$_{eff}$ = 1.08104261

Fig. 8. Most important contributor to uncertainty in k_{eff} ($\%\Delta k/k$) in PWRs.

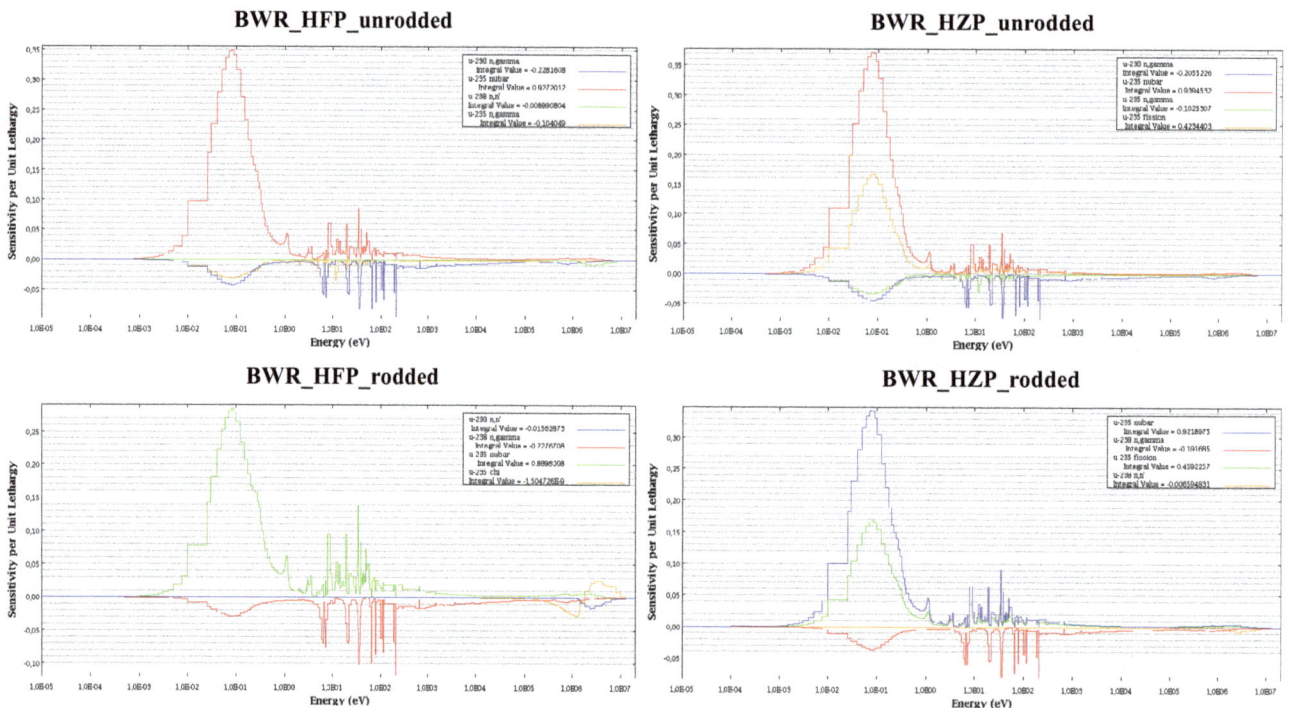

Fig. 9. Sensitivity profiles of most important contributor to uncertainty in k_{eff}, BWRs cases.

sensitivity per unit lethargy profiles looks similar, with peaks ranging from 0.28 to 0.35. Sensitivity at Hot Full Power state has emerged greater than sensitivity at Hot Zero Power state, for all cases of study. Moreover, sensitivity profile varies only slightly from case to case, and they do not lead to changes in the uncertainty of the k_{eff}. Figures 9 and 10 represent BWRs and PWRs calculations, respectively.

4.4 SAMPLER results

SAMPLER calculation provides not only estimates for the expected values of the data but also covariance data describing the correlated uncertainty. SAMPLER repeats perturbation steps for a specified number of samples to obtain a distribution of results that can be converted to a

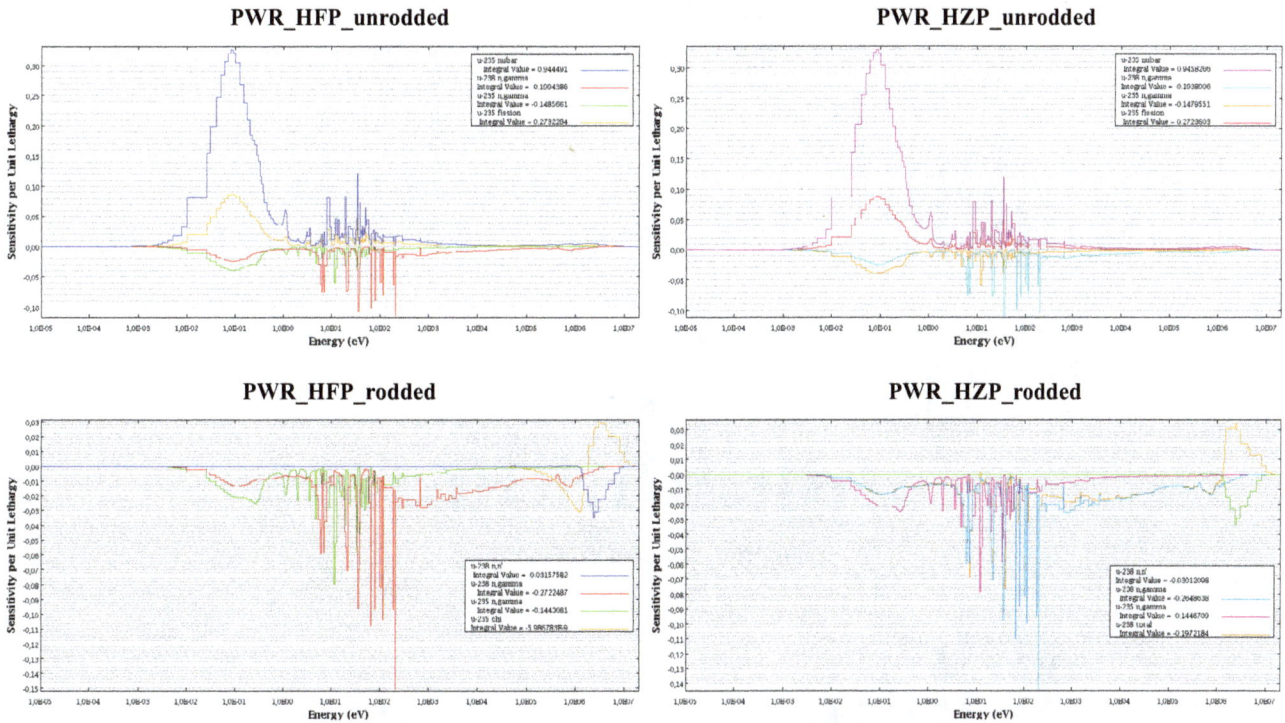

Fig. 10. Sensitivity profiles of most important contributor to uncertainty in k_{eff}, PWRs cases.

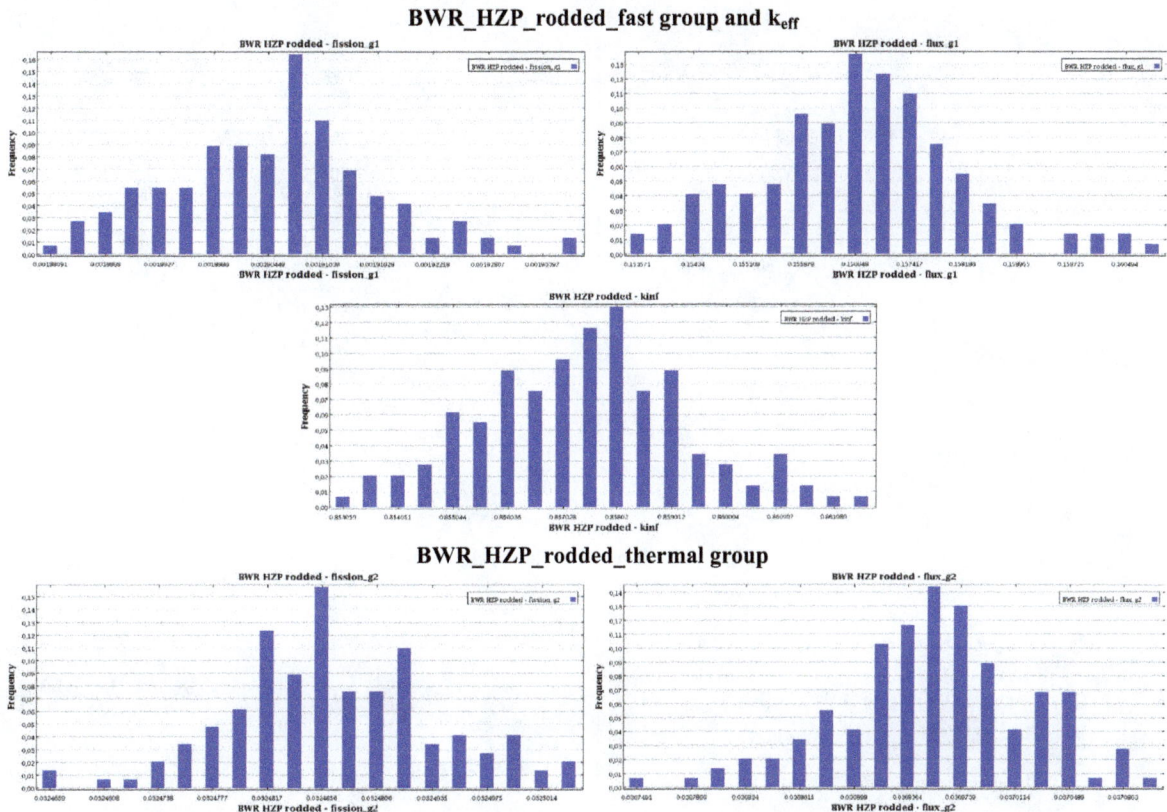

Fig. 11. Histogram plot for k_{eff} and two cross-section calculations, for BWR HZP rodded case.

PWR_HFP_unrodded_fast group and k_eff

PWR_HFP_unrodded_thermal group

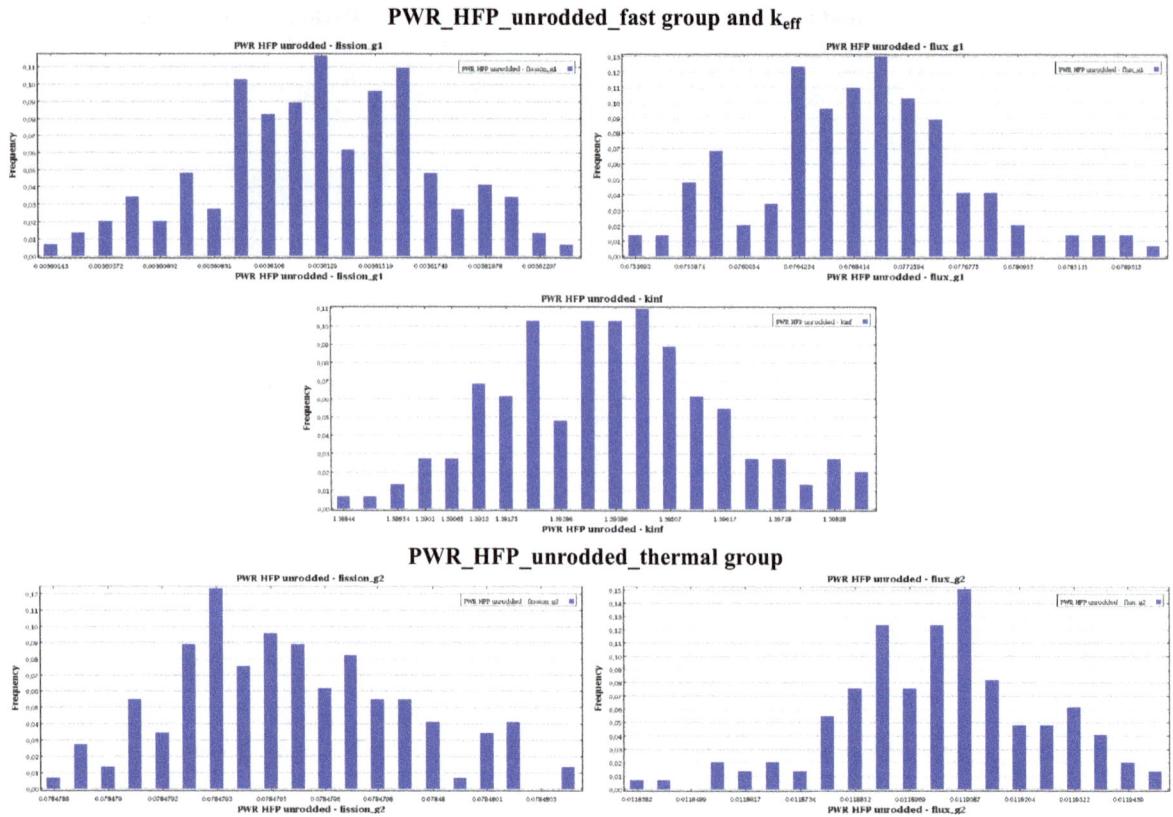

Fig. 12. Histogram plot for k_{eff} and two cross-section calculations, for PWR HFP unrodded case.

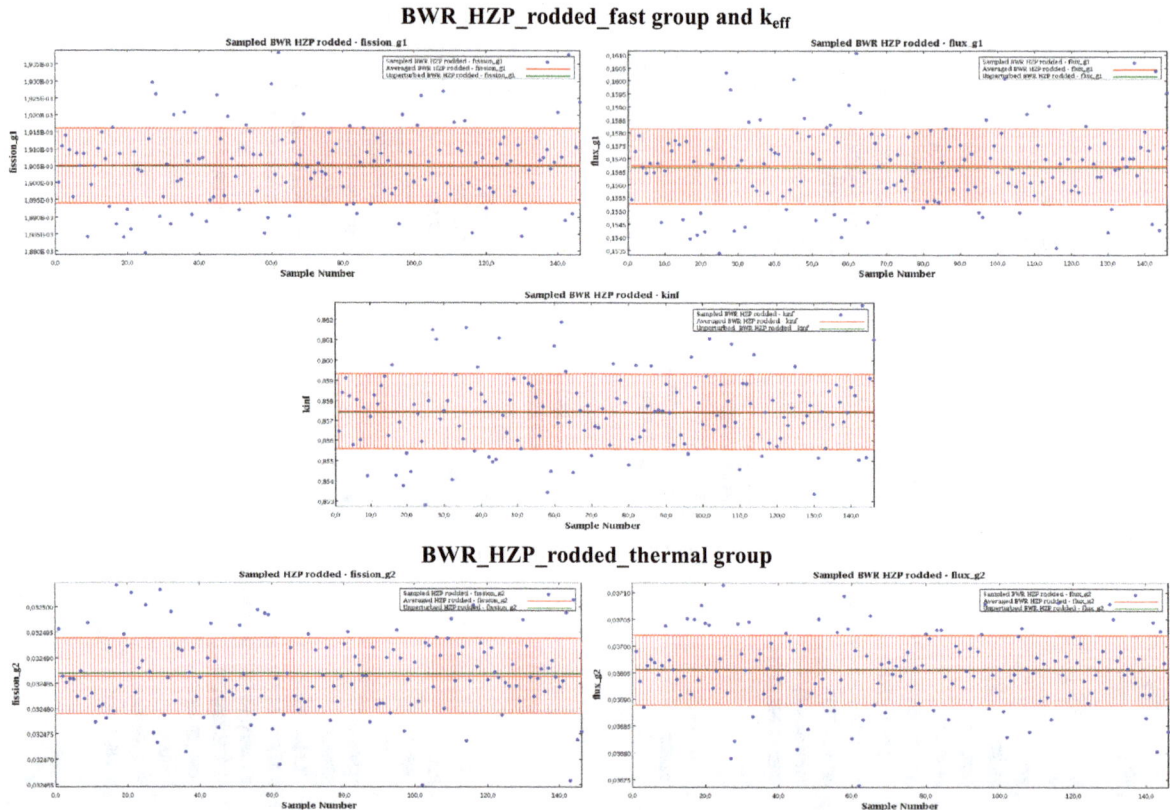

BWR_HZP_rodded_fast group and k_eff

BWR_HZP_rodded_thermal group

Fig. 13. Samples population for k_{eff} and two cross-section calculations (fission and flux), for BWR_HZP with control rod.

PWR_HFP_unrodded_fast group and k$_{eff}$

PWR_HFP_unrodded_thermal group

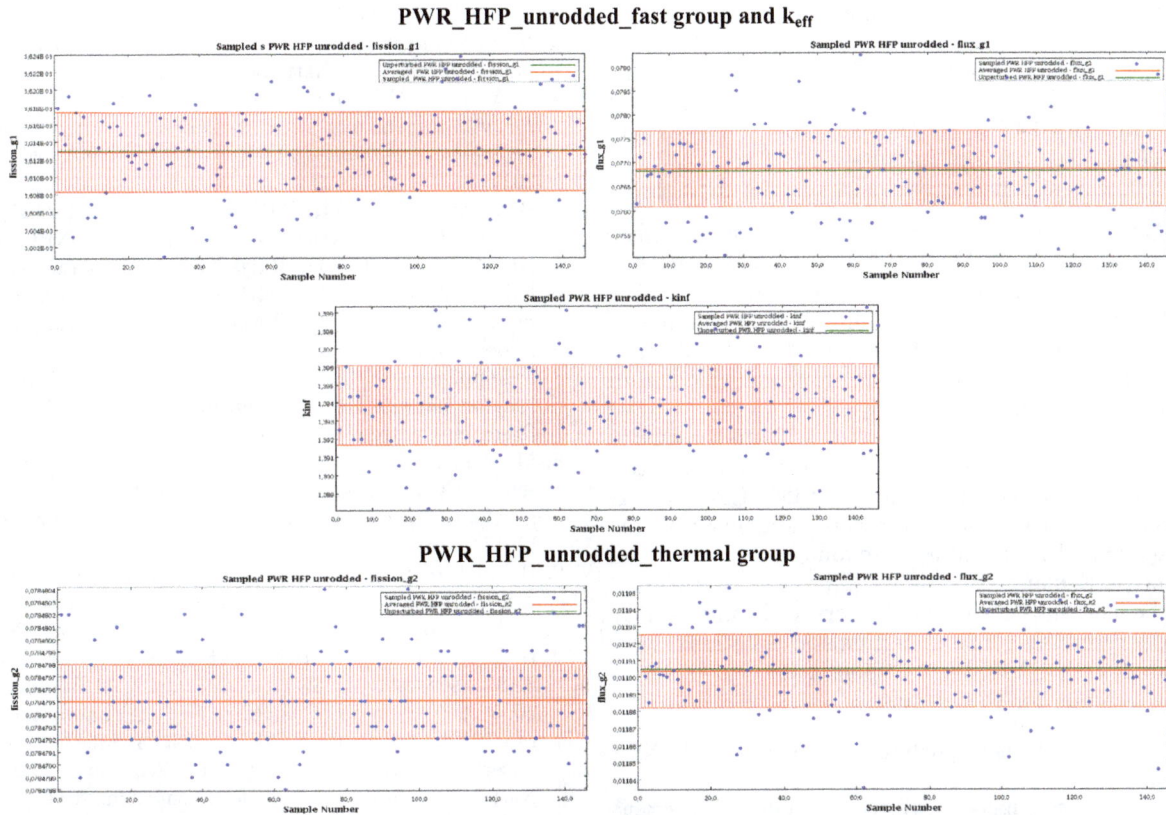

Fig. 14. Samples population for k_{eff} and two cross-section calculations (fission and flux), for PWR_HFP without control rod.

standard deviation and correlation coefficients. The SCALE Criticality Safety Analysis Sequence (CSAS) with the 238-group nuclear data library was used for the computations.

Based on the Wilks' approach [14], the sample size for double tolerance limits with a 95% of uncertainty and with 95% of statistical confidence for the output variables is equal to 146 samples [15], which is the number of runs performed in this work.

To provide information on sampling convergence, layout response of SAMPLER results is presented. For every case run within SAMPLER, any number of responses can be extracted. For instance, the histogram plots that indicate the distribution of the k_{eff} and some cross-section computed values for these benchmark exercises are listed below. Figure 11 shows histogram results of BWR with control rod, at hot zero power state (HZP). The k_{eff} and two cross-sections (fission and flux) values for the fast and thermal group are represented.

While Figure 12 shows PWR histogram results, in unrodded configuration and at hot full power state (HFP).

In the same way, are represented the k_{eff} and fission, and flux cross-section results for the fast and thermal group.

According to the print flag set by the user, SAMPLER also prints a list of tables with interesting information, like average values and standard deviation, correlation matrices, and covariance matrices and so on. Another of interesting results of perturbed variables in SAMPLER layout is the running average, which represent average values and standard deviation for samples of the population during simulation.

Moreover, in SAMPLER response it is possible to see that samples population are closer to the average value and almost entirely within its standard deviation, this is shown in the figures below. According to Wilks' theory, more than 95% of reliability is reached with 146 samples.

Figure 13 shows samples population with averaged values and standard deviation of BWR with control rod configuration, at hot zero power state (HZP). The k_{eff} and two cross-sections (fission and flux) values, for the fast and thermal group, are represented. While Figure 14 shows PWR results of samples population with averaged values and standard deviation, in unrodded configuration and at hot full power state (HFP). In the same way, are represented the k_{eff} and fission and flux cross-section results, for the fast and thermal group.

5 Conclusions

This work has been carried out in the framework of UAM-Benchmark Exercise I-1 Cell Physics and I-2 Lattice Physics. The two test cases (PB-2 BWR and TMI-1 PWR) have been analyzed in two different configurations and two different states, with the objective of quantifying the uncertainty in all step calculation and propagate uncertainties in the LWR whole system.

Transport calculations have been analyzed with the deterministic code TRITON/NEWT and stochastic code SERPENT-2.1.22 with the aim of comparing k_{eff} and cross-sections results between both codes and with UAM-Benchmark reference values.

Sensitivity calculations have been performed with TSUNAMI module, which uses Generalized Perturbation Theory, and SAMPLER for the perturbed cross-section with stochastic sampling techniques.

The following significant conclusions can be highlighted:

– TRITON/NEWT is a solid, validated code that has performed well the UAM-Benchmark calculations but spent more computational times comparing against the SERPENT-2 results. Even though the slight differences comparing SERPENT with validated and reference results, this code could be a good transport calculation code, ongoing testing and validation;
– TSUNAMI module was adopted to estimate sensitivity and uncertainty analysis, the impact of the uncertainties in the basic nuclear data on the calculation of the multiplication factor and microscopic and macroscopic cross-sections. Uncertainties were found to be $\approx 0.5\%$ on k_{eff}. The particular $^{238}U_{n,\gamma}$, $^{235}U_{nubar}$, $^{238}U_{n,n'}$ were found to be the most important contributors to the uncertainty in these exercises. The deterministic solutions were compared with SAMPLER response, and good agreement was found for this exercise.

This work contains findings produced within the OECD/NEA UAM-Benchmark.
This work has been supported by the Generalitat Valenciana under GRISOLIA/2013/A/006 (037) subvention and partially under Project PROMETEOII/008.

References

1. K. Ivanov et al., *Benchmarks for Uncertainty Analysis in Modelling (UAM) for the design, operation and safety analysis of LWRs* (NEA/NSC/DOC, 2012)
2. SCALE: a comprehensive modelling and simulation suite for nuclear safety analysis and design, ORNL/TM-2005/39, Version 6.1, Oak Ridge National Laboratory, 2011
3. M.A. Jessee, M.D. DeHart, *TRITON: a multipurpose transport, depletion, and sensitivity and uncertainty analysis module* (Oak Ridge National Laboratory, 2011)
4. M.A. Jessee, M.D. DeHart, *NEWT: a new transport algorithm for two-dimensional discrete-ordinate analysis in non-orthogonal geometries* (Oak Ridge National Laboratory, 2011)
5. B.J. Ade, SCALE/TRITON Primer: a primer for light water reactor lattice physics calculations, U.S. NRC Report NUREG/CR-7041, Oak Ridge National Laboratory, 2012
6. D.F. Hollenbach, L.M. Petrie et al., *KENO-VI: a general quadratic version of the KENO program* (Oak Ridge National Laboratory, 2009)
7. J. Leppänen, *SERPENT: a continuous-energy Monte Carlo reactor physics burnup calculation code* (VTT Technical Research Centre of Finland, 2013)
8. M. Aufiero et al., A collision history-based approach to sensitivity/perturbation calculations in the continuous energy Monte Carlo code SERPENT, Ann. Nucl. Energy **85**, 245 (2015)
9. B.T. Rearden, M.A. Jessee, M.L. Williams, *TSUNAMI-1D: control module for one-dimensional cross-section sensitivity and uncertainty* (Oak Ridge National Laboratory, 2011)
10. B.T. Rearden, L.M. Petrie, M.A. Jessee, M.L Williams, SAMS: sensitivity analysis module for SCALE, ORNL/TM-2005/39 Version 6.1, Oak Ridge National Laboratory, 2011
11. M.L. Williams et al., SAMPLER: a module for statistical uncertainty analysis with SCALE sequences, Oak Ridge National Laboratory, Draft documentation, 2011
12. M.L. Williams et al., A statistical sampling method for uncertainty analysis with SCALE and XUSA, Nucl. Technol. **183**, 515 (2012)
13. K. Ivanov et al., *Benchmarks for Uncertainty Analysis in Modelling (UAM) for the Design, Operation and Safety Analysis of LWRs. Volume I: Specification and Support Data for Neutronics Cases (Phase I)* (OECD Nuclear Energy Agency, 2013)
14. S.S. Wilks, *Mathematical statistics* (John Wiley & Sons, 1962)
15. I.S. Hong, D.Y. Oh, I.G. Kim, Generic Application of Wilk's Tolerance Limits Evaluation Approach to Nuclear Safety, in *OECD/CSNI Workshop on Best Estimate Methods and Uncertainty Evaluations, 2011* (NEA, CSNI, 2011)

12

Thermal-hydraulics/thermal-mechanics temporal coupling for unprotected loss of flow accidents simulations on a SFR

Cyril Patricot[1*], Grzegorz Kepisty[1], Karim Ammar[1], Guillaume Campioni[1], and Edouard Hourcade[2]

[1] CEA, DEN, DM2S, SERMA, 91191 Gif-sur-Yvette, France
[2] CEA, DEN, DER, CPA, 13108 Saint-Paul-Lez-Durance Cedex, France

Abstract. In the frame of ASTRID designing, unprotected loss of flow (ULOF) accidents are considered. As the reactor is not scrammed, power evolution is driven by neutronic feedbacks, among which Doppler effect, linked to fuel temperature, is prominent. Fuel temperature is calculated using thermal properties of fuel pins (we will focus on heat transfer coefficient between fuel pellet and cladding, H_{gap}, and on fuel thermal conductivity, λ_{fuel}) which vary with irradiation conditions (neutronic flux, mass flow and history for instance) and during transient (mainly because of dilatation of materials with temperature). In this paper, we propose an analysis of the impact of spatial variation and temporal evolution of thermal properties of fuel pins on a CFV-like core behavior during an ULOF accident. These effects are usually neglected under some a priori conservative assumptions. The vocation of our work is not to provide a best-estimate calculation of ULOF transient, but to discuss some of its physical aspects. To achieve this goal, we used TETAR, a thermal-hydraulics system code developed by our team to calculate ULOF transients, GERMINAL V1.5, a CEA code dedicated to SFR pin thermal-mechanics calculations and APOLLO3®, a neutronic code in development at CEA.

1 Introduction

The CFV (*Cœur Faible Vidange*, low void coefficient core) concept [1], which includes several innovations, is viewed as a way to improve the sodium void effect (reactivity effect of a core voiding) and the accidental behavior of large sodium fast reactors (SFRs). A scheme of this kind of core is given in Figure 1. A sodium plenum, with an upper absorbing protection, is positioned just above the core in order to increase the neutrons leakage in case of voiding. This effect is enhanced by the heterogeneities of the inner core, and by the height difference between the outer core and the inner core. These particularities increase the flux at the top of the core, and therefore in the plenum.

Loss of flow accidents are especially difficult for large SFRs and are therefore studied in depth in the frame of their designing. A detailed analysis of these accidents can be found in reference [2]. In order to clarify the explanations, our paper focuses on the unprotected loss of flow accident, during which primary pumps are lost, but not the secondary ones (we will call it ULOF/PP). The reactor is not scrammed, and the power evolution is driven by the neutronic feedbacks

(Doppler, sodium dilatation and dilatations of structures). During the accident, the coolant mass flow decreases until it reaches the natural convection equilibrium. It results in sodium heating in the upper part of the core, making the power decrease, thanks to CFV design. As a consequence, fuel temperature decreases and the Doppler effect is positive. Thus, the stabilization effect of the Doppler is, in this case, an obstacle to the power decrease.

An accurate evaluation of fuel temperature evolution during the transient is therefore necessary. It is usually derived from diffusion equation with given thermal properties. These properties are often homogenized over core zones and are usually constant in time. However, in reality, their spatial variations (mainly due to the heterogeneity of the core and to the mixing of sub-assemblies of different ages) and temporal evolutions (mainly due to differential thermal dilatations) can be quite important.

In this work, we propose an analysis of the impact of spatial variation and temporal evolution of thermal properties of fuel pins on a CFV-like core behavior during an ULOF/PP accident. Section 2 presents the evolution of the core under irradiation, calculated with APOLLO3® [3] and GERMINAL V1.5 [4]. In Section 3, ULOF/PP accidents are calculated with TETAR (developed in the frame of TRIAD [5]) and different spatial descriptions of

* e-mail: cyril.patricot@cea.fr

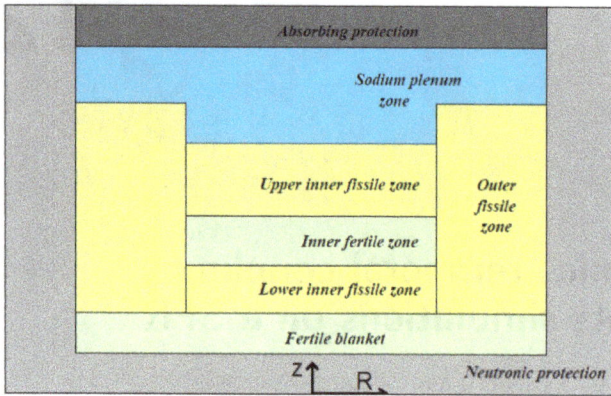

Fig. 1. Scheme of the CFV core concept.

thermal properties. In Section 4, we show the results of the temporal coupling. Section 5 provides some general conclusions.

Note that TETAR is not ASTRID reference tool and that the CFV-like core used is an academic model. As a consequence, the numerical results of this paper should not be considered as reference ones. They are given for the physical analysis of the phenomena.

2 Core evolution under irradiation

2.1 Neutronic evolution

We used APOLLO3® for the neutronic calculations with 33 energy groups. Cross-sections were computed by the module ECCO of ERANOS [6]. Control rods are withdrawn in every calculation.

The chosen reloading procedure uses four batches. As the sub-assemblies are not moved during the reloading, the core is a mixing of sub-assemblies with different burn-up. The resulting power distribution is quite heterogeneous, as shown in Figures 2 and 3. In Figure 2, the power distribution is given, for a cut in the center of the core, at beginning of cycle.

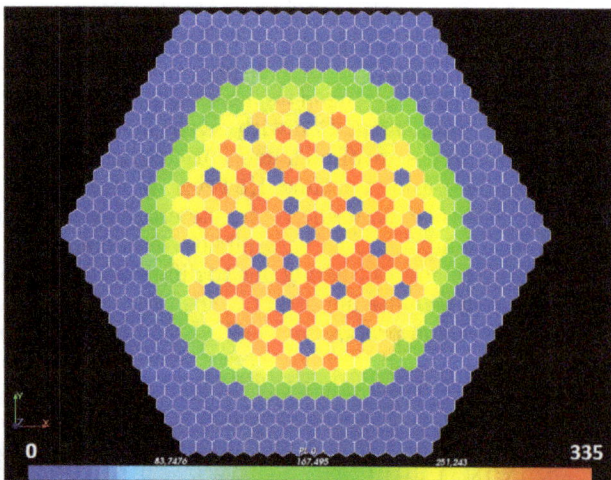

Fig. 2. Linear power distribution (W/cm by pin) in the center of the core at beginning of cycle.

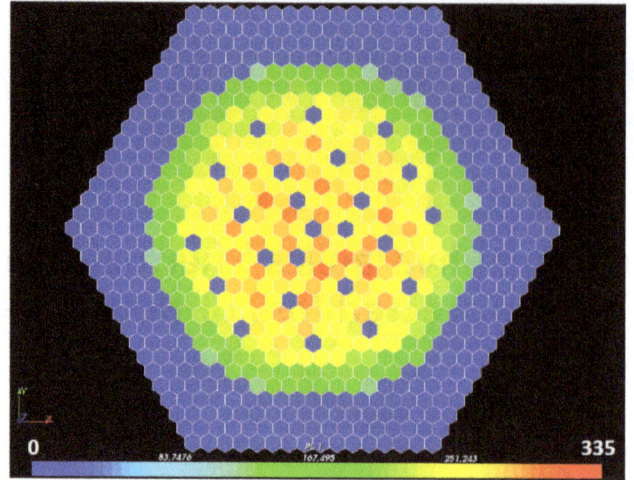

Fig. 3. Linear power distribution (W/cm by pin) in the center of the core at end of cycle.

Fresh sub-assemblies have high fissile content and have therefore a high linear power. At end of cycle, in Figure 3, the power distribution becomes more homogeneous. The color ranges are the same for both figures.

The same kind of flux and power redistributions occurs axially because of the combination of consumption of Pu in fissile zones and breeding in fertile ones (located at the bottom of the core).

2.2 Thermo-mechanical evolution

The evolution of thermo-mechanical properties of fuel pins is evaluated with GERMINAL V1.5. It uses simplify fuel description model based on mono-group neutron flux, linear power and irradiation damage distributions calculated by APOLLO3®. It also needs sodium inlet temperature and mass flow per pin.

The heat transfer coefficient between fuel pellet and cladding, called H_{gap}, has strong non-linear variations with irradiation. H_{gap} and gap size evolutions are given in Figures 4 and 5 respectively, at a fixed position (in fissile) of

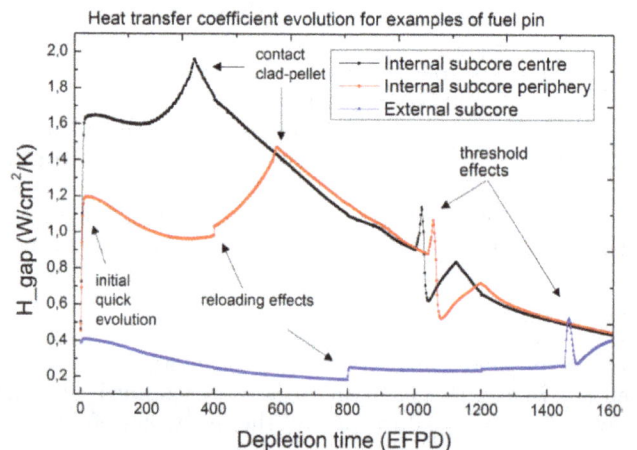

Fig. 4. Typical heat transfer coefficient evolution for chosen sub-assemblies (in top fissile zone).

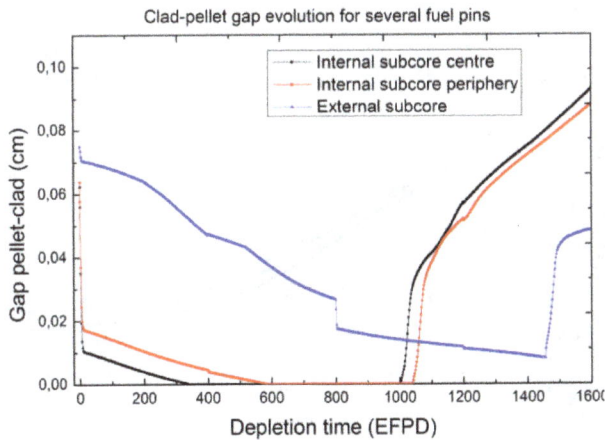

Fig. 5. Typical gap size evolution for chosen sub-assemblies (in top fissile zone).

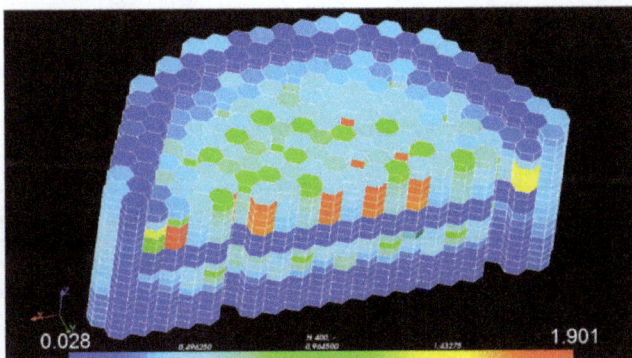

Fig. 7. Typical 3D map of λ_{fuel} (W/cm/K).

chosen sub-assemblies. One can see that the initial thermal dilatation of the pellet makes the H_{gap} increase, at the very beginning. A peak is then observed when the pellet comes in contact with the cladding (it does not occur here for the external subcore sub-assembly). A quite linear phase follows, with constant decrease of the H_{gap} due to the degradation of the contact surface. Finally, threshold effects occur, swelling of the cladding, creation of an oxide layer on its surface and strong gaseous fission products release. The discontinuities at 400, 800 and 1200 EFPD (equivalent full power days) are due to the reloading of a quarter of the core, which changes the linear power and flux in the studied sub-assemblies.

This non-linear behavior, together with the positioning of sub-assemblies in the core, and the axial heterogeneity of the fuel produce quite heterogeneous 3D maps of H_{gap}, as one can see in Figure 6. To build this 3D map, one mean pin per sub-assembly has been calculated. A 3D map of thermal conductivity of fuel (called λ_{fuel}) is given in Figure 7. The evolution of this quantity is much more linear: the irradiation degrades the ceramics and thus its conductivity. As a consequence, λ_{fuel} is maximal where the irradiation damages are minimal.

3 Impact of spatial descriptions of thermal properties and of neutronic feedbacks on the ULOF/PP accident

3.1 Calculations comparison with integrated neutronic feedback coefficients

We used TETAR (Transients Estimation Tool for nA-cooled Reactors) to calculate the ULOF/PP accident. It solves 1D thermo-hydraulic equations in each sub-assembly. We emphasize that each sub-assembly is calculated separately by a dedicated 1D thermo-hydraulic channel in all calculations presented in this paper. This ability of TETAR allowed us to perform our studies on spatial descriptions impacts. Mass flow in each sub-assembly is calculated to create a given pressure drop. Pin temperature is calculated through 1D diffusion. Point kinetic, fed with feedback coefficients (integrated or local) from APOL-LO3®, is used for the power estimation. The system is closed with sodium collectors and sodium-sodium heat exchangers simple models. The accident is driven by a given decrease of the pumps pressure. The overall pressure drop due to gravity (this term leads to natural convection) is calculated precisely.

In this section, the thermal properties are constant during the transient. Four models were used to estimate their initial value:

– *Exact*: one mean pin per sub-assembly is calculated by GERMINAL V1.5, and the results feed directly the TETAR calculation;
– *Global average*: we calculate, from the exact core calculation, the mean H_{gap} and λ_{fuel} of the core and use them everywhere in the TETAR calculation;
– *Zones average*: we calculate, from the exact core calculation, the mean H_{gap} and λ_{fuel} of the core main five zones (Fig. 1). They are used in the corresponding meshes in TETAR;
– *Groups*: we gather sub-assemblies in groups and calculate one mean pin per group (sub-assemblies of the same ring, from the same batch). In comparison with the exact model, the number of GERMINAL V1.5 calculations is reduced by almost a factor 10.

Sodium maximal temperature and power evolutions during the ULOF/PP accident are given in Figures 8 and 9

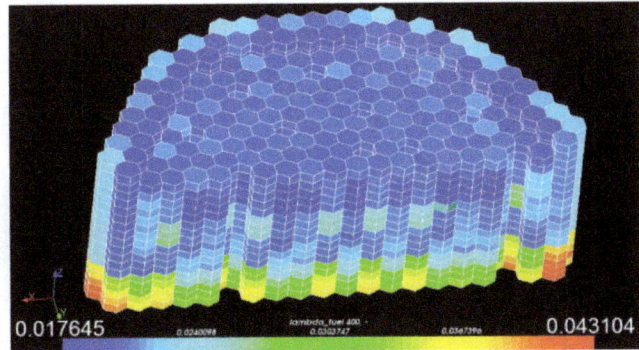

Fig. 6. Typical 3D map of H_{gap} (W/cm^2/K).

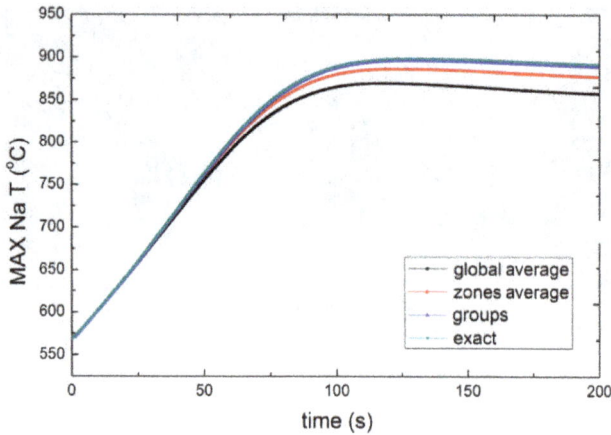

Fig. 8. Sodium maximal temperature during ULOF/PP accident for different thermal properties models.

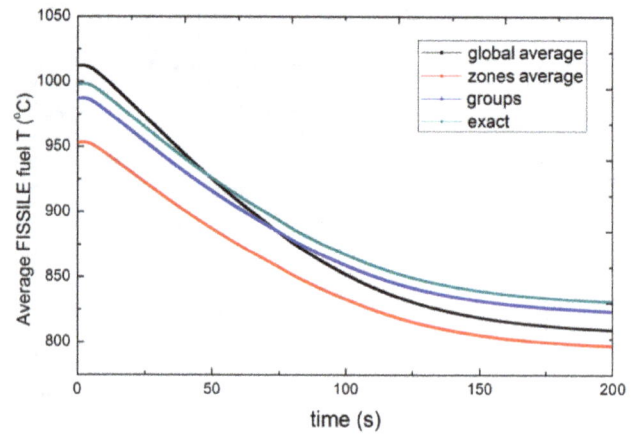

Fig. 10. Average fissile temperature during ULOF/PP for different H_{gap} treatments.

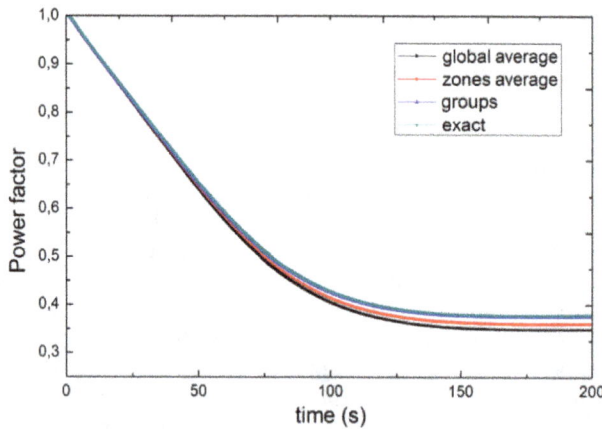

Fig. 9. Power during ULOF/PP accident for different thermal properties models.

for the models above. One can see that the exact and the group models are indistinguishable and that the maximal temperature they reached is slightly above the zone average model, which is slightly above the global average model.

Sodium maximal temperatures for some other models are given in Table 1. One can see that the zones average model is enough for λ_{fuel}, its results are very close to those of the exact model. In addition, non-linearities seem to be weak; the effect of a combination of models is the sum of effects of the models. Finally, the difference between the groups and the exact models is very small in all cases, about 3 °C.

3.2 Interpretation of the results with integrated neutronic feedback coefficients

All the presented calculations used two integrated Doppler coefficients, one for the fertile zones, and one for the fissile zones. The power is therefore affected by the average fissile and fertile temperatures. One can see their evolution in Figures 10 and 11, for the calculations of the second column of Table 1 (λ_{fuel} model is always global average). Except for the global average model which mixes fertile and fissile meshes, one can see that every H_{gap} averaging leads to a cooler fuel.

This observation can be explained. Let us consider two fuel meshes, i and j, in contact with the cladding. Because of the linearity of the diffusion equation, the temperature of i can be written as:

$$T_i = T_i^{Cl} + \frac{\alpha_i}{h_i}, \qquad (1)$$

with T_i^{Cl} the temperature of the cladding, h_i the H_{gap} coefficient and α_i a scalar depending on local power. The same equation can be written for mesh j. We introduce now the temperatures T_i^m and T_j^m obtained using average H_{gap} value, that is to say $\frac{h_i+h_j}{2}$. The difference between the average values with exact and average H_{gap} is equal to:

$$\frac{T_i^m + T_j^m}{2} - \frac{T_i + T_j}{2} = \frac{(h_i - h_j)\left(\frac{\alpha_i}{h_i} - \frac{\alpha_j}{h_j}\right)}{2(h_i + h_j)}. \qquad (2)$$

Table 1. Comparison of sodium maximal temperature (°C) during ULOF/PP accident for different thermal properties models.

Model	H_{gap} (λ_{fuel}: global average)	λ_{fuel} (H_{gap}: global average)	H_{gap} and λ_{fuel}
Global average	869.2	869.2	869.2
Zones average	877.4	877.8	886.4
Groups	887.2	878.3	896.1
Exact	889.9	878.6	898.9

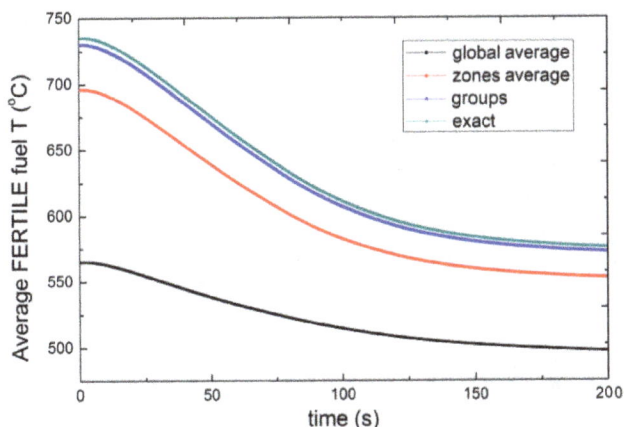

Fig. 11. Average fertile temperature during ULOF/PP for different H_{gap} treatments.

Table 2. Comparisons of sodium maximal temperature (°C) during ULOF/PP accident with and without local neutronic feedbacks.

Model	H_{gap} and λ_{fuel} (and integrated NF)	H_{gap} and λ_{fuel} (and local NF)
Global average	869.2	878.9
Zones average	886.4	898.0
Groups	896.1	902.4
Exact	898.9	904.3

decreases a little bit more. These results explain the impact of spatial description of thermal properties of fuel pins we observed in Section 3.1.

$\frac{\alpha_i}{h_i}$ is the temperature increase between fuel and cladding. This equation means therefore that using average H_{gap} reduces average fuel temperature if the H_{gap} of the hottest mesh is smaller than the one of the coolest mesh. The point is that it may be the reason why the hottest mesh is the hottest. Therefore, without strong positive correlation between power and H_{gap}, using average H_{gap} usually reduces fuel temperature.

In addition, we can prove that, starting with a cooler fuel, for the same power decrease, the Doppler effect is smaller. To show that, let us write the temperature of a given mesh in the situation i like:

$$T_i = T^{Cl} + \alpha_i P, \qquad (3)$$

with T^{Cl} the temperature of the cladding, P the local power and α_i a scalar depending on mesh state. The same equation can be written for the same mesh in the situation j by replacing α_i by α_j.

Now we consider that the power becomes, at time t, $P.f$ with f a given factor ($f < 1$ in the case of a ULOF). The mesh contribution to the Doppler effect is:

$$\frac{T_i(t) - T_i(t=0)}{T_i(t)} C = \frac{\alpha_i P(f-1)}{T^{Cl} + \alpha_i Pf} C, \qquad (4)$$

with C a given feedback coefficient (usually $C < 0$). Thanks to the form chosen for equation (4) (this is the usual one), the coefficient C has no dependence on temperature. We assume here that the cladding temperature is constant. The difference between the Doppler contributions of the mesh in both situations is:

$$\frac{T_i(t) - T_i(t=0)}{T_i(t)} C - \frac{T_j(t) - T_j(t=0)}{T_j(t)} C$$
$$= \frac{T^{Cl} P(f-1)(\alpha_i - \alpha_j) C}{(T^{Cl} + \alpha_i Pf)(T^{Cl} + \alpha_j Pf)}. \qquad (5)$$

This quantity is positive if $C < 0$, $f < 1$ and $\alpha_i > \alpha_j$. In other words, if the Doppler effect is negative and if the power decreases, we show that the Doppler effect is smaller for an initially cooler mesh. As a consequence, the power

3.3 Impact of local neutronic feedback coefficients

The previous analysis is based on the use of average fuel temperatures to calculate the Doppler feedback. One could wonder if it still stands if we use local neutronic feedbacks. Because this work is on the impact of the fuel pin thermo-mechanics on ULOF/PP accident, we focused our study on the Doppler effect. Comparisons of sodium maximal temperatures reached during ULOF/PP accident with and without local Doppler coefficients (the global Doppler effect is the same) are presented in Table 2. 3D maps of Doppler coefficients, derived from the perturbation theory, are given in Figure 12 (fissile) and Figure 13 (fertile).

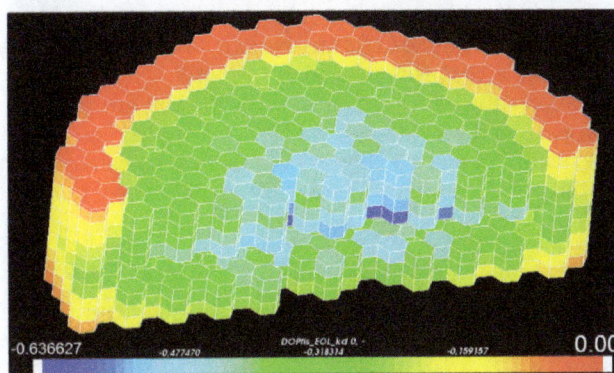

Fig. 12. 3D map of Doppler coefficients (pcm) in fissile zones.

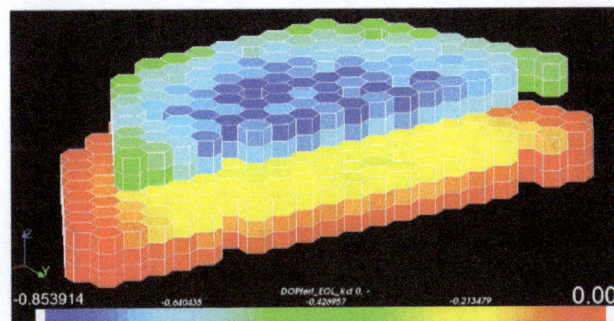

Fig. 13. 3D map of Doppler coefficients (pcm) in fertile zones.

One can see that, here, using local neutronic feedbacks always increases the sodium maximal temperature. The impact depends on the thermal properties model, but is pretty small (about 5 °C) for exact treatment of H_{gap} and λ_{fuel}. An analysis shows that this difference is mainly due to the heterogeneity of the fertile zones. Indeed, in the inner fertile there are in the same time a much stronger fuel temperature decrease and much stronger Doppler coefficients than in the fertile blanket (this is visible in Fig. 13). These two differences together create a bias when one uses integrated Doppler coefficients.

In Section 3, we saw the impact of spatial description of thermal properties. The more accurate it is, the hotter the sodium becomes during the ULOF/PP accident, whatever is the spatial treatment of the Doppler effect. We will now see the impact of the temporal evolution of the thermal properties.

4 Impact of temporal evolution of thermal properties during the ULOF/PP accident

4.1 Calculations comparison with integrated neutronic feedback coefficients

We used the simple explicit coupling scheme illustrated in Figure 14. GERMINAL V1.5 gives local H_{gap} and λ_{fuel} values to TETAR, which returns mass flow per pin and local power (through the global power factor calculated by the point kinetic). The coupling time step is set to 10 s. The already presented groups model for GERMINAL V1.5 is chosen in order to save calculation time.

It would be interesting to enhance this coupling scheme, and it should be done in future work. However, preliminary studies show that this scheme is correct.

The results of the coupled calculation are given in Figures 15 and 16 with equivalent non-coupled case. Here, integrated Doppler coefficients are used. One can see that the temporal coupling has a very strong impact, about −38 °C.

4.2 Interpretation of the results with integrated neutronic feedback coefficients

This very strong impact of the coupling is due to the opening of the gap during the transient: as the cladding is

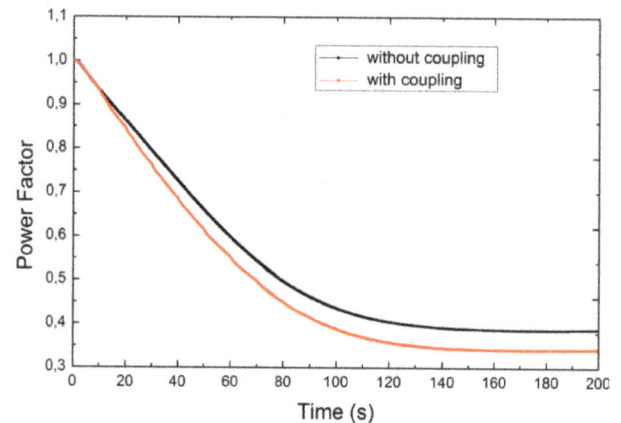

Fig. 15. Sodium maximal temperature during ULOF/PP accident with and without coupling.

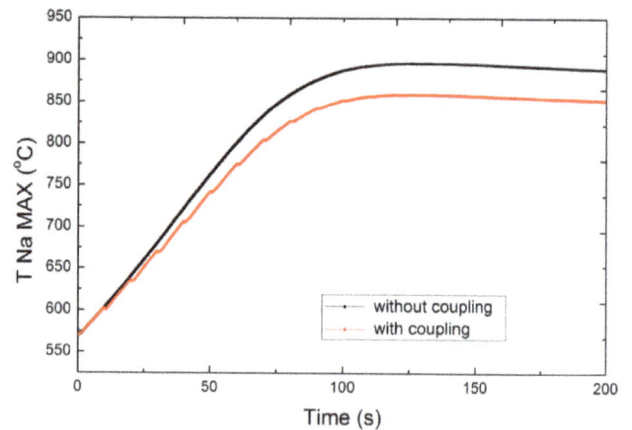

Fig. 16. Power during ULOF/PP accident with and without coupling.

getting hotter and the fuel cooler, the differential thermal expansion takes them away from each other. As a consequence, the H_{gap} decreases and the fuel temperature drop is reduced, leading to a smaller Doppler effect.

H_{gap} profile evolutions for one sub-assembly from the inner core is given in Figure 17, and for one sub-assembly from the outer core in Figure 18. The H_{gap} does decrease everywhere. It is especially important (divided by about 3) where the initial value was high: there was a contact between the cladding and the pellet at these locations. This contact is lost during the transient.

While H_{gap} changes a lot during the transient, λ_{fuel} is found to be almost constant.

One can note that the H_{gap} temporal evolution is rather smooth, and could be approximated by polynomial functions, as suggested in reference [2].

4.3 Impact of local neutronic feedback coefficients

We found, in Section 3, that local neutronic feedback coefficients have a small impact on ULOF/PP when used with a good spatial discretization of the thermal properties

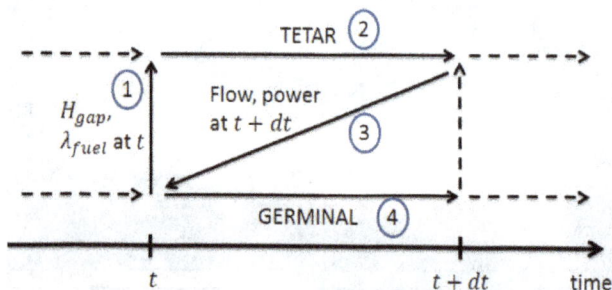

Fig. 14. The temporal coupling scheme used.

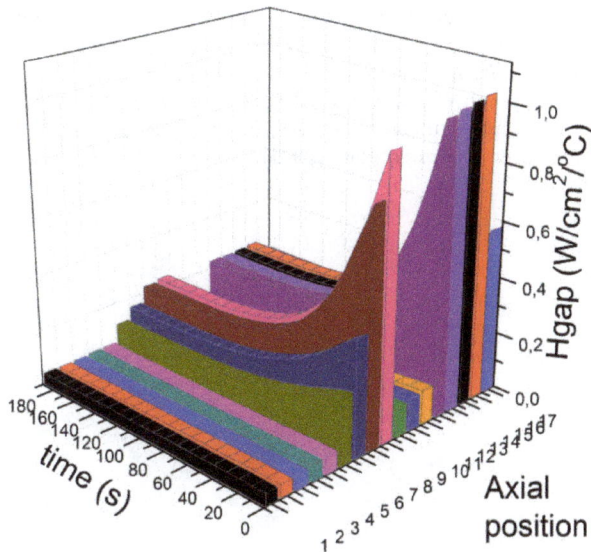

Fig. 17. H_{gap} profile evolutions during ULOF/PP accident for one sub-assembly from the inner core.

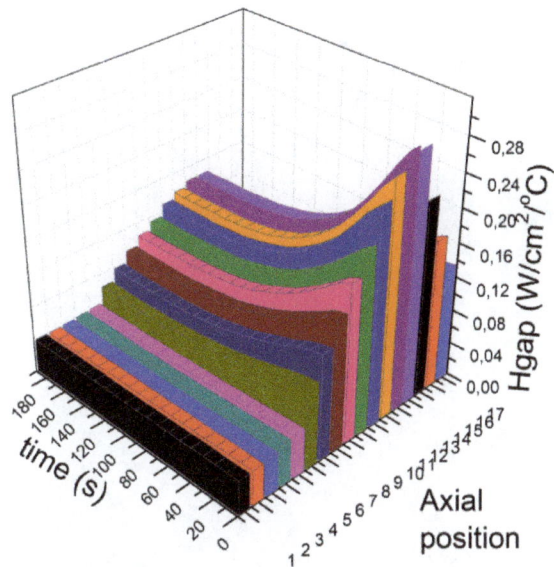

Fig. 18. H_{gap} profile evolutions during ULOF/PP accident for one sub-assembly from the outer core.

of the pins (see Tab. 2). The cause of the discrepancy has been identified to be the combined heterogeneities of H_{gap} and Doppler coefficients in the fuel zones. The temporal coupling, because it reduces the H_{gap} preferentially where it is high, that is to say in the center of the core, where the Doppler effect is the strongest, reduces these heterogeneities. As a consequence, the impact of the local feedback coefficients is reduced. This is visible in Figure 19. We used, here again, the groups model for H_{gap} and λ_{fuel}.

The comparison with the non-coupled equivalent calculation with local feedbacks coefficients leads to a reduction of the sodium maximal temperature of about 45 °C.

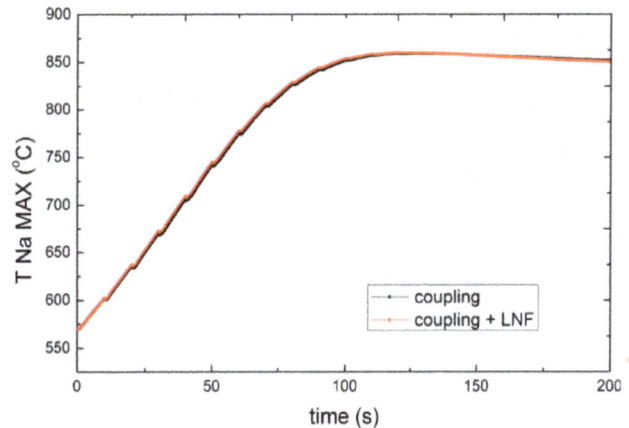

Fig. 19. Sodium maximal temperature during ULOF/PP accident with coupling and different Doppler effect treatments.

5 Conclusions

In this paper, we proposed an analysis of the impact of spatial variation and temporal evolution of thermal properties of fuel pins on the behavior of a CFV-like core during an ULOF accident.

Sources of spatial variations and temporal evolution of the main thermal properties of fuel pins were identified. The impact of their spatial variations was found to be about +30 °C on sodium temperature during ULOF/PP transient. It is mainly due to H_{gap}, and simple zones averages seem to be enough for λ_{fuel}. The combined effect of local thermal properties and local Doppler coefficients leads to an impact of about +35 °C. On the other hand, the temporal coupling, because of the opening of the gap, improves the reactor behavior during the ULOF/PP and leads to a decrease of about 45 °C of the sodium temperature. This improvement of the core behavior is very strong and could help greatly to demonstrate the safety of large SFRs.

From the above observations we can make the following comments:

– a static estimation of λ_{fuel} in the main zones of the core is sufficient;
– for a conservative calculation, the spatial variations of H_{gap} and of the Doppler effect should be taken into account;
– the temporal coupling between thermal-hydraulics and thermal-mechanics of fuel pins brings out substantial margins, because of the H_{gap} evolution.

References

1. M.S. Chenaud et al., Status of the ASTRID core at the end of the pre-conceptual design phase 1, in *Proceedings of ICAPP 2013, Jeju Island, Korea, 2013* (2013)
2. R. Lavastre et al., State of art of CATHARE model for transient safety analysis of ASTRID SFR, in *Proceedings of NUTHOS-10, Okinawa, Japan, 2014* (2014)
3. H. Golfier et al., APOLLO3: a common project of CEA, AREVA and EDF for the development of a new deterministic

multi-purpose code for core physics analysis, in *Proceedings of M&C 2009, New York, USA, 2009* (2009)

4. L. Roche, M. Pelletier, Modelling of the thermomechanical and physical process in FR fuel pins using GERMINAL code, in *Proceedings of MOX Fuel Cycle Technologies for Medium and Long Term Deployment IAEA-SM-358/25, Vienna, Austria, 1999* (1999), pp. 322–335

5. E. Hourcade et al., SFR core design: a system-driven multi-criteria core optimisation exercise with TRIAD, in *Proceedings of FR13, Paris, France, 2013* (2013)

6. G. Rimpault et al., The ERANOS code and data system for fast reactor neutronic analyses, in *Proceedings of PHYSOR 2002, Seoul, Korea, 2002* (2002)

13

Testing of high temperature materials within HTR program in Czech Republic

Jan Berka[1,2*] and Jana Kalivodová[1]

[1] Research Centre Rez Ltd., Husinec – Řež, Hlavní 130, 25068 Řež, Czech Republic
[2] University of Chemistry and Technology Prague, Technická 1905, 16628 Prague 6, Czech Republic

Abstract. Research institutes and also industrial companies in Czech Republic are involved in High Temperature Gas Cooled Reactor (HTGR) program and activities related to the study of advanced materials and HTGR technologies. These activities are supported by EC (within international projects, e.g. FP7-ARCHER, ALLIANCE, GoFastR can be mentioned) and also by Technology Agency of Czech Republic. Within these activities, degradation of metallic and ceramic materials in the high temperature helium atmosphere is investigated, and also new experimental facilities for material testing are built. As examples of tested materials, Alloy 800 H, ferritic steel P91, austenitic steel 316, Inconel 713 and 738 and corundum ceramics could be named. The selected results of exposure experiments in the high temperature helium environment are presented in this paper.

1 Introduction

Czech research organizations, universities and industrial companies are involved in High Temperature Reactor (HTR) and also Gas Fast Reactor (GFR) Research program. The examples of these organizations are listed in Table 1. The research used to be supported by the Ministry of Industry and Trade of Czech Republic, presently it is supported by the Technology Agency of Czech Republic. Some of Czech organizations also participate in the international projects aimed to HTR and GFR – as examples the ARCHER [1] and ALLIANCE projects could be named. One of the most important tasks of HTR program is testing and the evaluation of properties and degradation of materials for HTR and other high temperature applications. For these activities, several experimental facilities are used – one of the most significant facility is the High Temperature Helium Loop (HTHL) – the scheme of the device is shown in Figure 1. The main operational parameters of the device are: gas pressure 3–7 MPa, temperature in the test section 25–900 °C, gas flow 12–38 kg.h^{-1} (for limited time the gas flow could be even lower than the mentioned lower limit). The gas in the loop should consist of helium with only minor impurities (H_2, H_2O, CO, CO_2, N_2, O_2, CH_4) in concentrations up to approximately 500 vppm. See reference [2] for more details

about the device. Another large research infrastructure (and also two new helium loops) is planned to be built in Czech Republic within the SUSEN project [3]. The program of testing the compatibility of metallic alloys with high temperature HTR helium coolant refers to previous activities performed within HTR program in the world (mostly in the last century). Some results are summarized e.g. in references [4–6]. In reference [4], the results from material research within the HTR program in approximately 1960–1990 are summarized, the list of metallic alloys for possible use for HTR components is introduced in this reference. In references [5,6], the high temperature corrosion mechanism of nickel alloys in HTR helium environment is described, the results of corrosion tests of alloys Inconel 617 and Haynes 230 in impure helium at up to 950 °C are summarized.

2 Experimental

The high temperature testing program is focused on corundum and cordierite ceramics and special metallic alloys. The first experiments concerning ceramic materials were aimed at electrical properties at high temperature. Ceramics are usually used as an insulating material for heating elements in experimental devices produced in Research Centre Řež and ÚJV Řež. Previously, cordierite ceramics were used for this purpose, but during the test operation of HTHL temperature above ca. 670 °C, it was

Table 1. List of organizations involved in Czech HTR program.

Name	Type of organization	Alignment and activities
Research Centre Rez Ltd.	Research	Testing of materials, investigation of technologies, operation of test facilities
University of Chemical Technology Prague	Research	Chemical university, testing, development, experiments
MICo	Industry	Developing seals and heat exchangers for nuclear power engineering
EVECO	Industry	Gas cleaning technologies
ÚJV Řež	Industry, engineering	Tests and evaluation of material specimens, engineering
Prague Casting Services	Industry	Production of high temperature components by precision castings by the lost wax method
ESTCOM-oxidová keramika a.s.	Industry	Production of high temperature ceramics based on corundum

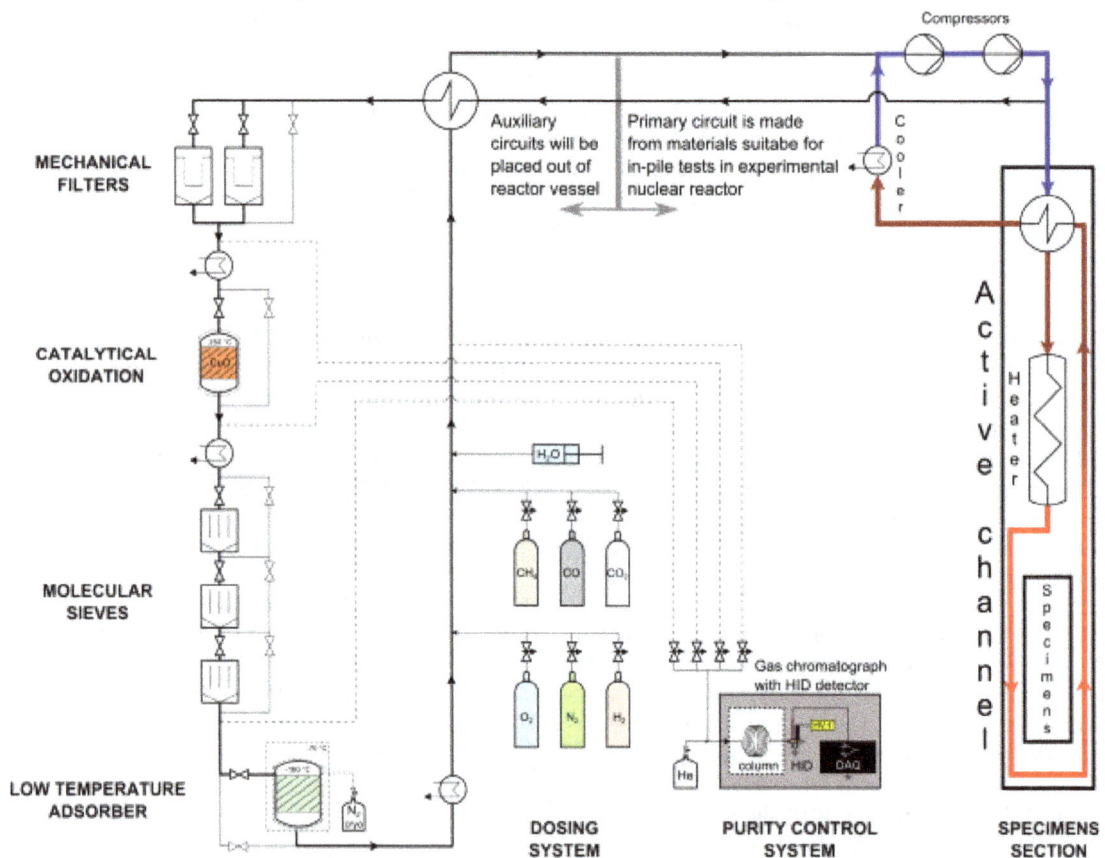

Fig. 1. Scheme of the High Temperature Helium Loop.

not possible to reach the higher temperatures with the heating elements insulated by cordierite ceramics, due to rapid decrease of electrical resistance. When the temperature reached 670 °C, the limit of leakage current given by the standard ČSN 33 1610 – see reference [7] for details – was exceeded. The standard ČSN 33 1610 gives the upper limit of leakage current 1 mA per 1 kW of output for devices with output higher than 3.5 kW. Therefore, the new ceramic insulator for heating elements was developed and its electrical resistance depending on temperature in helium environment tested. See references [8,9] for details about the testing of electrical resistance. For the basic properties of tested ceramic materials, see Table 2.

Table 2. Basic parameters of tested ceramics.

Parking according to ČSN EN 60672	Symbol	Unit	CORDIERIT C410	Corundum ceramics C799
Commercial name			TH 7/7 R12 BM	Luxal 203
Porosity	p_a	[%]	0.5	
Density	ρ_a	[g.cm^{-3}]	2.1	min. 3.8
Bending strength	σ	[MPa]	60	min. 300
Thermal expansivity coefficient	α_{30-600}	[10^{-6}K^{-1}]	2–4	7–8
Heat conductivity	λ_{30-10}	[Wm^{-1}K^{-1}]	1.2–2.5	
Resistance against sudden chase of temperature	ΔT	[K]	250	min. 150
Relative permittivity	ε_r	[–]	5	
Al$_2$O$_3$ content		% by weight	33	min. 99.5
Inner electric resistance at 30, 200 and 600 °C	$\rho_{v,30}$	[Ω.m]	10^{10}	
	$\rho_{v,200}$	[Ω.m]	10^6	10^{12}
	$\rho_{v,600}$	[Ω.m]	10^3	10^8

Table 3. Chemical composition of steel 316 L (wt.%).

Element	C	Si	Mn	P	S	Cr	Mo	Ni	Co	N	Fe
Min.							2.00	10.00			Bal.
Max.	0.021	0.34	1.73	0.027	0.025	16.50	2.03	10.03	0.120	0.0320	Bal.

Table 4. Chemical composition steel P91 (wt.%).

Element	C	S	Mn	Si	P	Cu	Ni	Cr	Mo	V
	0.12	0.002	0.36	0.39	0.011	0.041	0.034	10.06	0.88	0.22
Element	Ti	W	Co	Nb	As	Sb	Sn	Al	N	Fe
	0.007	<0.005	<0.003	0.052	0.003	0.001	0.002	0.005	0.065	Bal

Table 5. Chemical composition of alloy 800 H (wt.%).

Element	C	S	Cr	Ni	Mn	Si	Ti	Nb	Cu	Fe
	0.06	<0.002	20.5	30.5	0.7	0.50	0.34	0.01	0.10	R46.7
Element	P	Al	Co							
	0.010	0.28	0.1							

Within the test program also, testing of other types of ceramic materials has begun, e.g. investigation of mechanical properties of ceramics Lunit 73 (C610), Luxal 203 (C799), AG 202 (C795) after log-term exposure in high temperature helium environment (up to 900 °C) is planned. See reference [10] for details about the materials and their manufacturing.

Also the high temperature metallic alloys were tested during the Czech HTR program: Alloy 800 H (WIG welded by Nicrofer S7020), ferritic steel P91 and austenitic steel 316 L. The chemical composition of tested alloys is listed in Tables 3–5.

The specimens were exposed:

– in the quartz retort in the furnace at 750–760 °C (the test temperature was determined – among others – with regard of the submission of the projects within which the tests were performed) for up to 1500 hours at atmospheric pressure in the impure helium environment. The chemical composition of inlet helium gas is listed in Table 6 (composition of premixed gas mixture in pressure vessel, concentration of oxygen in the pressure vessel guaranteed by the producer). Concentration of residual moisture was measured by optical hygrometer in the inlet to the retort and ranged from 1–10 vppm;

Table 6. Chemical composition of premixed gaseous mixture used for experiment.

Component	Concentration [vppm]	Partial pressure [Pa]
CH_4	100	10
CO	500	50
H_2	100	10
O_2	<0.1	<0.01
Helium	Bal.	Bal.

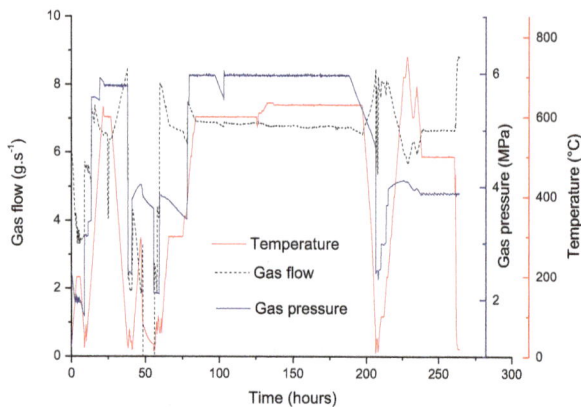

Fig. 2. Parameters during the first period of the test operation of High Temperature Helium Loop.

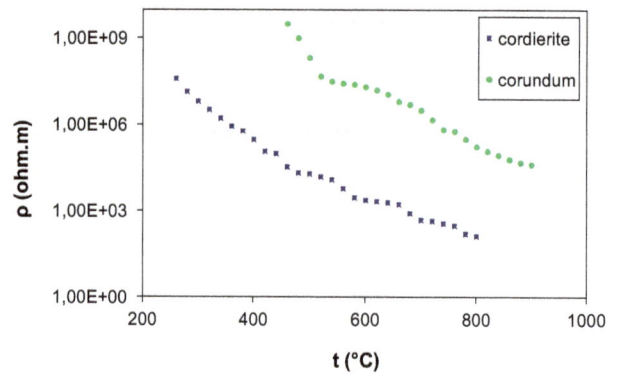

Fig. 3. Dependence of specific electric resistance of tested ceramic materials on temperature.

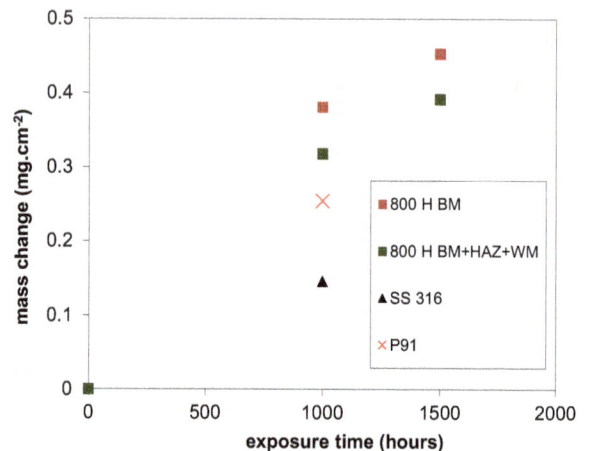

Fig. 4. Mass change of the specimens of alloys exposed in impure helium environment at 750–760 °C. BM: base metal, WM: weld metal, HAZ: heat affected zone.

– in High Temperature Helium Loop (HTHL) during the first period of the test operation of the device. The first period of the test operation lasted 264 hours, HTHL was filled with the pure technical helium (purity 4.6; producer determines that concentration of impurities in this helium contains less than 30 vppm of impurities, and less than 5 vppm of O_2). The temperature ranged from 25 to 750 °C, the average temperature was 500 °C. Residual moisture content measured by optical hygrometer in helium gas reached max. 250 vppm, several check sample analyzed by gas chromatography confirmed the concentration of impurities other than O_2 and H_2O was less than 2 vppm. During this period of the test operation of HTHL separate oximeter was not integrated to the gas circuit. See reference [2] for details about the system of helium purity control of HTHL. The gas pressure and gas flow ranged from 2 to 7 MPa and approximately 2–9 g.s^{-1} (7.2–32.4 kg.h^{-1}) respectively. The main parameters during this period of the test operation are illustrated in the graph in Figure 2.

After exposure, the degradation of specimens was investigated. Gravimetry, Scanning Electron Microscope (SEM) and theoptical microscope were used for this purpose. The cross-sections of specimens were prepared. The microstructure was further checked by etching by 10% oxalic acid solution. The change of hardness and micro hardness and fracture toughness (only for P91 and 316 L) was also investigated. More details about experiment could be found e.g. in reference [11].

Tests of other alloys, e.g. Inconel 713, 738 and austenitic steel N155, at 900 °C in impure helium environment are also planned. The corrosion test of Alloy 800 H, steel P91 and 316 L in HTHL is in progress.

3 Selected results

3.1 Selected results of tests of ceramics

The electrical resistivity of corundum ceramics at high temperature (up to 900 °C) was proven to be significantly higher than that of cordierite ceramics. The results of the tests are summarized in the graph in Figure 3. The limit of leakage current given by the standard ČSN 33 1610 will not be exceeded even at 900 °C if corundum ceramics is used for insulating the heating elements. Therefore, corundum ceramics C799 is convenient for this purpose.

3.2 Selected results of tests of metallic alloys

The mass changes of the specimens of alloys exposed in the furnace are summarized in the graph in Figure 4.

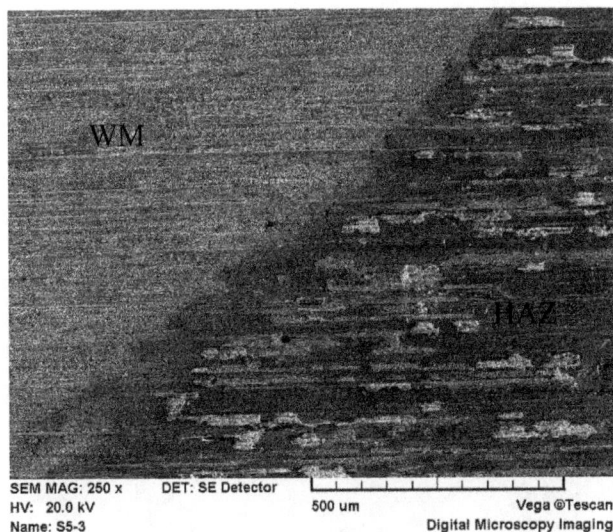

Fig. 5. Spalling of surface layer on welded specimen of Alloy 800 H after exposure of 1500 hours at 760 °C in impure helium.

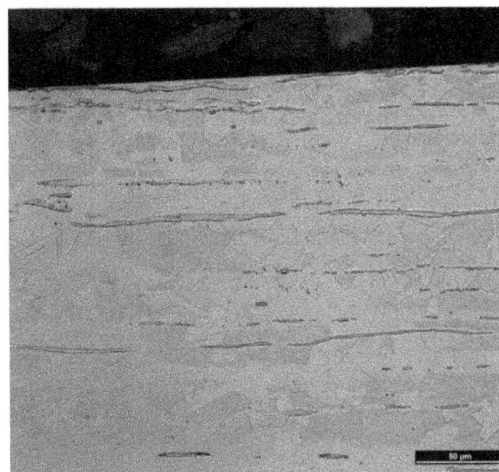

a

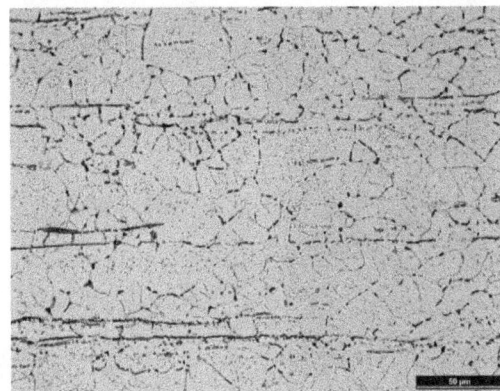

b

Fig. 7. Microstructure of steel 316 L (the cross-section of the specimen): (a) in as-received state, (b) after exposure in He 750 °C/1000 h.

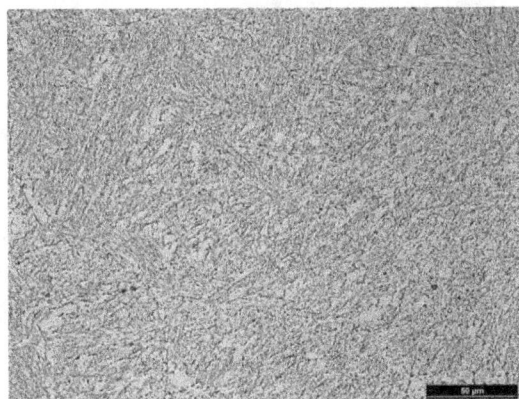

a

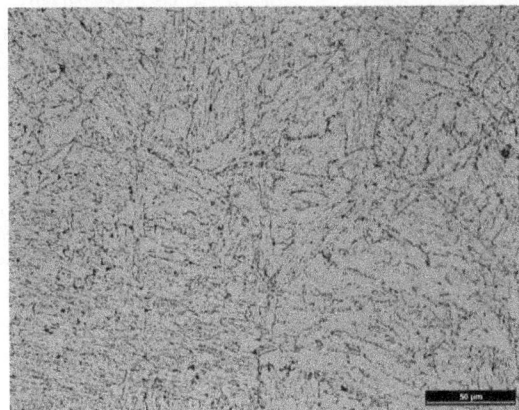

b

Fig. 6. Microstructure of steel P91 (the cross-section of the specimen): (a) in as-received state, (b) after exposure in impure helium 750 °C/1000 h.

The uncertainty of the results could be estimated to be about 10%. The mass gain of alloy 800 H was the highest of that of tested alloys.

The oxidation layer on the surface of exposed specimens was mainly formed by chromium and manganese oxides. The spalling of the oxide layer was observed mainly on the surface of the Heat Affected Zone (HAZ) – obviously visible on Figure 5. Thickness of the oxide layer on all tested metallic specimens was in the range 2–4 microns.

The images of microstructure of steels P91 and 316 L before and after exposure are shown in Figure 6 and Figure 7 respectively. Precipitation of particles (carbides) after exposure is apparent (see Figs. 6b and 7b). In case of P91, the subsurface layer without carbides (up to ca. 17 μm thick) appeared after exposure. The microstructure of base metal and weld join of Alloy 800 H before and after exposure is given in Figures 8 and 9. The precipitation of particles after exposure is visible by comparison of these figures. After exposure of 1000 hours, significant precipitation of particles ($M_{23}C_6$ and $\acute{\gamma}$) was noticed. Under the corrosive layer $\sim$20 μm undersurface layer without precipitates was observed. There are some differences in composition of the surface corrosive layer after exposure in the furnace and in HTHL during the test operation – except chromium although a significant percentage of iron and nickel was detected by SEM

a

b

c

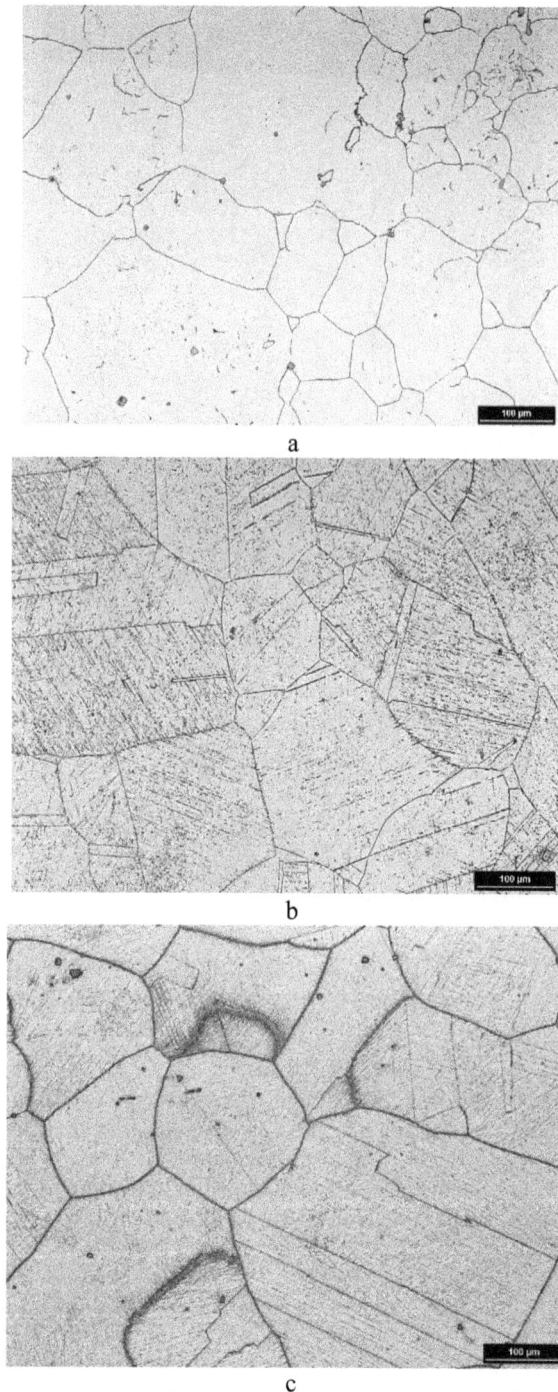

Fig. 8. Microstructure of base metal of Alloy 800 H on the cross-section of specimens: (a) in as-received state, (b) after exposure of 1500 hours in impure helium at 760 °C in the furnace, (c) after exposure in HTHL during 264 hours of the test operation.

a

b

Fig. 9. Microstructure of interface of weld metal and heat affected zone of alloy 800 H: (a) in as-received state, (b) after exposure of 1500 hours in impure helium at 760 °C.

The fracture toughness of austenitic steel 316 L decreased after exposure of 1000 hours at 750 °C in impure helium of 67% (from value of $J_{0.2}$ integral 62 J.cm^{-2} in as-received state to 20 J.cm^{-2} after exposure). The fracture toughness of ferritic steel P91 almost did not change after exposure. The change of fracture toughness of 316 L is probably caused by changes of material at high temperature independently of environment (e.g. due to the precipitation of the sigma phase after annealing at 750 °C during 1000 h) – however, to prove this assumption, other tests in different environments (e.g. in air) at the same temperature are needed.

4 Conclusions

The research organizations and industrial companies in Czech Republic participate in the research and development of materials and technologies for High Temperature gas cooled Reactors and other high temperature industrial applications. These activities are supported – among others – by the European Commission within the international FP7 projects and also by the Technology Agency of Czech Republic. The infrastructure for this investigation

analysis in the surface corrosive layer on Alloy 800 H after exposure in HTHL. Mass gain of specimen after exposure in HTHL was 0.06 mg.cm^{-2}. The dependence of hardness and micro hardness of tested materials on exposure time is illustrated in Figure 10.

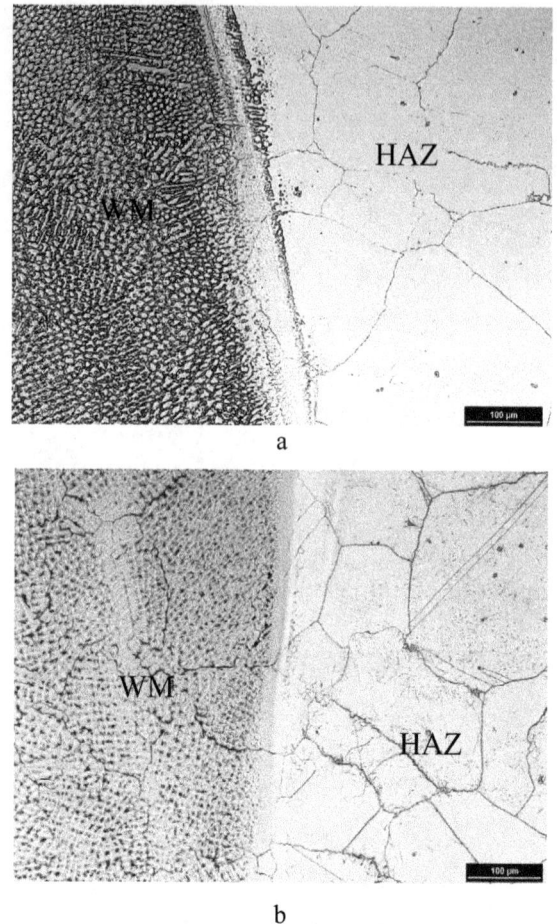

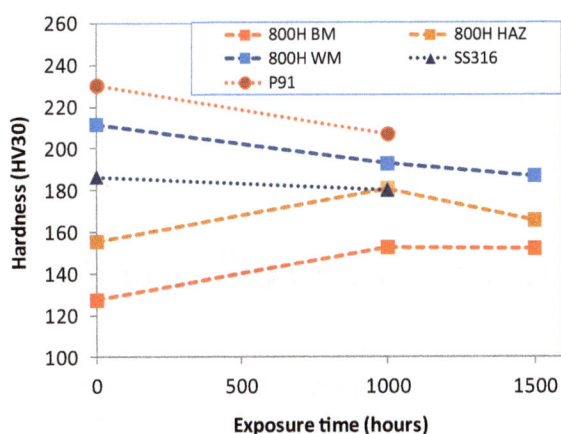

a

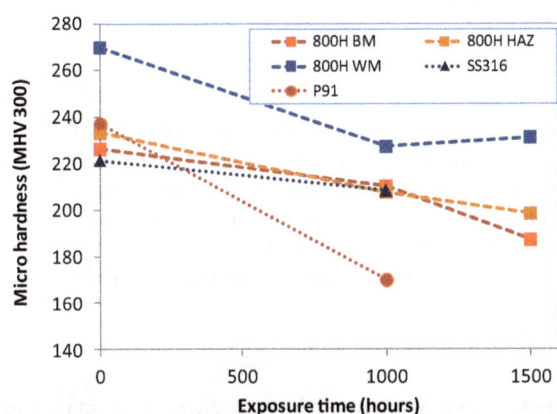

b

Fig. 10. (a) hardness, (b) micro hardness of tested alloys depending on exposure time in impure helium at 750–760 °C.

exists in Czech Republic and is being extended. Some results of investigation of high temperature degradation of metallic and non-metallic materials are already available, other tests and evaluation are still in progress and others are planned.

With regard to metallic materials which were tested so far: stainless steel 316 L proved the best corrosion resistance in impure helium at 750–760 °C, however mechanical properties of this steel changed after exposure at high temperature. Fracture toughness of steel P91 almost did not change after exposure at high temperature. These materials are not designed for long-term operations at such high temperatures; however these materials could be used e.g. for not mechanically loaded parts of experimental devices (for example sample holders) or colder parts. Mechanical properties of Alloy 800 H after

exposure at high temperature were not tested. According to obtained results, corrosion resistance of Alloy 800 H at high temperature in impure helium seems to be worse compared to other tested materials.

With regards to tested ceramic materials, corundum base ceramics seems to be a better material as an insulator of heating elements for high temperatures than commonly used cordierite ceramics.

The presented work was financially supported by the TACR (Alfa Project TA03010849 and TA03020850) and by the SUSEN Project CZ.1.05/2.1.00/03.0108 realized in the framework of the European Regional Development Fund (ERDF).
The presented work was also supported within the FP7-ARCHER project supported by the European Commission.
The authors also thank colleagues from ÚJV Řež and Institute of Plasma Physics of the Academy of Sciences of the Czech Republic for performing and evaluating some tests on exposed specimens.

References

1. http://archer-project.eu/, cited on 01/07/2014
2. J. Berka, J. Matěcha, M. Černý, I. Víden, F. Sus, P. Hájek, New experimental device for VHTR structural material testing and helium coolant chemistry investigation – High Temperature Helium Loop in NRI Řež, Nucl. Eng. Des. **251**, 203 (2012)
3. http://susen2020.cz/, cited on 01/07/2014
4. K. Natesan, A. Purohit, S.W. Tam, Report NUREG/CR-6824: Materials Behavior in HTGR Environments, Office of Nuclear Regulatory Research, Washington, 2003
5. C. Cabet, A. Terlain, A. Girardin, D. Kaczorowski, M. Blat, J. L. Séran, S. Dubiez Le Golf, Benchmark CEA – AREVA NP – EDF of the Corrosion Facilities for VHTR, in *Proceedings of ICAPP 2007, Nice, France, May 13–18, 2007* (2007), Paper 7192
6. C. Cabet, F. Rouillard, Corrosion of high temperature metallic materials in VHTR, J. Nucl. Mater. **392**, 235 (2009)
7. http://nahledy.normy.biz/nahled.php?i=71705J
8. J. Berka, A. Rotek, J. Vít, J. Kutzendorfer, Testing of ceramics for production of heating elements for usage in High Temperature Helium Loop, in *Proceedings of the 22th International Conference on Nuclear Engineering ICONE22, Prague, Czech Republic, July 7–11, 2014* (2014), paper No. ICONE22-31019
9. J. Berka, A. Rotek, J. Vít, Testing of electric properties of ceramic components for thermally stressed parts of High Temperature Helium Loop, Paliva **5**, 123 (2013)
10. http://www.estcom.cz/data/soubory/slozeni-vlastnosti-hmot. pdf, cited on 28/12/2014
11. J. Berka, J. Kalivodova, M. Vilemova, Z. Skoumalova, P. Brabec, Corrosion tests of high temperature alloys in impure helium, in *Proceedings of the HTR 2014, Weihai, China, October 27–31, 2014* (2014), Paper HTR2014-41235

14

A study of different cases of VVER reactor core flooding in a large break loss of coolant accident

Yury Alekseevich Bezrukov[*], Vladimir Ivanovitc Schekoldin, Sergey Ivanovich Zaitsev, Andrey Nikolaevich Churkin, and Evgeny Aleksandrovich Lisenkov

OKB GIDROPRESS, 21, Ordzhonikidze Street, Podolsk, Moscow Region, 142103, Russian Federation

Abstract. The paper covers the results of VVER core reflooding studies in fuel assembly (FA) mockup of 126 fuel rod simulators with axial power peaking. The experiments were performed for two types of flooding. The first type is top flooding of the empty (steamed) FA mockup. The second type is bottom flooding of the FA mockup with level of boiling water. The test parameters are as follows: the range of the supplied power to the bundle is from 40 to 320 kW, the cooling water flow rate is from 0.04 to 1.1 kg/s, the maximum temperature of the fuel rod simulator is 800 °C and the linear heat flux is from 0.1 to 1.0 kW/m. The test results were used for computer code validation.

1 Introduction

Loss of primary coolant accidents in pressurized water reactors (VVER reactors in Russia and PWR reactors in the West) belong to the most severe cases in the spectrum of accidents in the nuclear power industry. The MCP guillotine break is considered to be the maximum design basis accident. Different thermal-hydraulic processes take place in the reactor in the course of such an accident, namely, a sharp drop of pressure followed by coolant boiling up and loss of primary coolant mass, which leads to partial reactor emptying. At this point the fuel rods heat rapidly to high temperature, due to a sharp decrease in heat removal efficiency. After the emergency core cooling system has been activated, the coolant mass is replenished and the partially-dried out core is reflooded. In later-designed American PWRs water is transported from the ECCS system into the cold leg of the MCP, and in VVER-type reactors the water is uniformly supplied to the upper and lower reactor plenums. Water supply into the upper plenum is connected with the problem of countercurrent flow of the water poured down into the core and the flow of steam released out of the FA. Steam flow is assumed to counteract water penetration into the core from the top and actually "seal" the water level above the core. The paper covers a brief review of reflooding studies performed in different countries and the relevant tests performed in OKB GIDROPRESS are discussed in more detail.

2 Brief descriptions of experimental studies in western countries

This section contains a brief description of some experimental studies performed in western countries. The background of the studies of the structure of the steam-water flow in the rod bundle goes back to the investigations performed in the USA. In their papers, Lahey and Shiralkar presented the studies of General Electric in a heated 9-rod bundle [1,2]. The purpose of the studies was to determine the velocity fields and the distribution of the flow enthalpy across the rod bundle section. Pressure differentials and flow temperatures were measured in separate fuel rod bundle cells. The fuel rod simulators were not equipped with thermocouples.

At about the same time, investigations were performed on a full-scale 36-rod Marviken facility in Sweden and the results are covered in reference [3]. They were distinguished by the studies of the axial and radial distribution of the flow steam quality. The measurements were made with a gamma-transmission unit. References [1–3] describe studies dealing with boiling water reactors.

The first studies devoted to reflooding in pressurised water reactors date back to the mid-seventies. Reference [4] mentions the studies conducted under the FLECHT program in the USA. The main purpose of these experiments was to obtain data that could be useful for reflooding calculations during loss-of-coolant accidents in the USA. The experiments were performed with a 10×10 bundle with 91 heated fuel rod simulators and 9 simulators of guide tubes that housed the instrumentation. The bundle was

---

* e-mail: bezrukov@grpress.podolsk.ru

placed inside a square housing with a 19.05 mm thick wall and then heated up on the outside during the tests. The fuel rod simulators have an outside diameter of 10.72 mm and were located in a square grid with 14.3 mm pitch, and their heated length was 3.66 m. Electrically-heated fuel rod simulators had a cosine power distribution with peak power of 1.66 due to the different pitch of the internal heater wiring. One of the peculiarities of these experiments was a wide variation of the flooding rate. Also, in these experiments the initial temperature of the fuel rod simulator claddings was relatively low. Thermocouples were installed inside the fuel rod simulators to measure the cladding temperature. The heat flux from the cladding surface was determined by calculation. The temperature of the control rod guide tube simulators was determined with the thermocouples installed inside the tubes. The housing temperature was measured at several points along the height. The flow rate of the flooded water was measured as well as its temperature and pressure at the bundle outlet. Several pressure drop gauges were installed along the column height to measure the water mass in the bundle. One of the specific features was the installation of thermocouples that were built into the wall on the internal surface of the housing at heights of 2.137 m, 3.048 m and 3.810 m. They were used as indications of continuous steam in the given section. In the event that the thermocouple showed the housing temperature to be above the saturation temperature, it was considered to be located inside the superheated steam.

A series of FLECHT experiments was the first to calculate the mass and power balance at the outlet of the testing facility, in order to determine the local conditions and to divide the heat transfer by irradiation between the droplets, steam and housing. In subsequent studies under the FLECHT program, experiments were performed with another heat release profile along the fuel bundle height [5], and with the bundle flow area partially blocked (simulation of fuel rod cladding ballooning during the accident) [6].

The most complete information on all of the issues that deal with reflooding is presented in reference [7]. It identifies and ranks the phenomena that are typical of different modes of the steam-water flow during reflooding. A description of the experiments conducted at 12 different facilities is given with rod bundles of different scales available in the USA and Europe. In addition, many single-tube experiments are considered. Such deep analysis was performed in order to develop the technical requirements to carry out the tests under the Rod Bundle Heat Transfer (RBHT) program. The scope of information that was required to develop mathematical models of reflooding and introduce them into computer codes was determined. Requirements were offered for the RBHT experimental facility unit, and equipment modeling and requirements for the FA instrumentation were defined.

A large cycle of work to study reflooding phenomena was performed in Germany under the FEBA and REBEKA [8,9] programs; this studied the effect of such factors as the presence of a gas gap, internal structure of the fuel rod simulator, heating-induced cladding deformation and the availability of spacer grids. The experiments have shown that the presence of the gas gap, cladding ballooning and rupture contribute to quicker cool down fuel rod.

Since 1976, VTT Energy and the Lappeenranta University of Technology have cooperated in researching nuclear reactor thermal-hydraulics. During these years they have built a series of experimental test facilities (REWET-II, REWET-III and PACTEL). The REWET-II and REWET-III facilities were designed for investigation of the reflooding phase of a LOCA [10]. The main design principle was the accurate simulation of the rod bundle geometry and the primary system elevations. The rod bundle consists of 19 indirectly-electrically-heated simulator rods. The heated length, the outer diameter and the lattice pitch of the fuel rod simulators as well as the number (=10) and construction of the rod bundle spacers are the same as in the reference reactor VVER-440. The aim of the tests was to improve the understanding of the basic phenomena of accident situations and to provide experimental data for the development and verification of the LOCA and SBLOCA codes aimed for analysing pressurised water reactors in use in Finland.

3 Descriptions of experimental studies in Russia

The study of the reflooding processes in Russia began in 1974 in OKB GIDROPRESS, with the investigations using single-rod and 7-rod bundles. The purpose of the studies was to investigate the effect of different kinds of cooling water supply on the cladding temperature. Experiments on the heated 7-rod bundle at the OKB GIDROPRESS test facility were started in 1975. These tests are described in reference [11]. The test facility is a two-loop installation that schematically models the VVER-440 reactor. The facility had a reactor model, one simplified loop with a rupture device to simulate a MCP leak and one large loop with a circulation pump that models the remaining five operating loops. The facility was used to simulate circulation pipeline guillotine break, and also to simulate reflooding of the heated bundle with the cooling water from the ECCS. The above bundle consists of 7 fuel rod simulators 9.1 mm in diameter and the heated length of 2.13 m.

The experiments were implemented according to the procedure below: steam was supplied to the test section and simultaneously the bundle power was smoothly increased. The bundle heat-up was confined to the central rod simulator cladding with the temperature not above 600 °C. The steam was discharged from the circuit via the damaged loop. After the steady state was established, steam supply to the model was quickly interrupted, drainage was stopped, the instrumentation system measurement devices were switched on and the test section was fed with water at 40 °C. After the rod bundle was cooled down to a temperature below 200 °C, the flooding stopped and the power supply to the bundle was interrupted. All-in-all, the experiments covered 11 tests with different versions of flooding. The plots in Figures 1 and 2 show the thermocouple readings in the tests with the core flooding from the top and the bottom.

The plots show that for the 7-rod bundle with flooding from the bottom, the bundle cool down takes place far more quickly and without significant temperature pulses. At this point, the cooling down front goes from the bottom to the top.

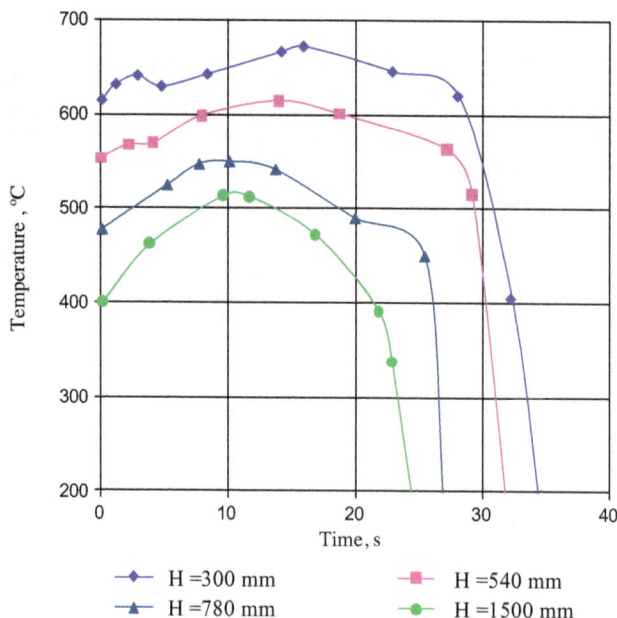

Fig. 1. Variation of cladding temperature in a 7-rod bundle in the case of flooding from the bottom.

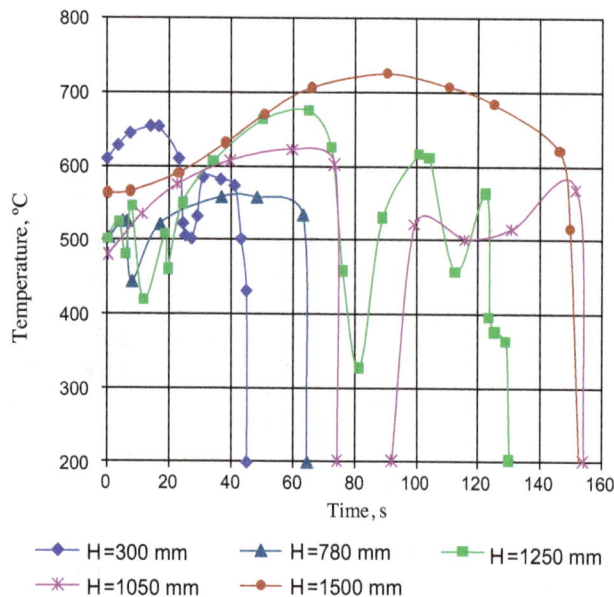

Fig. 2. Variation of cladding temperature in a 7-rod bundle in the case of flooding from the top.

In the case of core flooding from the top, the cooldown time increases significantly. The nature of the cooling front movement is also changed. The area where the outermost upper thermocouple is installed is the first to be cooled down. The thermocouples located below are cooled down later and with considerable fluctuations. This means that it is difficult for water to go inside the narrow bundle. The steam that is leaving the bundle impedes the water flow, i.e. the effect of the countercurrent flow of steam and water is quite significant.

In 1976, construction of the OKB GIDROPRESS test facility began [12]. The venture schematically modeled the

primary circuit of theVVER-440 reactor, with a full-scale FA mockup as the core simulator. As preparations for the work were underway, fuel rod simulators were designed, manufactured and tested, and their indirect heating up and thermal-physical characteristics were found to be close to a full-scale actual fuel rod. Such simulators were incorporated into a full-scale mockup FA for VVER-440 containing 126 heated rods 2.5 m long with uniform axial heating. The simulators were axially spaced with cell-type spacer grids 10 mm in height. One unheated rod was placed in the centre of the mockup. The grids were installed with a separation of 240 mm. The test facility was equipped with a large number of thermocouples to measure the cladding temperature, with a probe to measure the swell level along the fuel assembly height in the middle part of the bundle and the pressure values in the central part of the fuel assembly. A schematic diagram of the test facility is shown in Figure 3 and the FA mockup cross-section and the fuel rod simulator location pattern is given in Figure 4.

The procedure for the fuel assembly mockup testing is as follows. The valves in the damaged and operating loops were opened. Steam was supplied to the lower chamber of the test section at 0.3 MPa pressure.

The power supplied to the FA mockup kept increasing until the temperature of the most heat-powered simulator had reached 600 °C. Steam supply increased as the power increased. After the steady state was established, the power was increased spasmodically until it reached the assigned level. Simultaneously all of the recorders were switched on, the steam feed to the test section was stopped and the valve

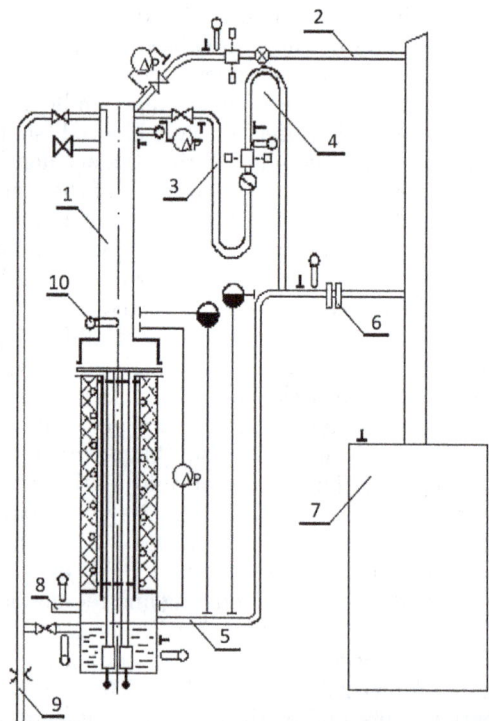

Fig. 3. Diagram of test facility with full-scale mockup of the VVER-440 FA. 1: test section; 2: tube of emergency leg; 3: tube of operable leg; 4: loop seal; 5: downcomer; 6: flowmeter; 7: discharge tank; 8: steam pipe; 9: tube of flooding water; 10: water-steam probe.

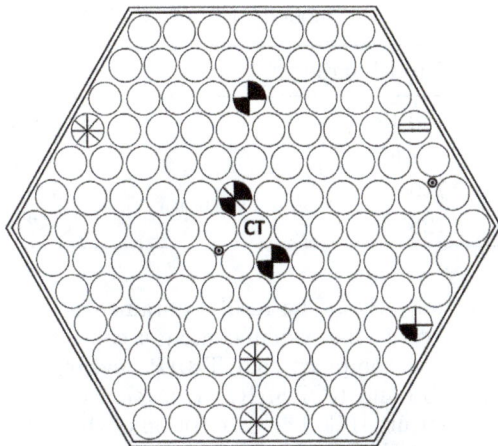

Fig. 4. Diagram showing the layout of imitators in mockup and their equipment by measuring sensors. CT: central tube with 5 submerged thermocouples; ⊕: simulator with 5 thermocouples; ◕: simulator with 4 thermocouples; ◕: simulator with 6 thermocouples; ◔: simulator with 3 thermocouples; ⊖: tube with 4 pressure taps; ⊙: the physical level sensor.

was opened on the cooling water supply line to the test section. In the course of the test, the assigned water flow rate was maintained.

The test was deemed to be over when the FA mockup was completely cooled down (to a temperature below 200 °C).

The steam flowrate was limited by pressure increases in the test section not exceeding 0.5 MPa. The plots of the fuel rod simulator cladding temperature variations in two tests with different types of flooding are shown in Figure 5. The thermocouple location points were indicated by their distance from the upper boundary of the simulator heating up.

Parameters associated with the most typical tests of the FA mockup are listed in Table 1, where G_w and t_w are flowrate and temperature of the flooding water respectively, q_{max} is the maximal heat flux, and t_{cl} is the fuel rod simulator cladding temperature.

The results of the FA simulator tests show that in the lower chamber flooding tests, gradual simulator cooldown from the bottom upwards can be observed. As the flowrate of the supplied water decreased, the cooldown time increased. In the tests with combined flooding and flooding from the top, significant pulsations of cladding temperature, especially in the upper and the middle parts of the FA, were observed. In the case of flooding from the top, there was no significant increase in the cooldown time, contrary to the phenomena observed in the 7-rod bundle. The only observation is that the middle part of the FA (the thermocouple is located at a height of 1510 mm) is actually cooled down simultaneously with the upper part of the FA. In addition, some temperature pulsations are observed in the middle part. It is likely that the water poured from the top goes down through the least heated parts of the mockup

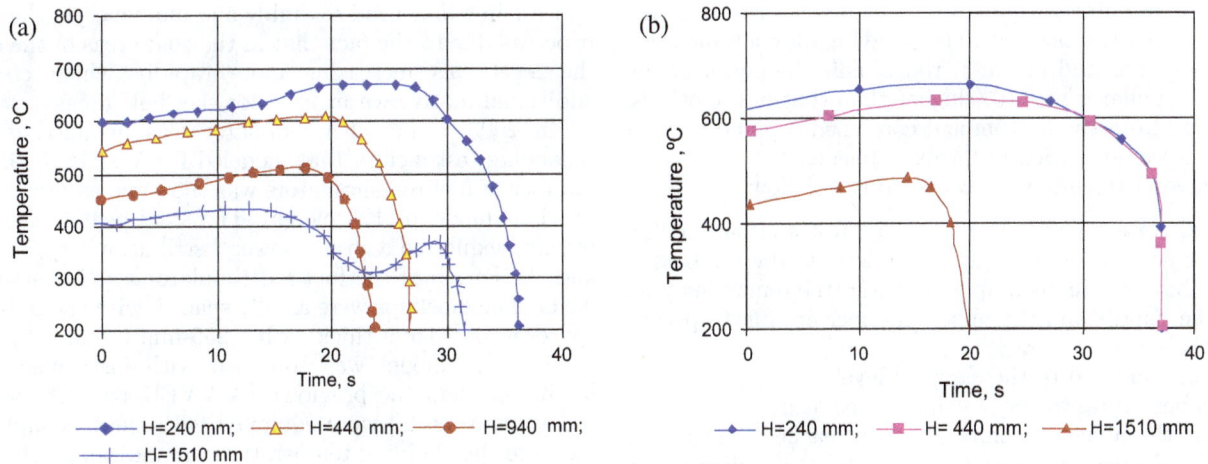

Fig. 5. Variation of cladding temperature depending on the kind of a reflood: (a) reflood from the top; (b) reflood from the bottom.

Table 1. Initial parameters of reflooding tests of the FA mockup.

Test No.	Flooding pattern	G_w, kg/s	t_w, °C	q_{max}, kW/m²	Max. t_{cl}, °C
1	To the lower chamber	1.4	40	30	570
2	To the lower chamber	2.4	46	24	610
3	To both chambers	1.4/2.4	41	31.2	630
4	To the upper chamber	2.06	46	28.5	640
5	To the upper chamber	1.9	46	31.4	690
6	To the upper chamber	1.9	40	29.2	630

Table 2. Comparison of the parameters of the tests of the 7-rod bundle and the mockup FA for VVER-440.

Parameter	7-rod bundle	FA mockup
Pressure, MPa	0.12	0.12
Specific power per one simulator, kW	2.65	2.24
Fuel rod cladding temperature before flooding, °C	720	690
Amount of flooding water per one simulator, kg/h	128	57

(over the edge of the periphery fuel rods, close to the central tube, along the hexahedral housing surface) and then the flooding proceeds from the bottom.

A comparison was made of the appropriate tests in the FA mockup and 7-rod bundle to investigate the effect of the scale factor on the process of water penetration in the lower chamber when the water is poured from the top. The parameters of the two comparable tests with flooding from the top are listed in Table 2.

It can be seen that for approximately the same mode parameters, even when there is greater water supply to a 7-rod bundle, it takes much longer for the small-scale bundle to be cooled down than to cool down the FA mockup. It is one demonstration of the fact that the efficiency of top flooding is influenced by the scale factor.

At the end of the nineties, SRC IPPE began investigating reflooding processes [13,14]. Testing facilities were created that modeled the primary circuit that contained the testing facilities with bundles of 7- and 37-rod simulators that simulated the geometry of VVER-1000 FAs. Axial heat flux profiling with a power peaking factor of 1.62 was envisaged for the mockups. The 7-rod bundle contained two unheated rods and in the 37-rod bundle the power of the central simulator was 10% higher than that of the others. Flooding from the bottom and combined top-and-bottom flooding were modeled in the experiments.

The experiments were carried out as follows:

– initial state – lower plenum of the test section and the lower part of the rods (up to the level where the heated area begins) are filled up with water, the remaining part of the bundle and the upper chamber are filled up with saturated steam;
– power increased to the assigned level;
– rods heated up to the starting temperature;
– when the assigned temperature is reached, the power begins decreasing under a set law and the cooling water flow is switched on;
– the experiment stops when the rod temperature decreases to the boiling temperature.

Four standard problems were arranged from the numerous tests of these bundles, of which we are going

to consider only two. The results obtained in these tests were used to verify the system computer codes KANAL-97 as a part of the computer code TRAP developed in OKB GIDROPRESS and KORSAR/V1 developed in the NITI Research Institute [15]. The description of the structure of the TRAP and KORSAR codes is given in reference [16]. Table 3 summarises the input parameters of the tests used in the calculations.

The time required for complete cooldown of the 7-rod bundle was about 700 s, and that of the 37-rod bundle was 300 s.

It is worth mentioning that both the experiment and the calculations were made in a one-dimensional statement. Therefore, the difference in the parameter behaviour in both bundles can hardly be attributed to the difference in the quantity of the rods. In a 37-rod bundle, the heat flux was smaller and the flooding rate was greater than in the 7-rod bundle.

A factor that accelerates the movement of the hot fuel rod wetting front in the case of flooding from the bottom appears in a multi-rod assembly and moreover, in the core. It occurs due to the fact that in the cold areas of the core, the level was increasing more rapidly which created additional motive water head for the "hot" areas.

In 2003, a new FA simulator was installed in the reflooding test facility that modeled FA VVER-1000. The number of fuel rod simulators was the same as that in the previous mockup (126 pieces). The axial heat in the new mockup exhibited a cosine power distribution ($K_z = 1.345$) and the length of the fuel rod simulator was increased to 3.5 m. The mockups were axially spaced, with the cell-type spacer grids 20 mm thick with a 255-mm pitch.

The FA model was equipped with instrumentation far better than the previous FA VVER-440. Of the fuel rod simulators, 20 were equipped with thermocouples to measure the cladding temperature. In addition, 6 thermocouples were installed inside the cladding of the instrumented simulator. The thermocouples were located at 10 levels along the simulator height beginning with the bundle bottom.

The initial experiments were performed with cold water supply to the upper chamber of the test section (top

Table 3. Parameters assigned in the calculations.

Bundle	Pressure, MPa	Maximum heat flux, kW/m	Flooding rate, cm/s
7 rods	0.278	2.94	2.0
37 rods	0.246	1.77	4.9

flooding). The subsequent tests were experiments with water boiling down (natural level decrease) and with subsequent cold water supply to the reactor downcomer (bottom flooding) at different flowrates and power supplied to the FA.

Methodologically, the experiments were performed in the following way. In the top flooding experiments, a small steam flow (up to 45 kg/h) was supplied from the steam generator within 10–15 minutes through the test section bottom inlet for the sake of heat-up. At the same time, power was supplied to the FA simulator where the maximum fuel rod simulator wall temperature did not exceed 400 °C. After this temperature was reached, the steam flow was quickly arrested, the reactor model was fed with water and within 2–3 seconds the power in the test section rose to the assigned value.

In the tests with water boiling down, the FAs were initially filled up with water, then the power was increased to the assigned level, then water boiled up, evaporated and the FA got uncovered. When the temperature of the fuel rod simulator claddings reached 650 °C, water was fed at the assigned flow rate to the reactor downcomer model. At this point, the upper part of the downcomer model was connected with the discharge line and the water excess streamed down to the discharge tank, i.e. the FA mockup makeup was realized with the free level. Several experiments were performed for each power value, with flood water flow rate varied to get the minimum flow rate at which the FA was cooled down.

During the experiment, the following parameters were recorded:

– the temperature of the fuel rod simulator claddings at all the points;
– flow rate of the supplied water;
– temperature of the flood water;
– pressure in the test section;
– water level in the FA channel and in the reactor downcomer model.

The experiment is complete once the FA has been completely cooled down or if the temperature of the fuel rod simulator cladding exceeds 800 °C.

Four tests were undertaken with the top of the test section flooding. The pressure was equal to 0.15 MPa. The

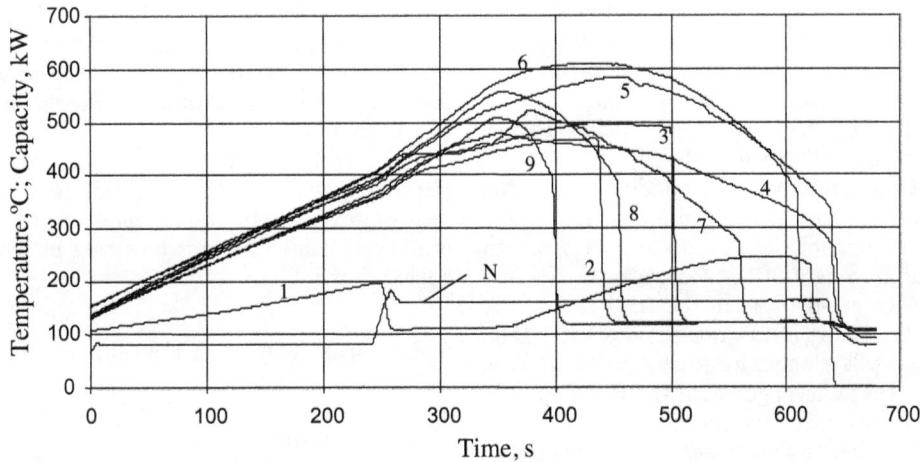

Fig. 6. Distribution of the cladding temperatures on the FA height in test No. 1. 1–9: numbers of sections of the thermocouples arrangement on the FA; N: capacity.

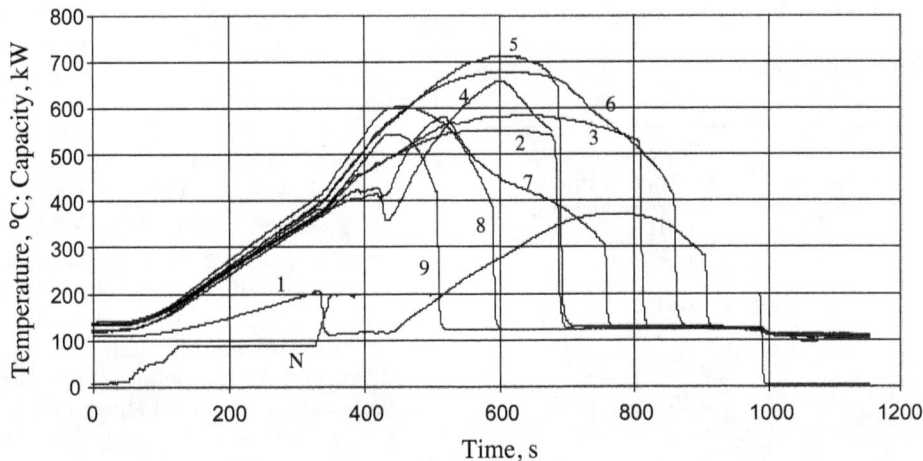

Fig. 7. Distribution of the cladding temperatures on the FA height in test No. 2. 1–9: numbers of sections of the thermocouples arrangement on the FA; N: capacity.

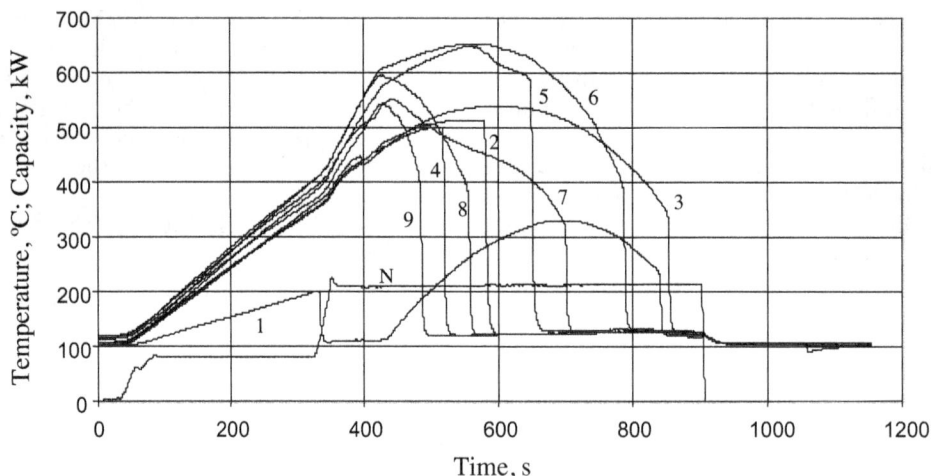

Fig. 8. Distribution of the cladding temperatures on the FA height in test No. 3. 1–9: numbers of sections of the thermocouples arrangement on the FA; N: capacity.

test results are given in Figures 6–8. The numbering of the cross-sections of the thermocouples takes place simultaneously with the lower boundary of the simulator heating. Detailed arrangement of thermocouples is listed in Table 4.

Cooling down of the FA takes place simultaneously from the FA top and bottom. The central axial parts of the fuel rod simulators remain hot for a longer time.

Full cooldown for top flooding is only realised after 400 s. Test parameters are listed in Table 5.

The maximum heat is released in the middle part of the fuel rod bundle, and therefore the maximum temperatures are also in the middle part of the FA mockup. It is worth mentioning that the flow rate of the fed water in the last experiments was twice as small as in the tests in the FA VVER-440, and far less than the flow rate in the VVER-1000 reactor. This is why it takes longer for the FA mockup to be cooled down. There was no level generation observed in the upper plenum. Thus, the upper flooding can be considered as efficient enough and all the supplied water quickly penetrates in the central part of the FA and cools down. Five tests were performed with level boiling down. The parameters of the boiling down tests are listed in Table 6.

Table 4. Coordinates of an arrangement of thermocouples on height of FA.

Number of section where installed thermocouples	Distance from the bottom of bundle, m
1	0.291
2	0.885
3	1.334
4	1.813
5	2.168
6	2.487
7	2.584
8	3.105
9	3.403

The results of the boiling down tests are given in Figures 9–11. It can be seen that at first, due to water boiling down, there was a level decrease in the mockup and after partial drying out and heat-up of the upper part of the fuel rod simulator cooling water supply into the lower chamber of the experimental model began. Due to water supply, the level in the model increased and the FA mockup was cooled down.

The model cooldown time depended on the flowrate of the cooling water and the value of the supplied power. Each test was repeated several times over in order to get the minimum value of water flow rate at which the FAs were cooled down.

4 Comparison with the system computer codes

As was mentioned above, SRC IPPE [14] has organised some standard problems for verification of the Russian codes. One of experiments has been simulated with the use of a code known as TRAP. The code package TRAP is intended for analysis of the variation of thermal and hydraulic parameters in the primary and secondary circuits and the core of NPP with VVER under conditions incorporating disturbances in operation of the primary and secondary equipment, such as accident conditions including LOCA. It is applied in the analysis of design basis accidents and beyond design basis accidents in substantiation of operability and safety of NPP with VVER and experimental facilities. The assumptions, common for the considered mathematical model, are given below:

– the equations to determine coolant parameters are put down as a one-dimensional approximation, not taking into account the power dissipation or metalwork strain;
– the process of surge of coolant boiling is assumed to be equilibrium from the point of view of thermodynamics;
– coolant movement in pipelines and in steam generator tubes is considered as an approximation to equilibrium steam-water mixture;
– the axial effect of thermal conduction in coolant and metalwork is not taken into account;

Table 5. The main results of the tests with top flooding.

Test No.	Supplied power in % from nominal heat transfer	Water flowrate, kg/s	Temperature of poured water, °C	Linear heat flux per one fuel rod, kW/m	Maximum temperature, °C	Cooldown time, s
1	2.55	1.10–0.9	88	0.46	610	400
2	2.7	1.10–0.9	80	0.61	700	600
3	2.95	1.1–0.9	87	0.67	650	500
4	3.1	1.0	75	0.70	870	Assembly was not cooled down

Table 6. The main results of the tests with level boiling down.

Test No.	Supplied power, kW	Water flowrate, kg/s	Temperature of flood water, °C	Heat flux density, kW/m	Maximum temperature, °C	Cooldown time, s
5	40	0.04	60	0.12	730	1200
6	80	0.07	60	0.24	800	1000
7	160	0.12	67	0.48	700	300
8	230	0.13	64	0.70	750	500
9	320	0.56	64	0.96	700	330

– the primary and secondary circuits of the plant represented in the model as a set of elementary cells (density, specific internal energy, etc.) are determined as average integrated per cells.

A comparison of the plots of cooling down the 37-rod bundle obtained from experiment and from the computations using the TRAP code is presented in Figure 12.

From the figure it can be seen that the peak of the cladding temperature in the experiment is a little more than in the calculation. However, the cooldown time of the bundle coincides for the experiment and for the calculation.

On the experiments OKB "GIDROPRESS" with initial evaporation of water from the test section and subsequent bottom reflooding, calculations with the use of code KORSAR/V1 [15] have been executed. The KORSAR is a code for the analysis of the non-stationary processes in NPP systems. It deals with water-cooled water-moderated reactor systems in stationary, transient and accident regimes, as well. The modeling of the thermal-hydraulic processes in RK KORSAR is performed on the basis of a non-equilibrium two-fluid model in one-dimensional approximation. The neutron kinetics calculation is performed in a quasi-three-dimensional approximation on the basis of the point kinetics model of the reactor.

Test No. 8 from Table 6 has been chosen to represent the results of these calculations. The initial and boundary conditions were set so that to comply fully with the scenario

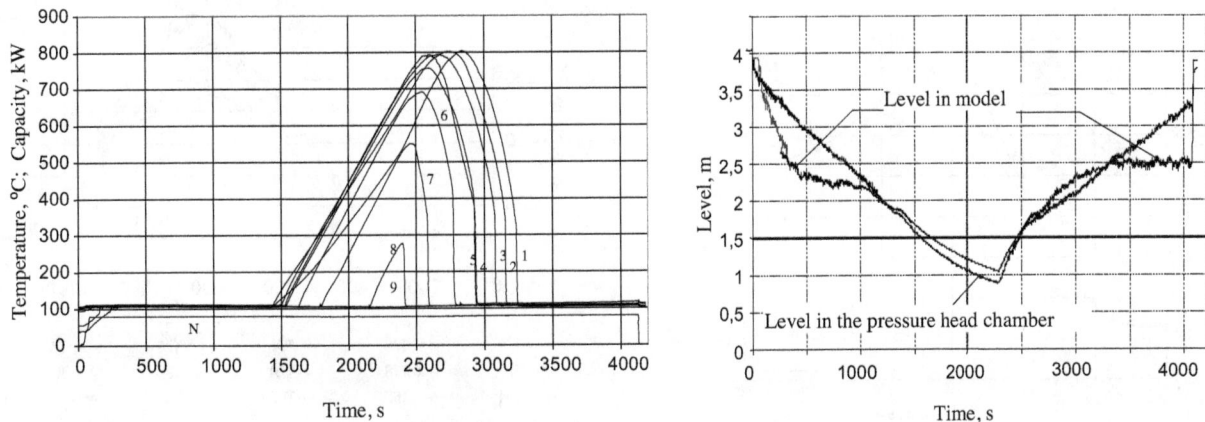

Fig. 9. Test 6. Distribution of the cladding temperature on the FA height. Variation of the level in the FA and the pressure head chamber. 1–9: numbers of sections of the thermocouples arrangement on the FA; N: capacity.

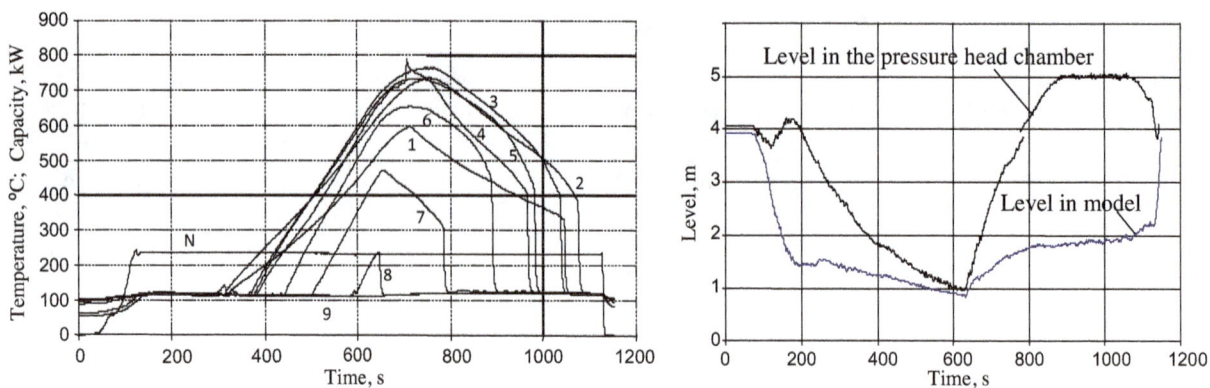

Fig. 10. Test 8. Distribution of the cladding temperature on the FA height. Variation of the level in the FA and the pressure head chamber. 1–9: numbers of sections of the thermocouples arrangement on the FA; N: capacity.

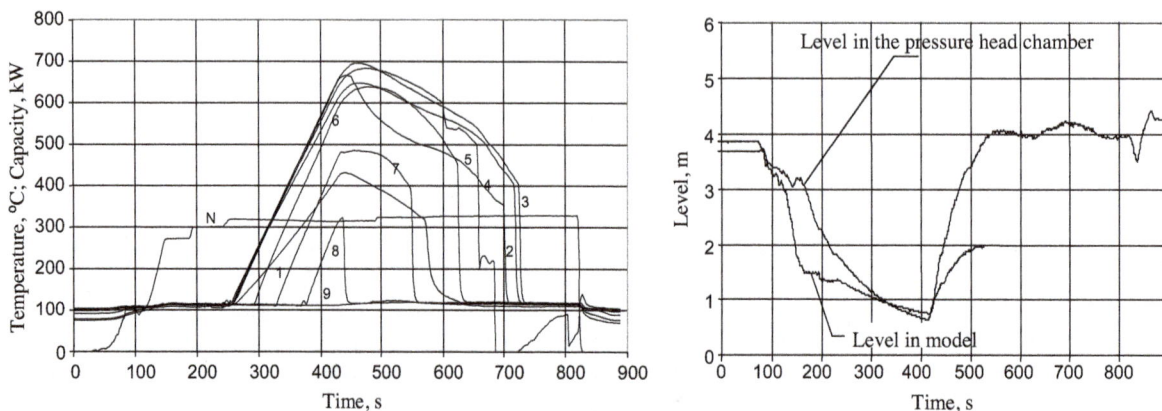

Fig. 11. Test 9. Distribution of the cladding temperature on the FA height. Variation of the level in the FA and the pressure head chamber. 1–9: numbers of sections of the thermocouples arrangement on the FA; N: capacity.

of the experiment. During the initial moment, the pressure in the test section was equal to 0.1 MPa, FA mockup was filled with water and then the capacity was switched on to the rod bundle. The boundary conditions used in the calculations were the given changes in time of values of capacity of simulators, the flowrate, the enthalpy of coolant at the inlet in the test section, and the pressure in the inlet

and outlet of a bundle. The comparison of the calculation with the experiment is presented in Figure 13.

The temperature curves are shown for the heat-intensity part of the FA mockup.

It can be seen that the results of the calculation are in reasonable agreement with the experimental data. The peak

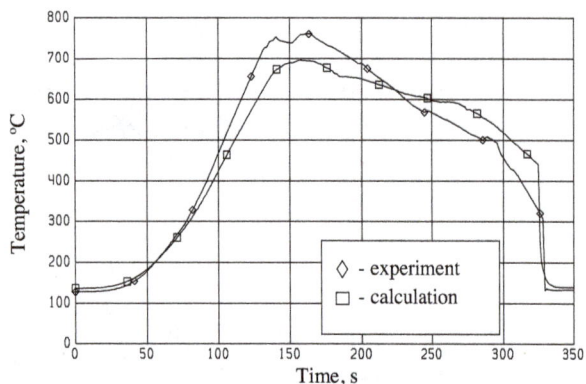

Fig. 12. Comparison of calculations using the TRAP code with experiment from SRC IPPE.

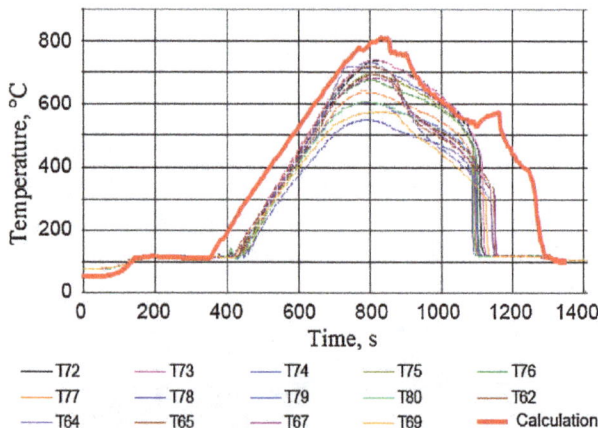

Fig. 13. Comparison calculation with experiment No. 8 from Table 6. Thermocouples located 2.487–2.584 m from the bottom of the bundle.

of the cladding temperature in the calculation is a little more than in the experiment. The code conservatively predicts the cooling down process of the experimental model.

5 Conclusions

The majority of the studies performed in OKB GIDRO-PRESS were devoted to top flooding. The reason is clear, as the flooding applied in the VVER is made immediately into the upper and the lower chambers of the reactor. The tests have shown that scale factor, i.e. the number of rods in the FA mockup influences the effectiveness of coolant supply from the top.

The experiments of OKB GIDROPRESS show that as the transverse dimension of the FA mockup increases, the flow choking of the water supplied from the top by the steam flow significantly decreases. This agrees well with the conclusions of the experiments in the UPTF facility [17,18] in Germany, where no flow choking was observed.

The experiments in the bundles with lower number of rods were performed at the end of the nineties in SRC IPPE. From the results of these experiments, several standard problems were solved and the Russian codes TRAP and KORSAR were verified.

It is worth mentioning that all of the experiments in OKB GIDROPRESS were performed in the rod bundles with strain-free fuel rod simulators equipped with cell-type spacer grids of small height. These grids, apart from the contemporary spacer grids applied in the VVER had small pressure loss coefficient and did not actually interfere with the cooling front movement. In the new designs mixing grids were additionally introduced to the spacer grids of increased height, which have considerable hydraulic resistance. Foreign researchers [19] have observed in experiments with bundles equipped with spacer grids of greater height with deflectors that the cooling front is passing, flood water accumulation is sometimes observed upstream of the grids and the cladding temperature increases downstream of the grids. However, this was not observed at low flooding rates; it only happened at high water flow rates with water supplied from the bottom.

The results of the studies in the RBHT facility in the USA [20] show that in case of bottom flooding, the spacer grids located along the bundle height quickly get wetted with the droplets of water flying in the steam flow, and have a temperature far lower than the cladding temperature in the same cross-section. No temperature increase of simulator claddings was observed in the places where the spacer grids were installed.

The introduction of the mixing grids into the new RP VVER designs and also for the operating NPPs with power increased to 107% requires experimental studies of the effect of the mixing grids for core reflooding in loss of primary coolant accidents.

Nomenclature

VVER water-cooled and water-moderated power reactor
SRC IPPE State Research Centre RF "Institute for Physics and Power Engineering"
NITI Technology research institute
MCP main circulation pipeline
PWR pressurised water reactor
RP reactor plant
ECCS emergency core cooling system
FA fuel assembly

References

1. R.T. Lahey, F.A. Schraub, A mixing flow regimes and void fraction for two-phase flow in rod bundles, Two-phase flow and heat transfer in rod bundles, ASME, 1969
2. R.T. Lahey, B.S. Shiralkar, D.W. Radcliff, A two-phase flow and heat transfer in multirod geometries: subchannel and pressure drop measurements in a nine-rod bundle for diabatic and adiabatic conditions, GEAP-13049, AEC, 1968
3. K.M. Becker, J. Flinta, O. Nylund, A dynamic and static burnout studies for the Full Scale 36-Rod Marviken fuel element in the 8 MW Loop FRIGG, in *Symposium on Two-Phase Flow Dynamics, Eindhoven, September 1967* (1967)
4. E.R. Rosal et al., A FLECHT low flooding rate cosine test series. Data report, WCAP-8651, 1975
5. E.R. Rosal et al., FLECHT low flooding rate skewed test series. Data report, WCAP-9108, 1977
6. M.J. Loftus et al., APWR FLECHT SEASET 21-rod bundle flow blockage task data and analysis report, NUREG/CR-2444, EPRI NP-2014, ECAP-992, Vol. 1 and Vol. 2, 1982
7. L.E. Hochreiter et al., Rod bundle heat transfer test facility test plan and design, NUREG/CR-6975, 2010
8. K. Rust, P. Ihle, F.J. Erbacher, Reflood heat transfer tests for PWR safety evaluation, in *Proceedings of Thermophysics-90 Obninsk, USSR, September 25–28, 1990* (1990)
9. F.J. Erbacher, H.J. Neitzel, K. Wiehr, The role of thermal-hydraulic in PWR fuel cladding deformation and coolability in a LOCA. Result of the REBEKA Program, in *Proceedings of Thermophysics-90 Obninsk, USSR, September 25–28, 1990* (1990)
10. T. Kervinen, H. Purhonen, T. Haapalento, *REWET-II and REWET-III facilities for PWR LOCA experiments.* (Espoo: Technical Research Centre of Finland, 1989) 25 p. + app. 8 p. (VTT Tiedotteita – Meddelanden – Research Notes 929)
11. S.A. Logvinov, Y.A. Bezrukov, Y.G. Dragunov, *Experimental justification of thermal-hydraulic reliability of VVER reactors* (Akademkniga Moscow, 2004) (in Russian)
12. Y.A. Bezrukov, S.A. Logvinov, S.V. Levchuk, V.D. Nakladnov, V.P. Onshin, A.S. Sokolov, Creation of a full-scale VVER-440 fuel assembly mockup to study the temperature mode in the core at the stage of reflooding, in *Proceedings of CMEA seminar "Thermal physics-82" Karlovy Vary, Czechoslovakia, 1982* (1982) (in Russian)
13. V.V. Lozhkin, O.A. Sudnitzin, B.I. Kulikov, Results of experimental studies of the VVER reactor the FA mockup reflooding with water fed from the bottom. Thermal-physical aspects of VVER safety, in *Proceedings of international conference "Thermal physics-98" Obninsk, May 26–29, 1998* (1998) Vol. 1, p. 389 (in Russian)
14. V.N. Vinogradov, V.V. Lozhkin, V.V. Sergeyev, S.I. Zaitsev, Y.V. Yudov, Verification of Russian thermal-hydraulic codes against standards problems of the VVER reflooding, in *Proceedings of the 2nd All-Russian Scientific and Technical Conference "Safety Assurance of NPP with WWER" V.5, Podolsk, November 19–23, 2001* (2001) (in Russian)

15. V.A. Vasilenko, Y.A. Migrov, Y.G. Dragunov, M.A. Bykov, E.A. Lisenkov, Thermal-hydraulic Code KORSAR. Development status and application experience, in *Proceedings of the 3rd All-Russian Scientific and Technical Conference "Safety Assurance of NPP with WWER" V.6, Podolsk, May 26–30, 2003* (2003) (in Russian)

16. A. Del Nevo et al., Benchmark on OECD/NEA PSB-VVER project Test 5A: LB-LOCA transient in PSB-VVER facility. UNIPI (Italy), DIMNP NT 638(08) Rev.2, Pisa, 2009

17. P.A. Weiss, R.J. Hertline, UPTF test results: first three separate effect tests, Nucl. Eng. Des. **108**, 249 (1988)

18. H. Glaeser, Downcomer and tie plate countercurrent flow in the Upper Plenum Test Facility (UPTF), Nucl. Eng. Des. **133**, 259 (1992)

19. Z. Koszela, Effect of spacer grids with mixing promouters on reflood heat transfer in a PWR LOCA, Nucl. Technol. **123**, 156 (1998)

20. L.E. Hochreiter et al., RBHT reflood heat transfer experiments data and analysis, NUREG/CR-6980, 2012

15

Comparison of CATHARE results with the experimental results of cold leg intermediate break LOCA obtained during ROSA-2/LSTF test 7

Piotr Mazgaj[1,2*], Jean-Luc Vacher[1], and Sofia Carnevali[3]

[1] EDF, SEPTEN, 12-14 avenue Dutriévoz, 69628 Villeurbanne, France
[2] Institute of Heat Engineering, Warsaw University of Technology, 21/25 Nowowiejska, 00-665 Warsaw, Poland
[3] CEA-Saclay, DEN, DM2S/STMF, 91191 Gif-sur-Yvette, France

Abstract. Thermal-hydraulic analysis is a key part in support of regulatory work and nuclear power plant design and operation. In the field of Loss Of Coolant Accident, evolutions of the regulations are discussed in various countries taking into account the very unlikely character of a double-ended guillotine break and questioning the necessity to study such an event with Design Basis Conditions assumptions. As a consequence, the consideration of intermediate size piping rupture becomes more and more important. The paper presents the modeling of the Test Facility ROSA-2/LSTF in the calculation code CATHARE 2.V2.5. OECD/NEA ROSA-2 Project Test 7 was conducted with the Large Scale Test Facility on June 14, 2012. The experiment simulated the thermal-hydraulic responses during a PWR 13% cold leg Intermediate Break Loss Of Coolant Accident (IBLOCA). The break was simulated by a cold leg upwardly mounted long break nozzle. The facility and the experiment conditions are modeled in CATHARE. The vessel is modeled by using a 3D module. A thermal-hydraulic analysis is conducted and the obtained results are subsequently compared with the experimental results from ROSA-2/LSTF Test 7. Evaluation of the differences between experimental and calculated results is discussed.

1 Introduction

The OECD/NEA ROSA-2 Project aimed to investigate key PWR thermal-hydraulics issues. It consisted in various experiments at ROSA/LSTF facility [1] operated by the Japan Atomic Energy Agency (JAEA). Among these experiments, three were devoted to the Intermediate Break Loss Of Coolant Accident (IBLOCA). The goal of the OECD/NEA ROSA-2 Project Test 7 [2] was to simulate the thermal-hydraulic responses during a PWR 13% cold leg intermediate break loss of coolant accident. The double-ended guillotine break (DEGB) of the Emergency Core Cooling Systems (ECCS) nozzle was simulated by using a 36.0 mm inner-diameter nozzle which was upwardly mounted on the cold leg.

The paper shows the comparison of the experimental results with the calculation results obtained from CATHARE 2.V2.5 [3].

2 Historical LOCA modeling approaches

In the field of LOCA Studies, the modelization choices may differ according to the break size and the involved physical phenomena, especially with respect to the vessel downcomer nodalization.

2.1 For large break LOCA (LBLOCA)

3D effects in the vessel downcomer during the blowdown stage of the accident are well known and have been widely investigated. It is usually recommended to use a 3D nodalization to represent the counter-current flow between the ascending steam exiting the core and the descending ECCS flow, although a 1D nodalization is considered to be conservative.

2.2 For small break LOCA (SBLOCA)

The dynamics of the accident is slow and the Peak Cladding Temperature is usually obtained before or at the very beginning of the ECCS accumulators injection. In such situations, 3D phenomena in the vessel downcomer have a very low influence, and a 1D nodalization of the downcomer is relevant.

2.3 For intermediate break LOCA (IBLOCA)

The dynamics of the accident is intermediate. The Reactor Coolant System (RCS) minimum mass inventory and the Peak Cladding Temperature are usually obtained shortly after the beginning of the ECCS accumulators injection (as it is observed during the ROSA test 7). Nevertheless, the overall reactor behavior during the accumulator injection is worth being studied with respect to the mass inventory distribution in the different parts of the vessel. More particularly, we have focused on the vessel downcomer area where three-dimensional effects are expected due to the asymmetric injection of subcooled ECCS water.

In France, the LOCA Intermediate Break Methodology (for the IBLOCA and SBLOCA break size range), developed by EDF and AREVA in the frame of the evolution of the LOCA regulation, retains a 2D nodalization of the vessel downcomer (theta, z, with one single mesh in the radial direction), using the 3D module and capability of the CATHARE code.

The work presented in this paper contributes to the understanding of this issue and to the related CATHARE validation.

3 CATHARE Code

CATHARE 2.V2.5 [3] is a multi-purpose multi-reactor concept system code that can describe several kinds of different circuits with various fluids either in single-phase gas or liquid or in two-fluid conditions possibly with non-condensable gases. The code is capable of simulating any kind of reactor concept and any kind of accidental transient. CATHARE 2 uses flexible structures for thermal-hydraulic modeling. The main hydraulic components or elements are pipes (1D), volumes (0D), a 3D module and boundary conditions, connected to each other by junctions. Apart from the modeling of the hydraulic components, the code modules can model pumps, turbomachines, control valves, T-junctions, sinks, sources, breaks and many other ones. The basic set of equation is based on a six-equation two-fluid model (mass, energy and momentum equations for each phase). Additionally, the code has

possibilities to take into account optional equations for non-condensable gases and radio-chemical components.

4 ROSA-LSTF modeling

The ROSA-LSTF (Large Scale Test Facility) is located in Japan and it was constructed to simulate the full-scale height and 1/48 volumetrically-scaled down reactor of a Westinghouse four-loop PWR with thermal power of 3423 MW. It is composed by two primary loops that correspond to four primary loops of the reference Westinghouse PWR. Figures 1 and 2 show the schematic view of the LSTF.

In order to compare the results, two nodalizations of the ROSA-LSFT vessel were developed with the CATHARE

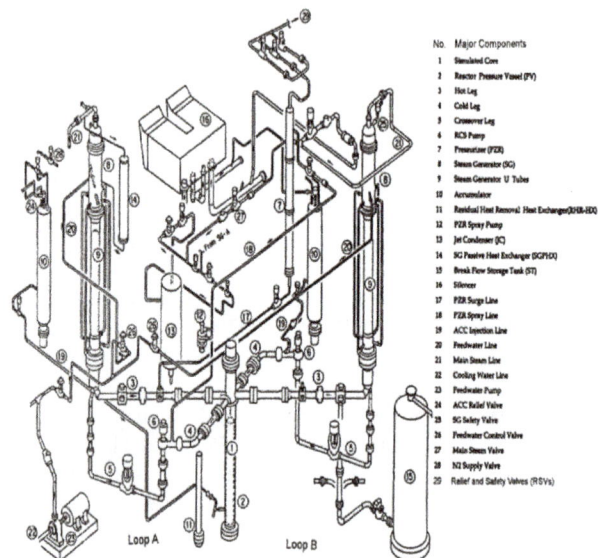

Fig. 1. General view of LSTF [1].

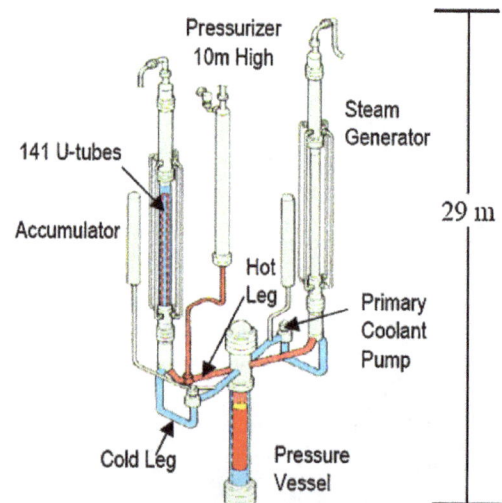

Fig. 2. The view of hot and cold legs in LSTF [2].

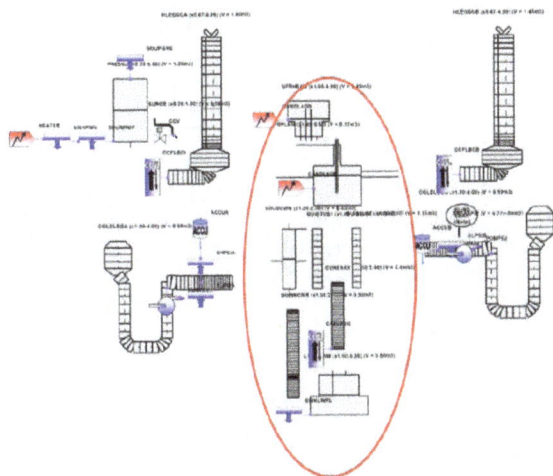

Fig. 3. 1D nodalization of ROSA-LSTF. In the 3D nodalization, the pressure vessel (the nodalization in the red circle) was modeled using 3D modules from CATHARE code.

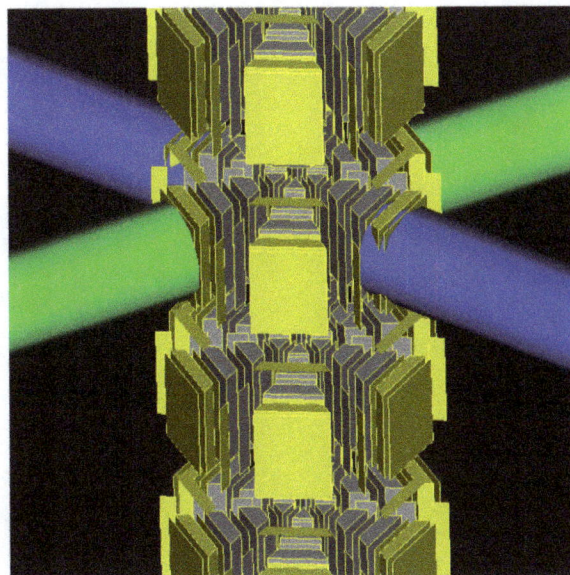

Fig. 5. The detailed view of the azimuthal and radial nodalization.

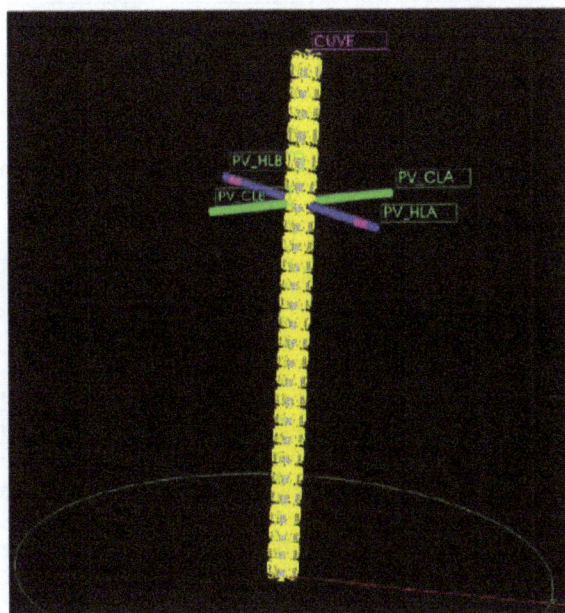

Fig. 4. 3D nodalization of the pressure vessel of ROSA-LSTF.

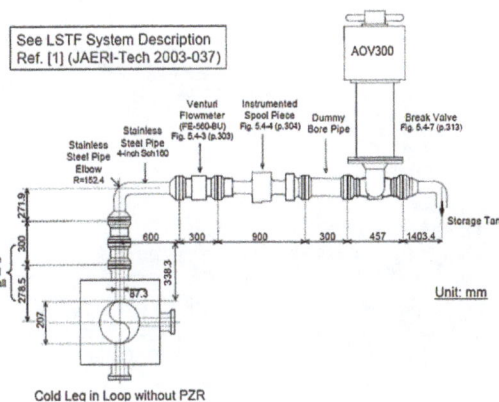

Fig. 6. ROSA-LSTF break unit [2].

shown below, which was upwardly mounted on the cold leg. The nozzle flow area corresponds to 13% of the volumetrically-scaled cross-sectional area of the reference PWR. The break is located in Cold Leg B.

5 Results of the calculations

CATHARE predicts the break flow rate and pressure in pressurizer rather well as it can be seen in Figures 7–9. In particular 3D model gives a pressure evolution which is very close to the experimental one. The major chronological events are listed in Table 1.

5.1 Downcomer mass inventory

The prediction (Fig. 10) of the overall pressure difference in the vessel downcomer (representative of the mass inventory)

code. The 1D nodalization is depicted in Figure 3. To better model the phenomena in LSTF, the 3D nodalization of reactor pressure vessel was introduced, using the 3D module from the CATHARE code.

The 3D nodalization of the pressure vessel is shown in Figures 4 and 5. As one can see, the pressure vessel has 26 meshes in vertical direction, six nodes in azimuth and radially four rings (the core region consists of 9 meshes in vertical direction, the lower plenum of 6 meshes and the upper plenum consists of 7 meshes). The most outer ring is used to model the downcomer, and the three inner rings are used to model the core (as shown in Fig. 4).

Figure 6 shows the ROSA-LSTF break unit which is used to model the Double-Ended Guillotine Break (DEGB) of the ECCS pipe by using a 36.0 mm inner-diameter nozzle

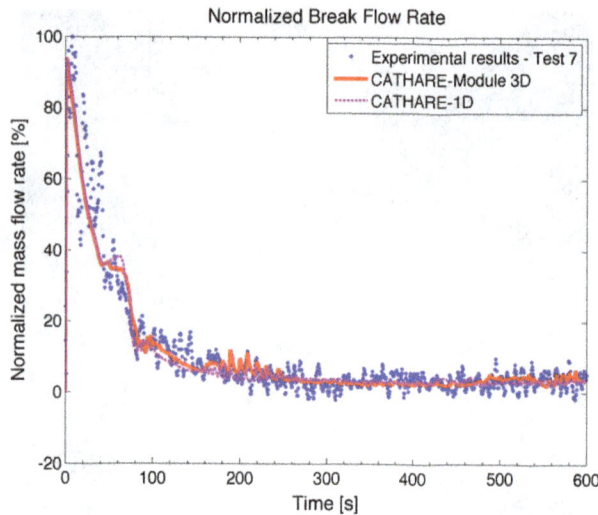

Fig. 7. Break flow rate.

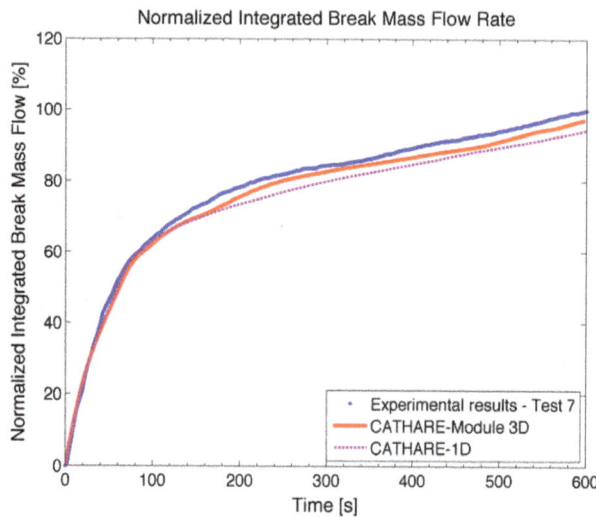

Fig. 8. Integrated break flow rate.

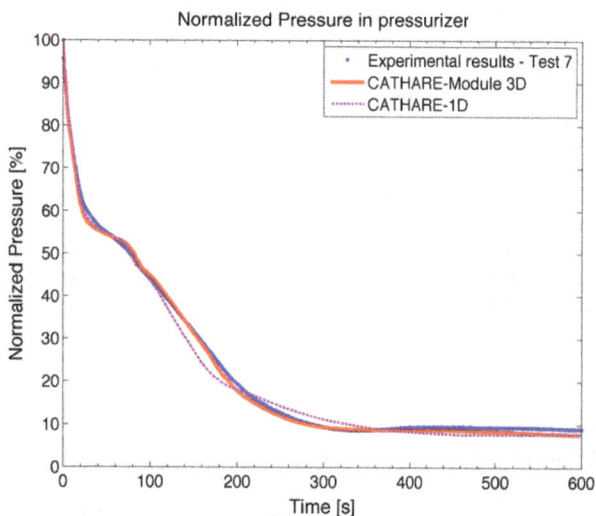

Fig. 9. Pressure in pressurizer.

Table 1. Chronology of major events.

Event	Time [s] – Experiment	Time [s] – CATHARE 3D
Break valve open	0	0
Initiation of coastdown of primary coolant pumps	11.5	12
Initiation of HPI system in loop with PZR (loop-A) only	26	27
Initiation of core power decay	30	30
Initiation of ACC system in loop-A only	154	158
Core reflooding	182	176
Upper plenum filling	195	190
Primary coolant pumps stopped	261	262
Termination of ACC system in loop-A only	350	340

is much better using the 3D model, for which the pressure difference behavior is very close to the experimental one, whether it is before or after the beginning of the accumulator injection (around 160 s). It can be thus noticed that the 3D effects that are present in IBLOCA modeling are much better modeled by a 3D nodalization of the pressure vessel.

Such a large difference between the results coming from 1D and 3D nodalization has an origin in the behavior of the ECCS water while passing through the pressure vessel downcomer. Once the level in the downcomer has reached the cold legs, the 1D modeling cannot predict the counter-current flow and the mixing between the cold ECCS water entering the top of the downcomer with the hotter water present in its lower part. Thus, the ECCS flow is mostly overpassing the downcomer and directly flows to the break.

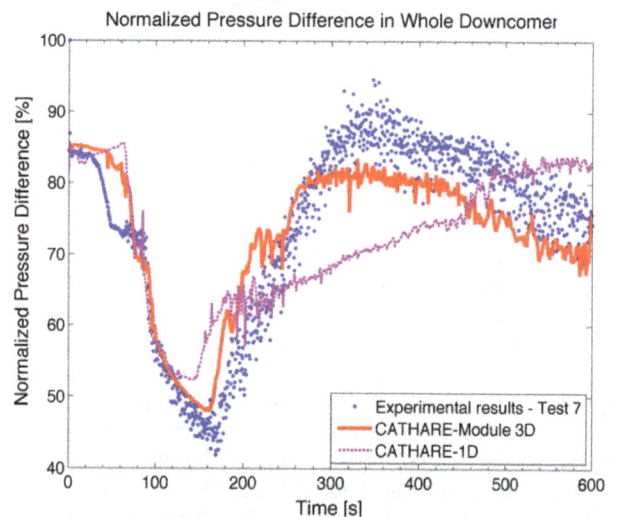

Fig. 10. Pressure difference in whole downcomer.

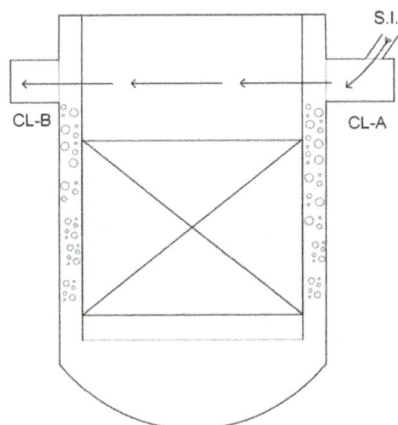

Fig. 11. 1D modeling effects during LOCA.

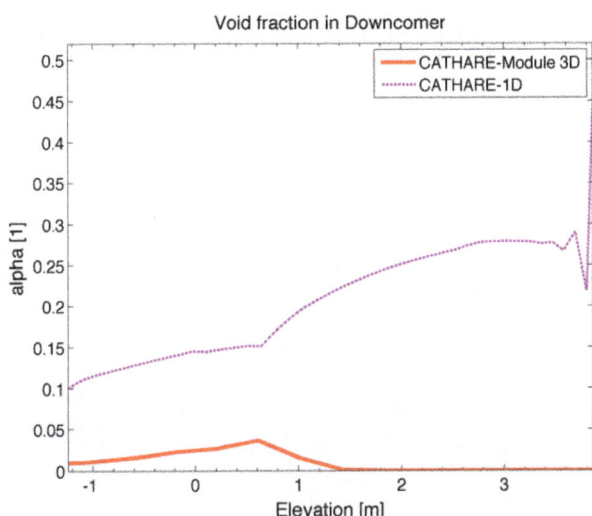

Fig. 12. The void fraction in vertical direction close in downcomer at the Cold Leg A (at 350 s).

Moreover, as the temperature in the lower part of the downcomer remains close to the saturation temperature, the fast RCS depressurization (combined with wall heat release) induces flashing and steam production. This irrelevant behavior obtained with the 1D calculation is depicted in Figure 11, while the calculated void fraction profile in Figure 12 clearly shows the steam production.

5.2 Overall vessel mass inventories in the 3D calculation

As the 3D model shows a good downcomer mass inventory prediction, we may expect also a good prediction of the mass inventories in the different other parts of the vessel and of the hot legs. The related pressure differences in different parts of the Pressure Vessel are shown in Figures 13–16. The results compare rather well with the ones of the experiment. In particular, in Figure 14 at 200 s, it can be pointed out that the beginning of the upper plenum filling is well predicted by CATHARE.

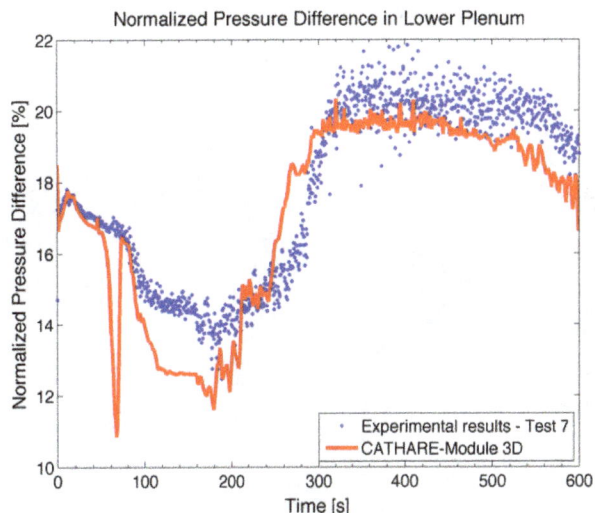

Fig. 13. Pressure difference in lower plenum of the pressure vessel.

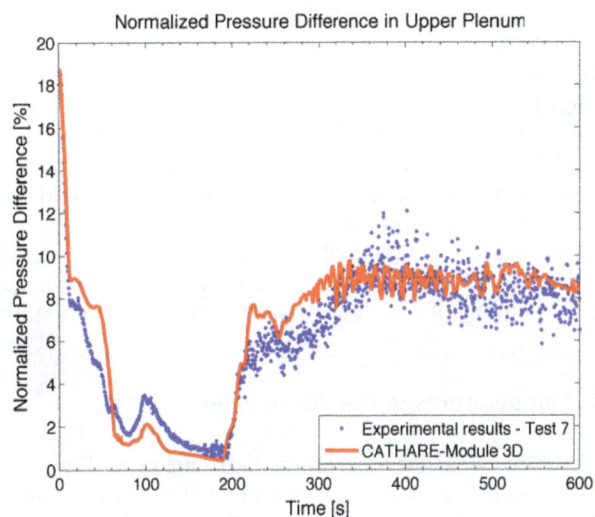

Fig. 14. Pressure difference in the upper plenum of the pressure vessel.

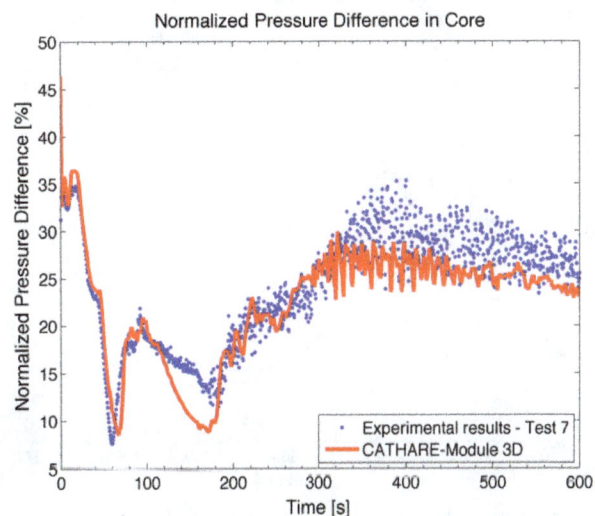

Fig. 15. Pressure difference in the core.

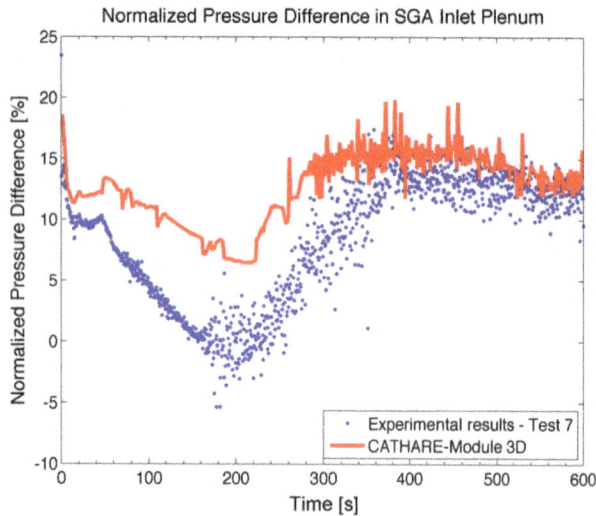

Fig. 16. Pressure difference in Steam Generator A (SGA) Inlet Plenum.

In Figures 15 and 16, before the accumulator injection until around 180 s, one can observe that the pressure difference in the core region is underestimated by the 3D model. This could be explained by an overestimation of the water retained in the Steam Generator inlet plenum (possibly due to a more severe counter-current flow limitation in the calculation). This observation could lead to a further investigation.

Figure 17 shows the fuel rod surface temperature in the fuel bundle located 5 cm under the top of the core. The curve is illustrating well the phase of the core reflooding.

5.3 Temperatures in the downcomer

The experimental temperature distribution in Figure 18 illustrates the idea that a 1D modeling of the downcomer is not accurate enough. One can realize that the schematic

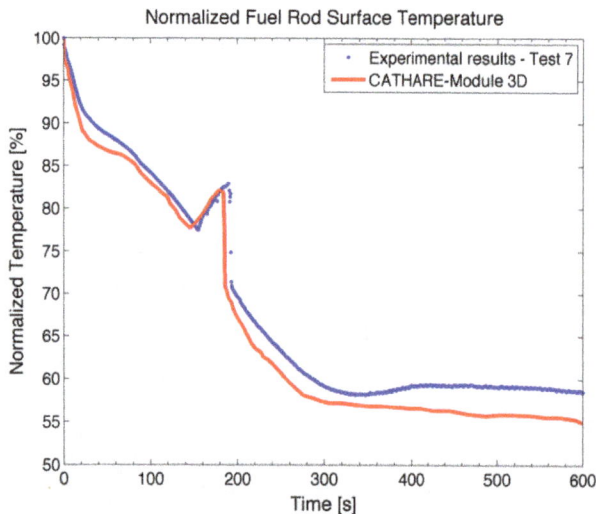

Fig. 17. Fuel rod surface temperature located in the fuel bundle 5 cm under the top of the core.

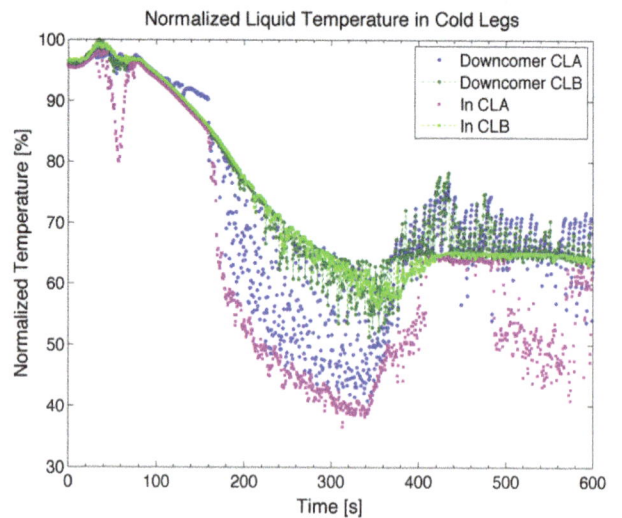

Fig. 18. Experimental results of temperature in Cold Leg A (CLA), Cold Leg B (CLB) and in the downcomer at the entrances of Cold Legs.

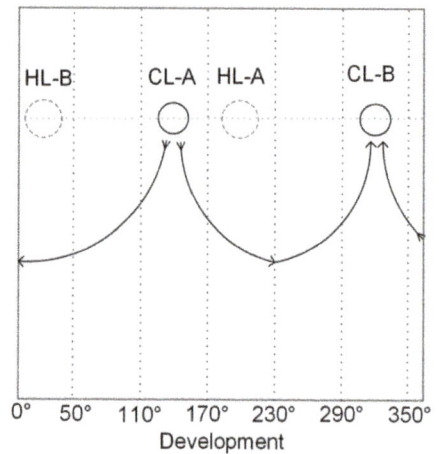

Fig. 19. The approximate flow of the coolant in the downcomer.

flow in downcomer would be as the one shown in Figure 19. Colder water is entering the circuit through the Safety Injection, then is flowing through the downcomer and core.

The temperature of fluids in Cold Leg A (intact leg) is lower than the temperature of fluids leaving through the break, located in Cold Leg B (as shown in Fig. 18).

The temperatures in Cold Legs are measured 1.6 m away from Pressure Vessel center. The temperature in the downcomer close to the Cold Leg A nozzle follows the Cold Leg A temperature, whereas the temperatures in Cold Leg B and in the downcomer close to the Cold Leg B nozzle follow the saturation temperature.

The 3D CATHARE model temperature predictions follow the experimental trends, though it can be observed that the code is underestimating the temperature in Cold Leg A and in the downcomer at the entrance of Cold Leg A (shown in Figs. 20 and 21). However, it can be noticed that the underestimation of the temperature in the downcomer

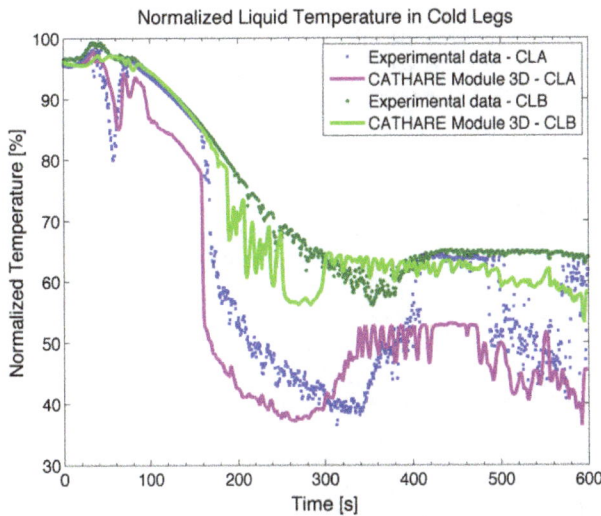

Fig. 20. Liquid temperature in Cold Leg A and B, comparison between experimental and numerical results.

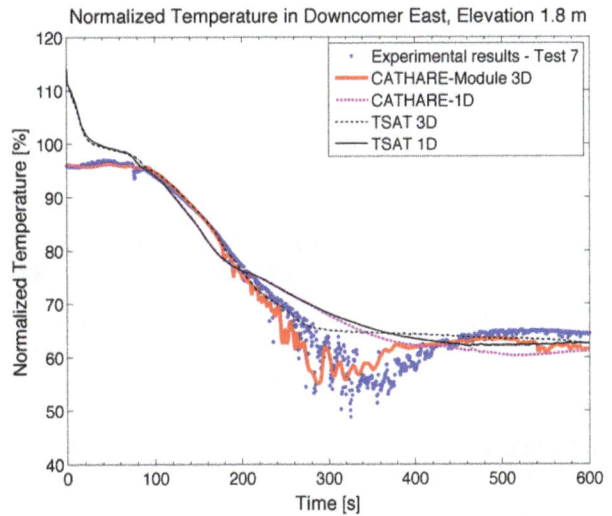

Fig. 22. The temperature in the downcomer at the elevation 1.8 m.

results from the underestimation observed in the cold leg. Further analysis of the Cold Leg A temperature could be performed, in connection with the condensation phenomena at ECCS injection.

To better visualize the physical conditions in the downcomer, the temperature distributions at two different positions (1.8 m and 3.6 m, above the core bottom) were depicted in Figures 22 and 23. It can be easily noticed that the 3D model predicts better the temperatures in these lower parts of the downcomer. After the accumulator injection (around 160 s), one can observe, as in the experiment, a slight subcooling, suggesting inflow of some ECCS water Cold Leg A to downcomer, whereas the 1D model predicts saturated conditions at the same locations.

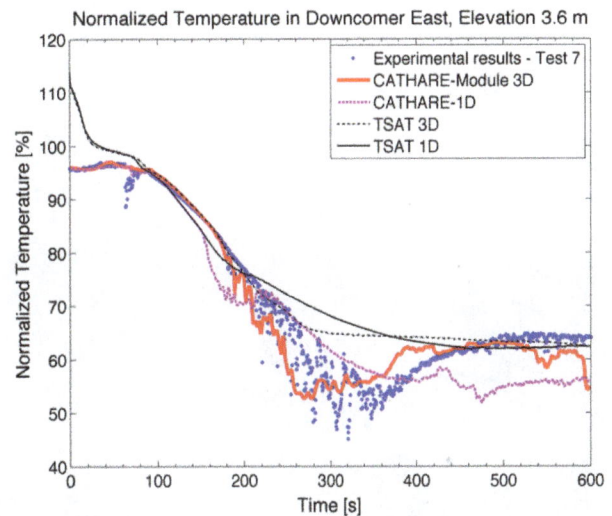

Fig. 23. The temperature in the downcomer at the elevation of 3.6 m. Comparison between experimental and numerical results.

6 Conclusions

The 3D vessel modeling of the LSTF facility for the calculation of the OECD/NEA ROSA-2 Project Test No. 7 gives much better results than the 1D modeling while comparing with experimental data related to the mass inventory and the temperatures in the vessel downcomer.

The ROSA-2 test No. 7 indicates that it exists strong 3D phenomena during the accumulator injection stage of an IBLOCA scenario. The calculations performed with the CATHARE code strongly suggest and justify the use of a 3D nodalization of the vessel downcomer, rather than a 1D nodalization, for such situations.

The present work contains findings that were produced within the OECD/NEA ROSA-2 Project. The authors are grateful to the

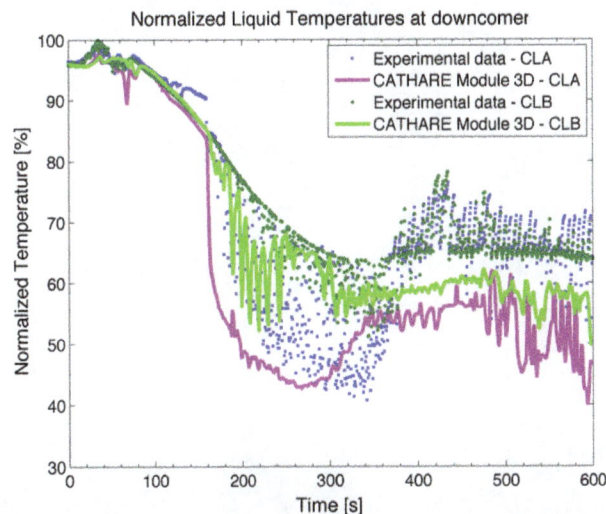

Fig. 21. Liquid temperature in the downcomer at the entrances of the Cold Legs. Comparison between experimental and numerical results.

Management Board of this project for its consent to this publication.

We would like to thank Pascal Bazin from CEA, for his guidelines and help in ROSA-LSTF modeling.

Thanks are extended to Camille Charignon for his advices and suggestions that have greatly helped to improve the modeling in CATHARE.

Piotr Mazgaj would like to extend his thanks and appreciation to his supervisor, Jean-Luc Vacher for his guidance, advices and motivation during writing this paper. He would like also to thank EDF, for giving him an opportunity to do the internship in EDF SEPTEN. The internship was funded by project co-financed by the European Union as part of the Human Capital Operational Program (INSPE - Innovative Nuclear and Sustainable Power Engineering).

Nomenclature

CL	Cold Leg
DEGB	Double-Ended Guillotine Break
ECCS	Emergency Core Cooling System
HL	Hot Leg
IBLOCA	Intermediate Break Loss of Coolant Accident
PWR	Pressurized Water Reactor
RCS	Reactor Coolant System
SI	Safety Injection

References

1. ROSA-V, Large Scale Test Facility (LSTF) - System Description for the Third and Fourth Simulated Fuel Assemblies, Japan Atomic Energy Research Institute, JAERI-Tech 2003-037, 2003

2. OECD/NEA ROSA-2, Project Experimental Data/Information Transfer Final Data Report of ROSA-2/LSTF Test 7, Thermohydraulic Safety Research Group, Nuclear Safety Research Center, Japan Atomic Energy Agency, 2013

3. G. Geffraye, O. Antoni, M. Farvacque, D. Kadri, G. Lavialle, B. Rameau, A. Ruby, CATHARE 2 V2.5_2: a single version for various applications, Nucl. Eng. Design **241**, 4456 (2011)

16

Safety operation of chromatography column system with discharging hydrogen radiolytically generated

Sou Watanabe*, Yuichi Sano, Kazunori Nomura, Yoshikazu Koma, and Yoshihiro Okamoto

Japan Atomic Energy Agency, 4-33, Muramatsu, Tokai-mura, Naka-gun, Ibaraki 319-1194, Japan

Abstract. In the extraction chromatography system, accumulation of hydrogen gas in the chromatography column is suspected to lead to fire or explosion. In order to prevent the hazardous accidents, it is necessary to evaluate behaviors of gas radiolytically generated inside the column. In this study, behaviors of gas inside the extraction chromatography column were investigated through experiments and Computation Fluid Dynamics (CFD) simulation. N_2 gas once accumulated as bubbles in the packed bed was hardly discharged by the flow of mobile phase. However, the CFD simulation and X-ray imaging on γ-ray irradiated column revealed that during operation the hydrogen gas generated in the column was dissolved into the mobile phase without accumulation and discharged.

1 Introduction

The extraction chromatography technology is one of the promising methods for the partitioning of minor actinide (MA: Am and Cm) from spent nuclear fuel [1], and Japan Atomic Energy Agency (JAEA) has been conducting research and development for the implementation. In those studies, we carried out design of an appropriate flow sheet [2], laboratory scale separation experiments on a genuine high level liquid waste [3], development of the engineering scale apparatus [4] and inactive repeated separation experiments using the large scale apparatus [5]. In order to progress the implementation, not only the performance of the column but also the safety of this system have to be guaranteed.

In respect of the safety, fire and explosion are one of the influential accidents which should be evaluated for nuclear chemical processing including the chromatography system. They are suspected to be caused by accumulation of hydrogen gas produced by radiolysis of adsorbents or mobile phase. Since radioactive nuclides in the aqueous solution are processed by adsorbents involving organic compounds, generation of hydrogen gas caused by radiolysis of water and the organic compounds is an unavoidable phenomenon. Consequently, the generated hydrogen gas has to be safely discharged from the column for the purpose of preventing fire or explosion.

Gas and heat are considered to be generated at the adsorption band of MA simultaneously. An increase in temperature of the mobile phase will lead to a decrease in the solubility of H_2 gas into it, thus heat from radioactive elements has also to be discharged as fast as possible. Our previous study has shown that flow of the mobile phase transports the decay heat to the outside of the column [4].

In this study, generation, accumulation and discharge behavior of hydrogen gas were investigated through experiments and Computation Fluid Dynamics (CFD) simulation.

2 Experimental

2.1 Behavior of gas in the engineering scale column

The large scale testing system consists of a column, tanks and pumps as shown in Figure 1. The column of ID $200\,mm\Phi$ with $650\,mm$ height was used for the experiments. The column has 18 ports for sensors for measuring the electric conductivity of the mobile phase, and a gas inlet was installed at the bottom of the column. The SiO_2-P support, which was prepared according to the article [6], was mixed with water in the slurry tank and transferred to the column by a mohno pump for packing.

N_2 gas was supplied into the packed bed through the gas inlet, and then N_2 gas discharged from the column was collected at downstream of the column as shown in Figure 2. In this measurement, amount of the supplied gas and flow

---

*e-mail: watanabe.sou@jaea.go.jp

Fig. 1. Overview of the large scale system.

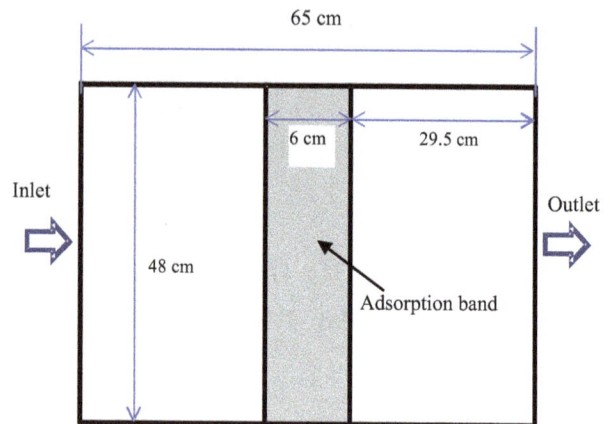

Fig. 3. Column configuration for the CFD calculation.

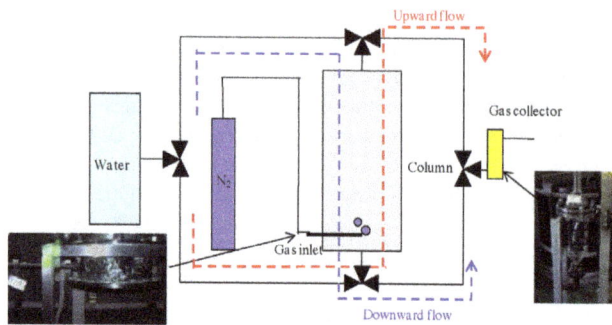

Fig. 2. Outline of the experiments for gas recovery.

2.2 CFD simulations

Simulation on two-dimensional side view of the column with 480 mm ID and 650 mm height was carried out to evaluate the accumulation behavior of heat and gas. Two-dimensional geometry was employed in order to evaluate influence of wall on distributions of velocity, temperature inside the column. The system consists of the bed, wall, inlet and outlet of the mobile phase as shown in Figure 3. Mobile phase was water, and outlet of the column was not pressured. Uniform and immobile adsorption band of MA was assumed at middle of the column. Heat from the adsorption band, which was calculated from the decay heat of ^{241}Am and ^{244}Cm, was 0.023 W/cm^3 and constant. Temperature of the wall was constant at 323 K which is one of the typical operational conditions of the extraction chromatography process [5]. In this simulation, H_2, O_2, NO_2 and CO_2 were considered as the products at the adsorption band by radiolysis. The generation rate of the gas was given by:

$$N_i = 3.73 \times 10^{-4} \times P \times G_i, \tag{2}$$

where N, P and G are amount of the generated gas [mol/h], heat from the adsorption band [W] and G value [molecules/100 eV] of component i, respectively. G values shown in Table 2 except for that of CO_2 are taken from an article [7], and G value of CO_2 was estimated from the results of γ-ray irradiation experiments on the adsorbents [8]. As shown in Figure 4, generated gas was assumed to stay at the original mesh unless it dissolves into the mobile phase. Dissolution of the gas into the mobile phase follows the Henry's law [9].

direction were parametrically changed as shown in Table 1. The average flow velocity in the column was determined by detecting the change in the electric conductivity of the mobile phase when certain amount of $Cu(NO_3)_2$ solution was mixed in the water carrier as a tracer [4]. The tracer profiles were analyzed by the same manner with deriving the height equivalent of the theoretical plate (HETP) according to the following equations:

$$N = 2\pi \left(\frac{t \times h}{A} \right)^2, H = \frac{L}{N}, \tag{1}$$

where N is the number of the theoretical plate, t is the retention time, h is the height of the profile, A is the area of the profile, H is the HETP, L is the length of the column.

Table 1. The experimental conditions for gas recovery.

No.	Amount of N$_2$ gas (mL)	Flow direction
(a)	200	Downward
(b)	50	Downward
(c)	200	Upward
(d)	50	Upward

Table 2. G values of the gas components.

Component	G value [molecules/100 eV]
H$_2$	1.6
O$_2$	0.20
NO$_2$	1.1
CO$_2$	3.9

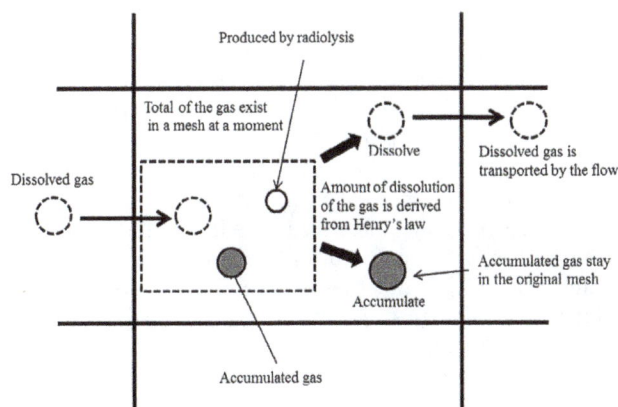

Fig. 4. Conceptual diagram of behavior of gas in the CFD calculation.

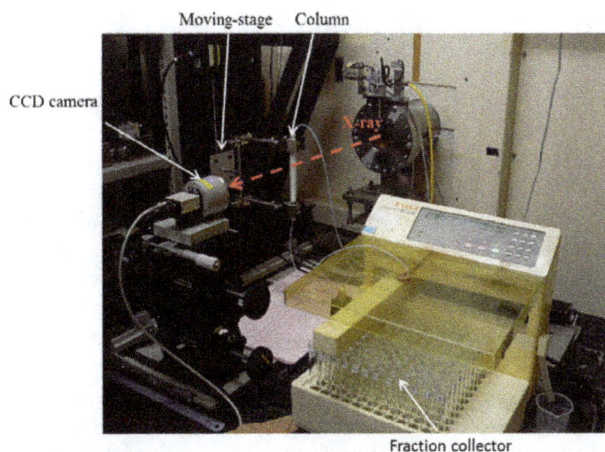

Fig. 5. Experimental setup for the X-ray imaging.

Geometry was produced by GAMBIT 2.4.6 [10] software and calculation was carried out by FLUENT 12.0 software [11]. The packed bed was simulated by water and porous media with porosity of 0.37, and the pressure drop of the bed was proportional to the velocity of the water. The thermal conductivity and heat capacity of the bed were experimentally measured to be $\lambda_{eff} = 0.525$ W/m·K and $Cp_{eff} = 7.40$ J/g·K, respectively. General features of CFD simulation are shown in Table 3. The flow velocity distribution was calculated with different mesh sizes, and an appropriate size was selected to eliminate dependence of results on the mesh.

2.3 X-ray imaging on γ-ray irradiated columns

The CMPO/SiO$_2$-P adsorbent contained CMPO (n-octyl (phenyl)-N,N-diisobutylcarbamoyl-methylphosphine oxide)

Table 3. General features of the CFD model.

Parameter	Model
Solver	Pressure based, double precision
Geometry	2-dimensional axisymmetric
Turbulence	Laminar flow
Discretization	Pressure: standardDensity: first order upwindMomentum: first order upwindTurbulent kinetic energy: first order upwindSpecific dissipation rate: first order upwind
Walls	No-slip
Temperature of the wall	323 K
Time step size	1 s
Mesh type	Uniform rectangle
Mesh size	1 mm × 1 mm
The number of mesh	156000
Pressure drop of the bed	$\Delta P/L$ [kPa/m] $= a \times v$ [m/s], $a = 1.45 \times 10^5$ [kPa·s/m^2]

as the extractant by impregnating it into the SiO$_2$-P support. The packed columns of 3 mmϕ-100 mmH (cylindrical bed) or 3 mm × 10 mm × 100 mmH (rectangular parallelepiped bed) containing the adsorbent were irradiated by γ-ray at ^{60}Co irradiation facility in Takasaki Laboratory of Japan Atomic Energy Agency. During the irradiation, mobile phase inside the column was continuously supplied with 0.9 mL/min or was stopped by closing the line. The irradiation dose rate was 3 kGy/h, and integrated irradiation dose was about 0.1 MGy.

Bubbles produced inside the bed by the irradiation were observed by X-ray imaging. The experiment was carried out at BL27B beamline of Photon Factory in High Accelerator Research Organization, Japan. Experimental setup for the imaging is shown in Figure 5. The incident X-ray obtained from synchrotron radiation was mono-chromaterized by Si(3 1 1) double crystal to 18.1 keV and then guided inside of the experimental hatch. Intensity of the X-ray passing through the column was measured by the CCD camera. The resolution of the X-ray imaging was about 25 μm. The column was set at moving-stage, and X-ray image of whole of the column was obtained by 2 min scanning. A pump for supplying solution and a fraction collector for sampling effluent were set at upstream and downstream of the column, respectively. Pump and tanks for the solutions were located outside the experimental hatch.

In order to evaluate influences of the bubbles on the separation performance, column separation experiments using the rectangle columns before and after the irradiation were also carried out. A feed solution (3 M HNO$_3$ containing Y(III), Sr(II) and Zr(IV)), wash solution (3 M HNO$_3$), eluents (H$_2$O and 50 mM Diethylene Triamine Pentaacetic Acid [DTPA] solution at pH = 3) were sequentially supplied to the columns, and then effluents were fractionally collected at every 1.2 BV of the column. Concentrations of the cations in the effluents were analyzed by ICP-AES measurements. During the separation experiments on the γ-ray irradiated column, distributions of Y(III) and Zr(IV) inside the column were evaluated from the X-ray absorption intensities in the same way to that described in reference [12].

3 Results and discussion

3.1 Behavior of gas in the engineering scale column

Figure 6 shows amount of the discharged N_2 gas plotted as time after the injection of the gas, where the broken line shows total amount of the supplied gas. Although almost all the supplied gas was accumulated inside the column when the flow direction is upward, the downward flow succeeded in discharging large part of the supplied gas. Since the gas inlet is located at the bottom of the column, the distance from the location of the gas to the outlet rather than the direction of the flow even when it is opposite to gravity must be essential for the difference in the results. If gases generate at close to the outlet of the column, almost all of them would be discharged through normal operation.

Although upward flow could not discharge the supplied gas, the accumulated gas was discharged when the upward flow was restarted after stopping the flow. The stop and restart of the upward flow was considered to change the distribution of gas, and then the gas accumulating inside the bed must be discharged. Therefore, switching of the feed pump is expected to be one of the effective methods to discharge the accumulated gas.

The flow velocity distribution inside the column and height equivalent of the theoretical plate (HETP) at the conditions of (a) and (c) were shown in Figure 7, where the flow velocities and the HETPs for the columns without supplying gas were also shown. The HETP at the condition of (c) shows greater value than that evaluated for the column without supplying the gas, whereas HETP of (a) showed little difference from that of without supplying gas. Therefore, accumulated gas may disturb the flow inside the bed. There is distinct difference in the flow velocity between at the center of the column and at close to the wall. The accumulation of gas must be impediment for obtaining the uniform flow. The gases generated by radiolysis have to be discharged with respect to not only the safety but also the separation performance of the column.

3.2 CFD simulations

The amount of the gas and increased temperature due to radioactive nuclides were calculated under the condition of $T = 323\,\mathrm{K}$ for the initial and ambient temperature and $v = 4\,\mathrm{cm/min}$ for the mobile phase. The generated products were properly dissolved into the mobile phase, and the gases did not accumulate. Temperature inside the column was almost constant, and the heat from adsorption band was transported to the downstream by the flow. Generation rate of the hydrogen from the adsorption band is $3.2 \times 10^{-5}\,\mathrm{mol/dm^3{\cdot}s}$ and the

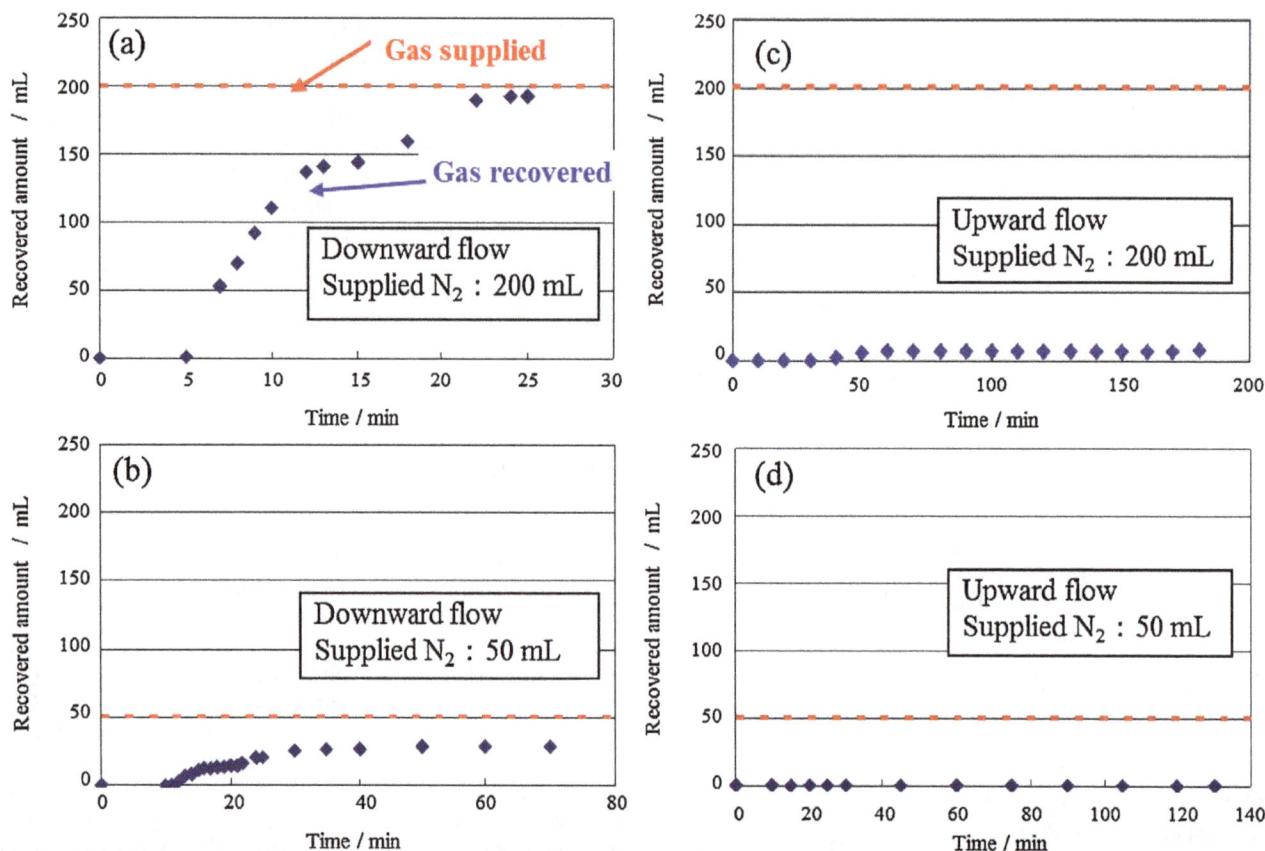

Fig. 6. Amount of recovered gas from the column. (a) Flow direction was downward and amount of supplied N_2 was 200 mL; (b) Flow direction was downward and amount of supplied N_2 was 50 mL; (c) Flow direction was upward and amount of supplied N_2 was 200 mL; (d) Flow direction was upward and supplied N_2 was 50 mL. Flow velocity was controlled at 4 cm/min.

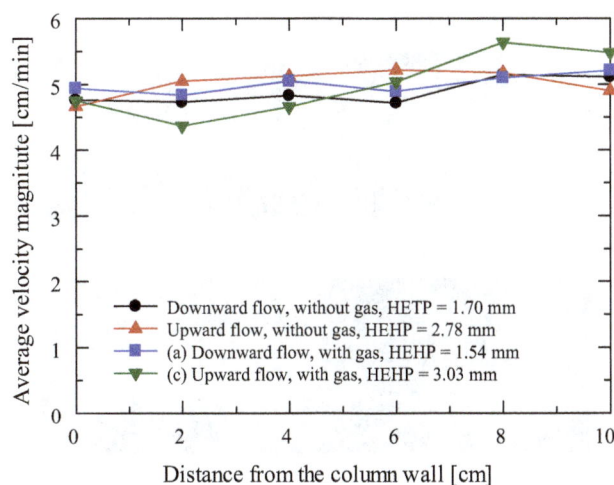

Fig. 7. The flow velocity distribution inside the column (ID = 20 cm) and HETP.

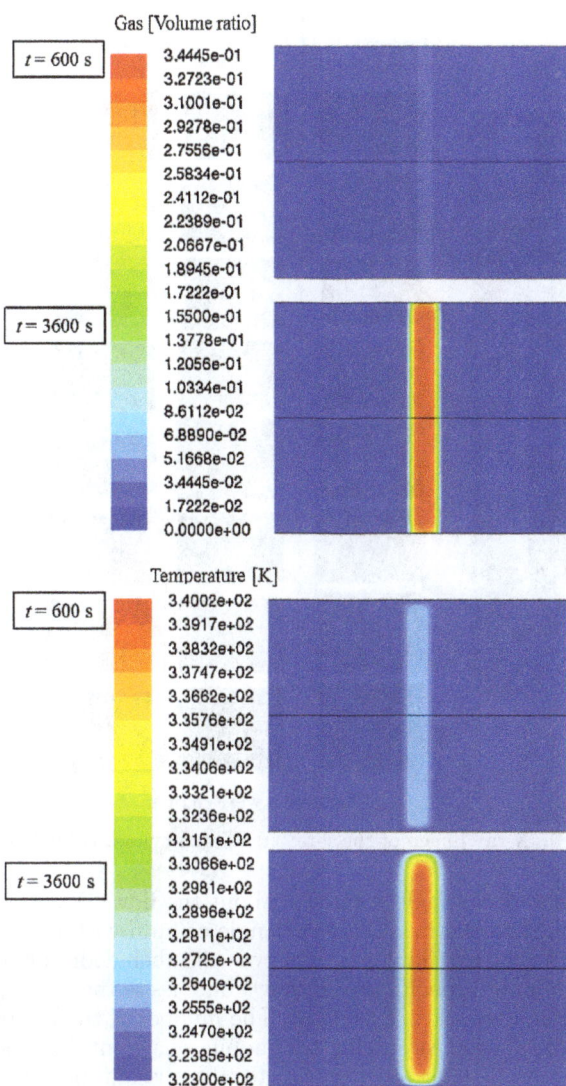

Fig. 8. Volume ratio of gas and temperature inside the column.

solubility of the hydrogen gas into the water at 1 atm and 300 K is ca. 7×10^{-4} mol/dm^3, then the generated products are considered to dissolve into the mobile phase immediately. In the case of O_2, CO_2 and NO_2, ratios of G values to the solubility of them into water are smaller than that of hydrogen, so they should dissolve in water as well. Therefore, hydrogen and oxygen do not accumulate inside the column but dissolve into the mobile phase and are discharged with an eluent during the operation.

Figure 8 shows the distribution of the accumulated gas and temperature inside the column at $t = 600$, 3,600 s, where the flow was stopped at $t = 0$ s. The gas began to accumulate before $t = 600$ s, and amount of the accumulated gas increased with proceed with time. About 1,700 mL (0.15 mL/1 mL bed) of gases at the standard condition was accumulated at $t = 3,600$ s. Composition of the gas is 93% of H_2 and 7% of O_2. Generated CO_2 and NO_2 were properly dissolved into the water. Since the mixture of hydrogen and oxygen shows explosive nature, the accumulated gases should be discharged from the column. The decay heat also accumulated at the adsorption band after the stop of the flow, and wall cooling was effective only at close to the wall. Thermal conductivity of the adsorbents must be too small to remove the decay heat inside the bed only by the wall cooling.

In order to evaluate the performance of chilled eluent for discharging the accumulated gas, the amount of the accumulated gas inside the column after the restart of the flow was calculated. This calculation was started from the state of 3,600 s after the stop of the operation as shown in Figure 8. The flow velocity and temperature of the coolant were $v = 16$ cm/min and $T = 278$ K, respectively. In this simulation, flow velocity and temperature of the mobile phase were changed from those for the normal operation in order to enhance the dissolution of the gas into the mobile phase. The accumulated gas and gas generated from the adsorption band were gradually dissolved into the coolant, and they were entirely discharged from the column at $t = 1,020$ s. The accumulated heat was simultaneously discharged from the column by the coolant. Since a part of gas accumulating at the lower part of the column could

be pushed out by the mobile phase as seen in the previous section, it must be required shorter time to discharge the gas accumulated. An equipment for supplying the emergency coolant which consists of pumps, tanks and pipes is important for the safety of the system.

3.3 X-ray imaging on γ-ray irradiated columns

Figure 9 shows X-ray image of the cylindrical columns. Bubbles generated by the external irradiation inside the bed were not confirmed in the images of the unirradiated column and of the irradiated column with the flow of the mobile phase. This result agrees with those obtained by the CFD simulation described in the previous section, and radiolytically generated hydrogen and oxygen should be dissolved in the mobile phase and be discharged. On the other hand, small bubbles with the size of ~0.3 mm ununiformly distributed inside the bed of the column irradiated without the flow of the mobile phase. As well as in the cylindrical column, bubbles were observed in the

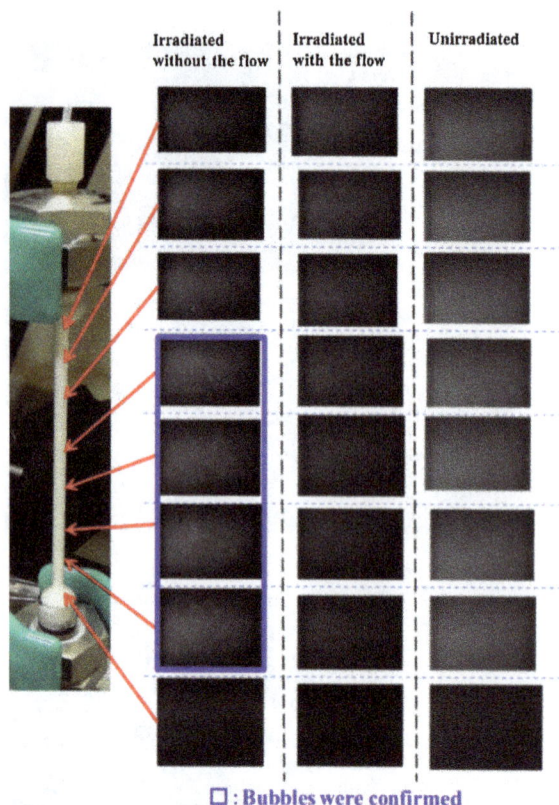

Fig. 9. X-ray image of the γ-ray irradiated cylindrical columns.

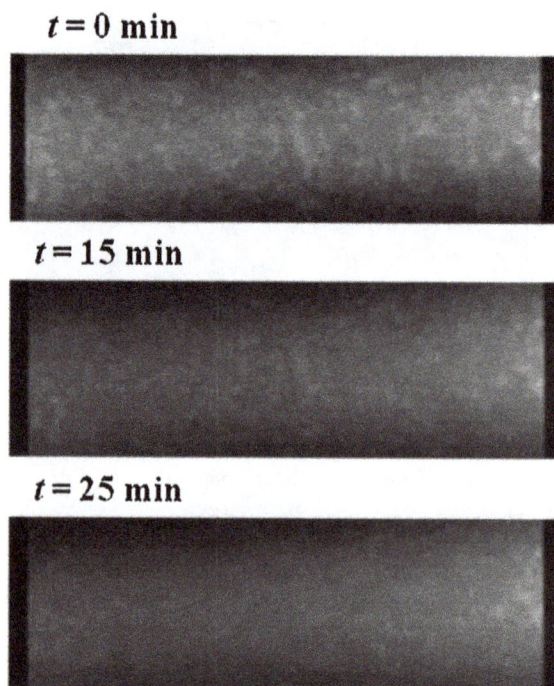

Fig. 10. X-ray image of the γ-ray irradiated rectangle column after restart of the flow of 3 cm/min.

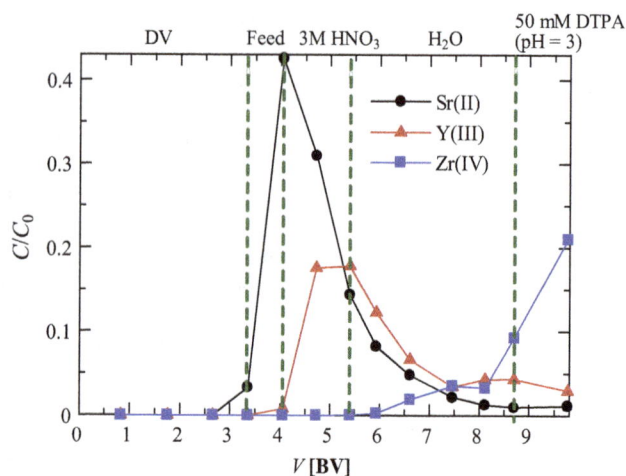

Fig. 11. Chromatogram obtained for the unirradiated rectangle column.

irradiated rectangle column without the flow. Water was supplied to the irradiated rectangle column with 0.9 mL/min ($v = 3$ cm/min) to observe the behavior of the accumulated gas. Figure 10 shows bubbles in the rectangle column after start of the flow. The number of the bubbles decreased after supplying the mobile phase into the bed. This result must correspond to the dissolution of the accumulated gas into the coolant observed in the CFD simulation. Although the CFD employed a quite simple model and carried out a conservative estimation, accumulation and dissolution behavior of the gas must be qualitatively reproduced in the simulation. In addition to that, bubbles moving with the effluent were also observed at the downstream of the column. This result supports the experimental results obtained for the large scale column system. Therefore, supplying the mobile phase is effective not only to dissolving the gas inside the column but also to pushing the gas away from the column. However, the bubbles at near the wall of the column stayed even after the start of the flow. Decrease in the flow velocity near the corner of the rectangle column is suspected to lead the remaining bubbles.

Elution curves obtained for the unirradiated rectangle column packed with CMPO/SiO$_2$-P adsorbent are shown in Figure 11. As shown in the previous report [3], Sr(II) was not extracted by CMPO and was eluted in the effluents. Y(III) is adsorbed and eluted by supplying H$_2$O and the wash solution (3 M HNO$_3$) [13]. Zr(IV) adsorbed by CMPO was retained and then eluted with the DTPA solution. Those not general elution behaviors are considered to be caused by mixing solutions in the relatively long flow channel between the

column and tanks located at the outside of the experimental hatch.

Figure 12 shows the elution curves obtained for the γ-ray irradiated rectangle column with the deposited gas. Elution behavior of Sr did not seem to be affected by the irradiation. Elution of Y(III) and Zr(IV) began faster than those observed for the unirradiated column. Degradation of CMPO/SiO$_2$-P adsorbent would not be significant for 0.1 MGy irradiation [14], however, the change was observed as shown in Figure 12 and this was attributed to the influence of degradation of CMPO, where adsorption of Y(III) which shows weak interaction with CMPO was apparently suppressed.

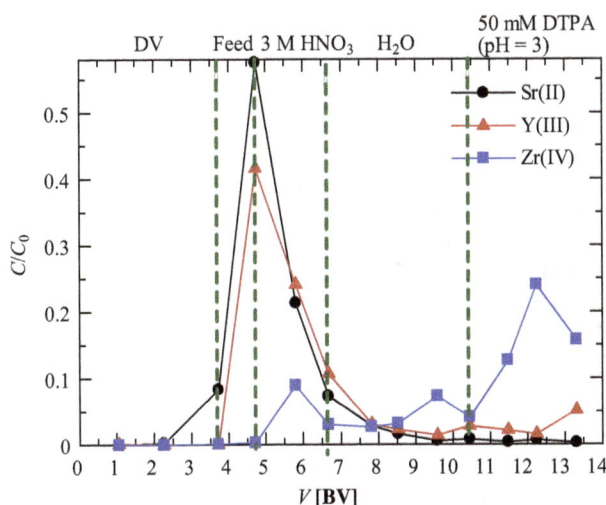

Fig. 12. Chromatogram obtained for the γ-ray irradiated rectangle column.

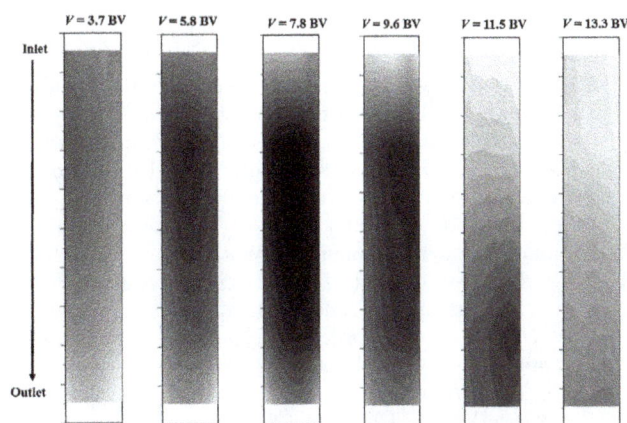

Fig. 14. Concentration profile of Zr(IV) during the separation operation with the rectangular column irradiated with γ-ray in advance.

Concentrations of Y(III) and Zr(IV) during the separation operation using the γ-ray irradiated rectangular column are shown in Figures 13 and 14, respectively. Color strengths correspond to the concentrations of cations, and V in the figure corresponds to V in Figure 12. For the ideal column operation, rectangle shape adsorption band is expected to move to the downstream of the column as the progress of time. However, distributions of Y(III) and Zr(IV) inside the column are not uniform in the laterally direction of the column as suggested from the elution curves. This indicates that flow velocity inside the bed was not uniform. Obstruction of the flow by the bubbles is considered to induce flow paths of the mobile phase inside the bed.

Consequently, the bubbles radiolytically generated disturb the uniform flow inside the bed as seen in Section 3.1. The non-uniform flow inside the bed may lead the non-uniform distribution of acidity or of DTPA concentration inside the bed. Besides degradation of CMPO extractant, those non-uniform distributions of them are considered to result in the

faster elution of Y(III) and Zr(IV). Accumulated gas is revealed to influence on the separation performance of the column, thus preliminary operation to discharge the gas and to recover the uniform flow is important even after a short period of unexpected stop of the system.

4 Conclusions

Generation, accumulation and discharge behavior of hydrogen gas radiolytically generated inside the extraction chromatography column were investigated through experiments with large scale column system, Computation Fluid Dynamics (CFD) simulation and X-ray imaging experiments on γ-ray irradiated column. Although both heat and gas accumulate at the adsorption band after the stop of the operation, supply of a coolant was revealed to be effective to discharge them. Bubbles inside the bed obstruct uniform flow inside the bed due to formation of flow paths. The accumulated gas should be discharged not only to secure safety of the system but also to guarantee the column performance. In the practical system, an equipment for feeding a coolant is effective.

This work was financed by the Ministry of Education, Culture, Sports, Science and Technology of Japan (MEXT) under the framework of "The Development of Innovative Nuclear Technologies". X-ray imaging experiments were carried out under the proposals 2010G047 of the Photon Factory, KEK.

Nomenclature

v	flow velocity inside the column
HETP	height equivalent of a theoretical plate
BV	volume of the packed bed
SiO_2-P	porous silica support coated by styrene divinyl benzene co-polymer
CMPO	n-octyl(phenyl)-N,N-diisobutylcarbamoyl-methylphos phine oxide
DTPA	diethylenetriaminepentaacetic acid

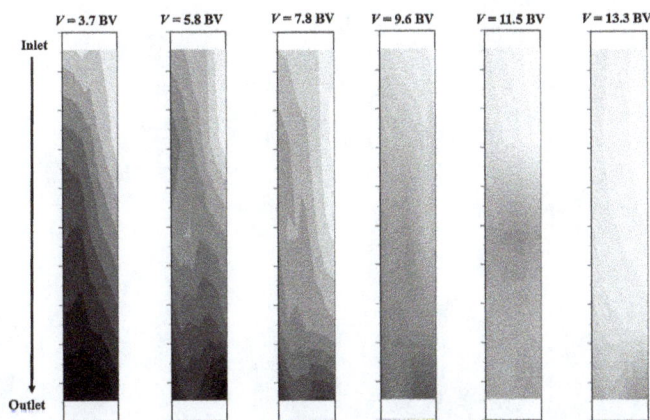

Fig. 13. Concentration profile of Y(III) during the separation operation with the rectangular column irradiated with γ-ray in advance.

References

1. E.P. Horwitz, M.L. Dietz, R. Chiarizia, H. Diamond, S.L. Mazwell III, M.R. Nelson, Separation and preconcentration of actinides by extraction chromatography using a supported liquid anion exchanger: application to the characterization of high-level nuclear waste solutions, Anal. Chim. Acta **310**, 63 (1995)

2. Y. Koma, Y. Sano, K. Nomura, S. Watanabe, T. Matsumura, Y. Morita, Development of the extraction chromatography system for separation of americium and curium, in *Proceedings OECD Nuclear Energy Agency 11th Information Exchange Meeting on Actinide and Fission Product Partitioning and Transmutation, San Francisco, USA, November 1–4, 2010*, IV-4, OECD/NEA (2010)

3. S. Watanabe, T. Senzaki, A. Shibata, K. Nomura, Y. Koma, Y. Nakajima, MA recovery experiments from genuine HLLW by extraction chromatography, in *Proceedings Global 2011, Makuhari, Japan, December 11–16, 2011*, paper 387433, Atomic Energy Society of Japan (CD-ROM) (2011)

4. S. Watanabe, I. Goto, Y. Sano, Y. Koma, Chromatography column system with controlled flow and temperature for engineering scale application, J. Eng. Gas Turbines Power **132**, 102903 (2010)

5. S. Watanabe, I. Goto, K. Nomura, Y. Sano, Y. Koma, Extraction chromatography experiments on repeated operation using engineering scale column system, Energy Procedia **7**, 449 (2011)

6. Y.-Z. Wei, K.N. Sabharwal, M. Kumagai, T. Asakura, G. Uchiyama, S. Fujine, Studies on the separation of minor actinides from high-level wastes by extraction chromatography using novel silica-based extraction resins, Nucl. Technol. **132**, 413 (2000)

7. Research Group for Aqueous Separation Process Chemistry, Nuclear Science and Engineering Directorate, Japan Atomic Energy Agency, *JAEA-Review2008-037 Handbook on process and chemistry of nuclear fuel reprocessing*. Version 2 (Japan Atomic Energy Agency, 2008), p. 530

8. Y. Koma, S. Watanabe, Y. Sano, T. Asakura, Y. Morita, Extraction chromatography for Am and Cm recovery in engineering scale, in *Proceedings ATALANTE 2008: nuclear fuel cycles for a sustainable future, Montpellier, France, May 19–23, 2008*; O1-19 (CD-ROM) (2008)

9. R. Sander, *Compilation of Henry's law constants for inorganic and organic species of potential importance in environmental chemistry*. Version 3, http://www.henrys-law.org (1999)

10. ANSYS Inc, GAMBIT 2.4 user's guide, 2007

11. ANSYS Inc, FLUENT 12.0 user's guide, 2009

12. S. Watanabe, Y. Sano, M. Myouchin, Y. Okamoto, H. Shiwaku, A. Ikeda-Ohno et al., In-situ analysis of chemical state and ionic distribution in the extraction chromatography column, J. Ion Exchange **21**, 73 (2010)

13. Y. Wei, A. Zhang, M. Kumagai, M. Watanabe, N. Hayashi, Development of the MAREC process for HLLW partitioning using a novel silica-based CMPO extraction resin, J. Nucl. Sci. Technol. **41**, 315 (2004)

14. S. Watanabe, S. Miura, Y. Sano, K. Nomura, Y. Koma, Y. Nakajima, Alpha-ray irradiation on adsorbents of extraction chromatography for minor actinides recovery, in *Proceedings OECD Nuclear Energy Agency 12th Information Exchange Meeting on Actinide and Fission Product Partitioning and Transmutation, Prague, Czech Republic, 24–27 September 2012, V-31*, OECD/NEA (2013)

17

PWR circuit contamination assessment tool. Use of OSCAR code for engineering studies at EDF

Moez Benfarah[1*], Meddy Zouiter[1], Thomas Jobert[1], Frédéric Dacquait[2], Marie Bultot[2], and Jean-Baptiste Genin[2]

[1] EDF, SEPTEN, 12-14 avenue Dutrievoz, 69628 Villeurbanne, France
[2] CEA Cadarache, 13108 Saint-Paul-Lez-Durance, France

Abstract. Normal operation of PWR generates corrosion and wear products in the primary circuit which are activated in the core and constitute the major source of the radiation field. In addition, cases of fuel failure and alpha emitter dissemination in the coolant system could represent a significant radiological risk. Radiation field and alpha risks are the main constraints to carry out maintenance and to handle effluents. To minimize these risks and constraints, it is essential to understand the behavior of corrosion products and actinides and to carry out the appropriate measurements in PWR circuits and loop experiments. As a matter of fact, it is more than necessary to develop and use a reactor contamination assessment code in order to take into account the chemical and physical mechanisms in different situations in operating reactors or at design stage. OSCAR code has actually been developed and used for this aim. It is presented in this paper, as well as its use in the engineering studies at EDF. To begin with, the code structure is described, including the physical, chemical and transport phenomena considered for the simulation of the mechanisms regarding PWR contamination. Then, the use of OSCAR is illustrated with two examples from our engineering studies. The first example of OSCAR engineering studies is linked to the behavior of the activated corrosion products. The selected example carefully explores the impact of the restart conditions following a reactor mid-cycle shutdown on circuit contamination. The second example of OSCAR use concerns fission products and disseminated fissile material behavior in the primary coolant. This example is a parametric study of the correlation between the quantity of disseminated fuel and the variation of Iodine 134 in the primary coolant.

1 Introduction

In a PWR, the release, the activation and the transfer of corrosion products generate radiation fields which cause occupational dose rates. In addition, the cases of fuel failures can cause the dissemination of actinides and fission products in the primary coolant. The fuel damages are the sources of the contamination of the PWR circuits by alpha emitters. To optimize reactor design and to reduce risks during reactor operations, it is essential to understand the behavior of corrosion products, fission products and actinides in PWR circuits.

In France, since 1970s, many R&D studies have been carried out by using test loops to simulate the behavior of contaminant species in PWR conditions. Furthermore, many engineering studies of PWR contamination have been based on the examination of data from plant measurements. The test loops and the data are key to understanding the contamination phenomena in PWR. Nevertheless, to thoroughly understand and control the mechanisms of the PWR contamination, it is strongly advised to develop and use tools for the numerical simulations of the contamination of the PWR circuits.

The simulation of PWR contamination is an important challenge for the following reasons:

- PWR contamination is the consequence of many physical and chemical phenomena impacted by a large number of design and operation parameters. It is difficult to clearly identify the individual impact of a specific parameter by just analyzing the plant data;
- the concentrations of the species generating the contamination are very low. Some species, which could represent a very low concentration, could paradoxically generate significant activities. However, it is very difficult to reproduce and control the behavior of species at very low concentrations in the test loops;
- PWR primary circuit presents high operating conditions regarding temperature, pressure, neutron flux and fluid velocity. Measuring chemical and physical data in these conditions is not an easy matter.

* e-mail: moez.benfarah@edf.fr

Since 1970s, CEA, EDF and Areva have been cooperating for the development of a contamination transfer code [1]. The OSCAR v1.3 code is a new version that incorporates the most recent advances on corrosion products, fission products and actinides modeling.

The code has been qualified by comparing the simulation results to measured contamination data from EDF fleet [2,3]. It is used by CEA, EDF and Areva to assess contamination of operating PWRs and to optimize new plant design.

The aim of this paper is to describe the OSCAR code and to illustrate its use at EDF through two examples of EDF engineering studies:

- the study of the impact of restart conditions on contamination by activated corrosion products following a reactor mid-cycle shutdown;
- the study of the connection between the quantity of disseminated fuel and the evolution of Iodine 134 in the primary coolant in the case of fuel damage.

2 Code description

For corrosion products, the source term is the result of the corrosion of the base metals. The corrosion phenomenon leads to the formation of oxide layers and induces the release of ions in the primary coolant. The metallic elements taken into account are those composing the main alloys found in PWR primary system: Ni, Co, Fe, Cr and Mn.

In the case of fuel failure, the source term of disseminated fissile material is defined by the dissemination rate specified by the user in the input file. The code takes into account U, Pu, Am, Cm isotopes as alpha emitter. The fission products taken into account are I, Xe, Kr, Cs, Rb, Ba, La, Ru, Sr and Te isotopes.

The OSCAR modeling is based on the subdividing of the PWR circuits into elementary regions:

- each region is defined by its geometric, thermal, neutron and hydraulic characteristics and by its base metal. These characteristics are the main input data required for an OSCAR simulation;
- each region is characterized by six media: the base metal, the oxide layer, the deposit layer, particles, ions and purification media. These media have different concentrations of corrosion products, fission products and actinides.

The OSCAR calculation is based on the resolution of the mass balance equations for each isotope in each medium of each region using the following equation:

$$\frac{\partial m_i}{\partial t} + (\dot{m}_{out} - \dot{m}_{in}) = \sum_{source} J_m - \sum_{sink} J_m,$$

with m_i the mass of the isotope (i) in a given medium [kg], t the time [s], $(\dot{m}_{out} - \dot{m}_{in})$ the convection term [kg·s^{-1}] and J_m the mass flux between two media [kg·s^{-1}].

The variations of the concentrations of the species in the six media result from corrosion, release, diffusion,

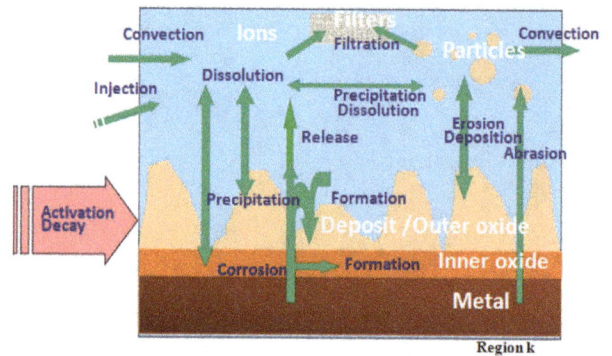

Fig. 1. Mass flux between the different media in an elemental region.

convection, activation, purification, radioactive decay mechanisms and the exchange flux between the media (dissolution/precipitation and erosion/deposition). Figure 1 describes the different media and flux in an elemental region. The main mechanisms involved in the transfers between the six media are dissolution/precipitation and erosion/deposition. A detailed description of these mechanisms has been reported by Dacquait et al. [2]. The dissolution of a deposit occurs when the concentration of a soluble species in the coolant is less than its equilibrium concentration. Soluble species precipitate when their concentration in the coolant reaches their equilibrium concentration. The dissolution and the precipitation flux are calculated using the following equations:

$$J_{dissol}^{elt} = \frac{S}{1/h + 1/V_{dissol}} \cdot \left(C_{equil}^{elt} - C^{elm}\right),$$

$$J_{precip}^{elt} = h.S.\left(C^{elm} - C_{equil}^{elt}\right),$$

with S the wet surface [m^2], h the mass transfer coefficient of ions in the fluid [m·s^{-1}], V_{dissol} the dissolution surface reaction rate coefficient [m·s^{-1}], C_{equil}^{elt} the equilibrium concentration of the element elt [kg·m^{-3}] and C^{elm} the bulk concentration of the element elt [kg·m^{-3}].

It is important to note that the dissolution and the precipitation phenomena depend on the equilibrium concentration. The equilibrium concentration of each element and the oxide speciation are calculated by an OSCAR chemistry module: PHREEQCEA, a version of PHREEQC code [4], associated to a thermodynamic database developed by CEA [5].

For insoluble species, the deposition flux, J_{depos} [kg·s^{-1}], is calculated by:

$$J_{depos} = \frac{4.V_{depos}}{D_h}.m^{part},$$

with V_{depos} the deposition velocity of particles [m·s^{-1}], D_h the hydraulic diameter [m] and m^{part} the mass of particles in the fluid [kg].

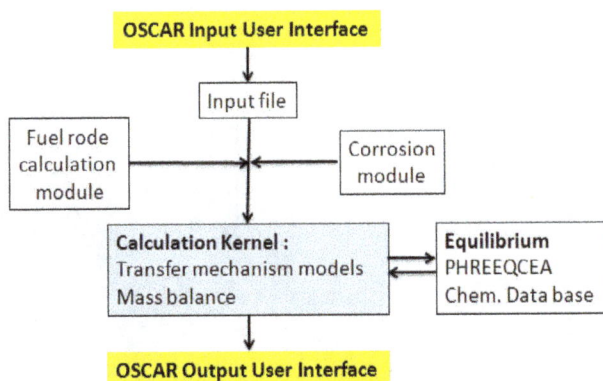

Fig. 2. Structure of the OSCAR code.

The erosion flux, resulting from the coolant friction forces, is calculated by:

$$J_{erosion} = \frac{m^{dep} - e_{lim} \cdot \rho^{dep} \cdot S^{dep}}{T_{erosion}},$$

with m^{dep} the mass of the deposit [kg], e_{lim} the thickness of the laminar sub-layer, ρ^{dep} the density of the deposit [kg·m^{-3}], S^{dep} the surface of the deposit [m^2] and $T_{erosion}$ the erosion characteristic time [s].

The source term of the corrosion products results from the base metal corrosion leading to the formation of oxide layers and the release of ions in the primary coolant. In the OSCAR code, the release of corrosion products is modeled by a parametric law which has been determined from the results of test loops.

In the case of fuel damage, the dissemination rate must be specified by the user in the input file. The isotopic distribution of the disseminated fissile material is the same as at the surface of the fuel pellet. The isotopic distribution depends on the fuel burn-up of the damaged fuel and is calculated by a specific module integrated in the code.

The global code structure is schematically described in Figure 2. The code validation was reported in previous papers [2,3].

3 The use of the OSCAR code for engineering studies at EDF

3.1 First example: study of the impact of the primary coolant activities at reactor restart after a mid-cycle shutdown

The radiochemical specifications applied at EDF plants indicate two requirements concerning the activity of the primary coolant before the reactor restart: 7 GBq/t for ^{58}Co activity and 14 GBq/t for the total gamma activity. These specifications have been designed to reduce the risk of recontamination by precipitation of corrosion products at high temperature. The objective of this study is to examine the impact of the ^{58}Co coolant activity on the contamination of the reactor circuits. The study only concerns the reactor restart occurring after a mid-cycle shutdown.

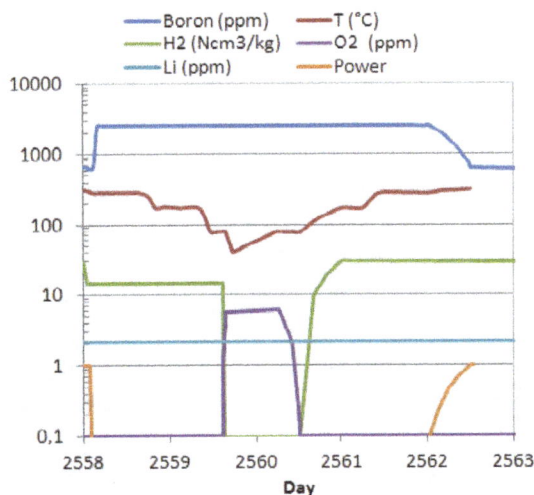

Fig. 3. Operating data used for the simulation of the mid-cycle shutdown and restart.

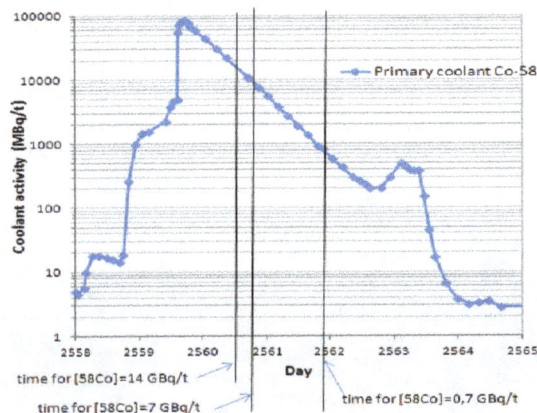

Fig. 4. Evolution of the activity of ^{58}Co in the primary coolant during a mid-cycle shutdown occurring 6 months before the end of cycle.

The input data used for the OSCAR simulation corresponds to the reference data for a 900 MWe PWR 3 loops reactor. Only the first 10 cycles have been simulated. The reference case refers to the case of 10 cycles without mid-cycle shutdown. The studied cases correspond to a mid-cycle shutdown and a restart occurring 6 months, 3 months or 1 month before the end of the 9th cycle. The operating data (power, temperatures, concentrations of boron, lithium, oxygen and hydrogen) from the beginning of the reactor shutdown to the end of the reactor restart are represented in Figure 3. The reactor restarts after oxygenation, during the purification phase, when the primary activity is decreasing (Fig. 4).

Figures 5 and 6 illustrate the ^{58}Co activities deposited on the surfaces of the hot legs, the steam generators and the letdown lines during cycle 9 and 10. These figures compare the reference case to the cases with mid-cycle shutdown occurring 6 months and 1 month before the end of the 9th cycle. These figures also describe the cases of a restart when the ^{58}Co activity in the coolant reaches 0.7, 7 or 14 GBq/t.

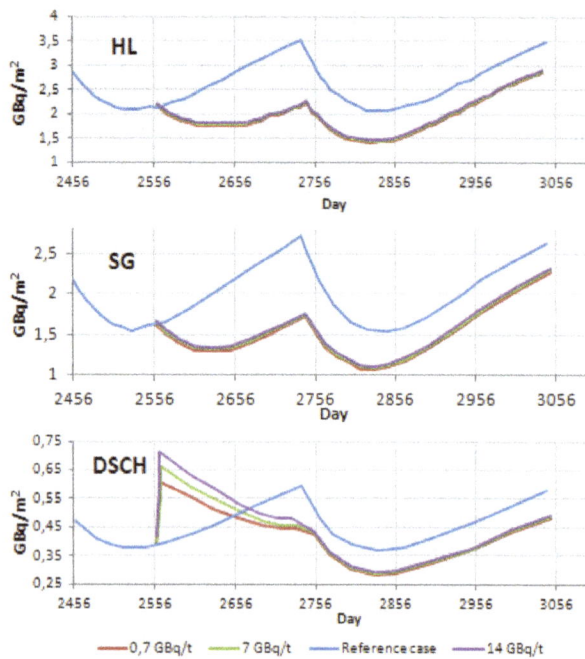

Fig. 5. ^{58}Co activities deposited on the hot legs (HL), the steam generators (SG) and the letdown lines (DSCH) in the case of a mid-cycle shutdown 6 months before the end of cycle.

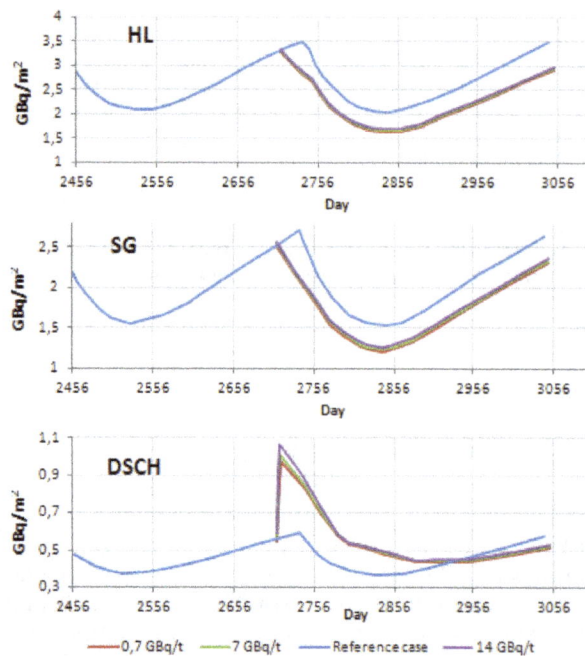

Fig. 6. ^{58}Co activities deposited on the hot legs (HL), the steam generators (SG) and the letdown lines (DSCH) in the case of a mid-cycle shutdown 1 month before the end of cycle.

The simulation results show that:

– the ^{58}Co activity in the primary coolant at reactor restart has no significant impact on the contamination of the main loop (legs and steam generators). The contamination of the main loop surfaces from the reactor

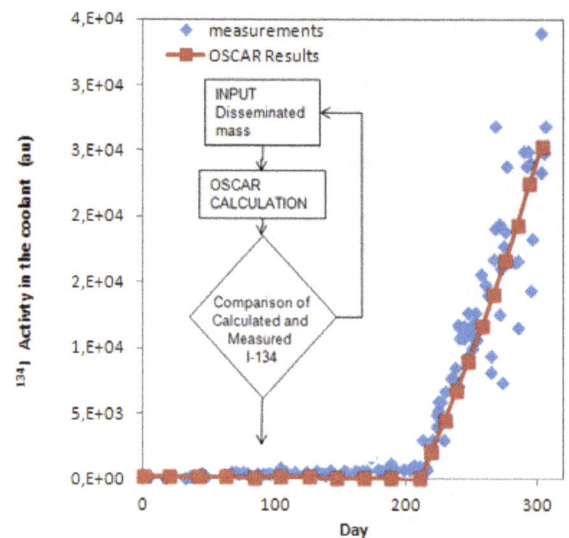

Fig. 7. Comparison of plant measurement and OSCAR calculation of ^{134}I variation in a case of fissile material dissemination.

restart to the end of the cycle remains lower than the reference case;

– for the letdown line, the ^{58}Co deposited activity increases at the reactor restart. The higher the ^{58}Co volume activity at the restart, the more significant the contamination. After the phase of the increase which occurs when the reactor restarts, the contamination of the letdown line decreases during the rest of the cycle. When the shutdown and the restart of the reactor occur 1 month before the end of the cycle, the contamination of the letdown line is notably higher than the reference case.

In conclusion, this simulation has shown that the criterion concerning the ^{58}Co coolant activity at the reactor restart after a mid-cycle shutdown has no significant impact on the contamination of the main loop. This criterion could have an effect on the contamination of the letdown line especially when the shutdown and the restart of the reactor occur near the end of the cycle.

3.2 Second example: study of the connection between the quantity of disseminated fuel and the evolution of Iodine 134 in the primary coolant

In the case of fuel rod damage, the release of a small amount of fissile material can cause a serious risk of contamination of circuits by alpha emitters. Furthermore, actinides disseminated in the primary system essentially have a particulate behavior and they depose easily on the primary circuit surfaces [6]. The detection of actinides disseminated in the primary fluid is very difficult when the reactor is operating.

The solution is the indirect monitoring of the fissile material dissemination using the evolution of ^{134}I activity. ^{134}I is a product of the fission reactions which occur in the fissile material deposited under neutron flux. The OSCAR code allows us to calculate the activity of ^{134}I in the primary fluid which is generated by the release of a given quantity of fuel. Thus, in the case of an increase in ^{134}I activity

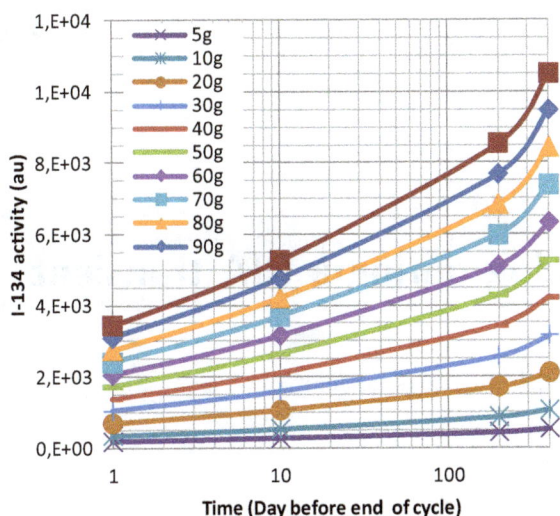

Fig. 8. Example of abacus developed using OSCAR code for the assessment of the quantity of disseminated fissile material in the case of fuel failure.

measured in the primary circuit, iterative calculations using the OSCAR code (Fig. 7) lead to the determination of the quantity of the released fissile material.

Thanks to the OSCAR code, we also managed to draw charts that may be easily used to estimate the amount of fuel disseminated in the PWR primary coolant in the case of fuel failure (Fig. 8). These charts aim at assessing the disseminated quantity by using the following criteria: (i) [134]I activity at the end of the cycle and (ii) the date of the [134]I increase.

4 Conclusions

OSCAR is a code that incorporates the French (CEA, EDF and Areva) scientific and industrial work focusing on the

PWR circuit contamination phenomena. In addition to the loop experiments and the plant measurement campaigns, the use of a simulation tool such as OSCAR is also essential in predicting the contamination of the PWR circuits by corrosion products, fission products and actinides. Indeed, the two examples described above illustrate the usefulness of the OSCAR code in the study of PWR contamination for the improvement of PWR operating parameters, as well as the optimization of new plant design.

References

1. P. Beslu, G. Frejaville, A. Lalet, A computer code PACTOL to predict activation and transport of corrosion products in PWR, in *Proceedings of the International Conference of Water Chemistry of Nuclear Reactors Systems 1, BNES, London* (1978), pp. 195–201
2. F. Dacquait et al., Simulation of corrosion product transfer with the OSCAR v1.2 Code, in *Nuclear Plant Chemistry Conference, Paris* (2012), P1-24-193
3. J.-B. Genin, M. Benfarah, C. Dinse, M. Corbineau, Simulation of alpha contamination in PWR with the OSCAR Code, in *Nuclear Plant Chemistry Conference, Sapporo* (2014), p. 10134
4. D.L. Parkhurst, C.A.J. Appelo, User's guide to PHREEQC (version 2) – A computer program for speciation, batch-reaction, one-dimensional transport and inverse geochemical calculations, Report 99-4259, US Geological Survey, Denver, Colorado, 1999
5. G. Plancque, D. You, E. Blanchard, V. Martens, C. Lamoureux, Role of chemistry in the phenomena occurring in nuclear power plants circuits, in *Proceedings of the International Congress on Advances in Nuclear Power Plants, ICAPP, Nice, France* (2011)
6. M. Benfarah, C. Dinse, M.-O. Sornein, J.-B. Genin, H. Marteau, Behavior of disseminated actinides in PWR primary coolant, in *LWR Fuel Performance Meeting, TopFuel 2013, Charlotte, North Carolina* (2013), p. 8323

18

2D simulation of hydride blister cracking during a RIA transient with the fuel code ALCYONE

Jérôme Sercombe[1,*], Thomas Helfer[1], Eric Federici[1], David Leboulch[2], Thomas Le Jolu[2], Arthur Hellouin de Ménibus[2], and Christian Bernaudat[3]

[1] CEA, DEN, DEC, Bâtiment 151, 13108 Saint-Paul-lez-Durance, France
[2] CEA, DEN, DMN, 91191 Gif-sur-Yvette, France
[3] EDF, SEPTEN, 69628 Villeurbanne Cedex, France

Abstract. This paper presents 2D generalized plain strain simulations of the thermo-mechanical response of a pellet fragment and overlying cladding during a RIA transient. A fictitious hydride blister of increasing depth (25 to 90% of the clad thickness) is introduced at the beginning of the calculation. When a pre-determined hoop stress is exceeded at the clad outer surface, radial cracking of the blister is taken into account in the simulation by a modification of the mechanical boundary conditions. The hoop stress criterion is based on Finite Element simulations of laboratory hoop tensile tests performed on highly irradiated samples with a through-wall hydride blister. The response of the remaining clad ligament (beneath the cracked blister) to the pellet thermal expansion is then studied. The simulations show that plastic strains localize in a band orientated at ~45° to the radial direction, starting from the blister crack tip and ending at the clad inner wall. This result is in good agreement with the ductile shear failures of the clad ligaments observed post-RIA transients. Based on a local plastic strain failure criterion in the shear band, ALCYONE simulations are then used to define the enthalpy at failure in function of the blister depth.

1 Introduction

The behavior of high burnup fuel during a Reactivity Initiated Accident (RIA) has been studied experimentally in the NSRR [1,2] and CABRI reactors [3,4]. It is now well established that the accumulation of hydrides beneath the thick outer zirconia layer that can form in Zircaloy-4 claddings during base irradiation is a key factor with respect to fuel rod failure during the Pellet Cladding Mechanical Interaction (PCMI) phase of a RIA [5]. In the extreme case of outer zirconia spalling, the local cold spot that appears triggers hydrogen diffusion in the cladding resulting in a massive hydride precipitation and eventually to a hydride blister (or lens).

Many experimental works have shown that precipitated hydrides result in a loss of ductility of zirconium alloys, especially at low temperatures. In the extreme case of a through-wall hydride blister, the failure can be brittle with no residual strains [6]. During simulated RIA transients on Zircaloy claddings, it has been reported that the rod failure proceeds in a mixed mode with a brittle fracture of the heavily hydrided periphery of the cladding and a ductile propagation in the remaining clad ligament [1–4]. Ductility is here associated to the change of direction of the through-wall crack, radially orientated in the hydride rim or blister and then bifurcating at ~45° until the clad inner wall.

In this paper, the failure of a fuel rod containing a fictitious hydride blister of varying thickness during a simulated RIA transient is studied with the 2D generalized plain strain scheme of the fuel code ALCYONE. The relationship between the blister depth and the maximum fuel enthalpy is seeked by multiple simulations of the CABRI REP-Na8 test [3,4].

2 The 2D model of the fuel code ALCYONE

ALCYONE is a multi-dimensional fuel code co-developed by the CEA, EDF and AREVA within the PLEIADES environment which consists of three different schemes [7]: a 1.5D scheme to model the complete fuel rod, a 3D scheme to model the behaviour of a pellet fragment with the overlying cladding, a 2D(r,θ) scheme to model the mid-pellet plane of a pellet fragment, see Figure 1. The different schemes use the same Finite Element (FE) code CAST3M [8] to solve the

* e-mail: `jerome.sercombe@cea.fr`

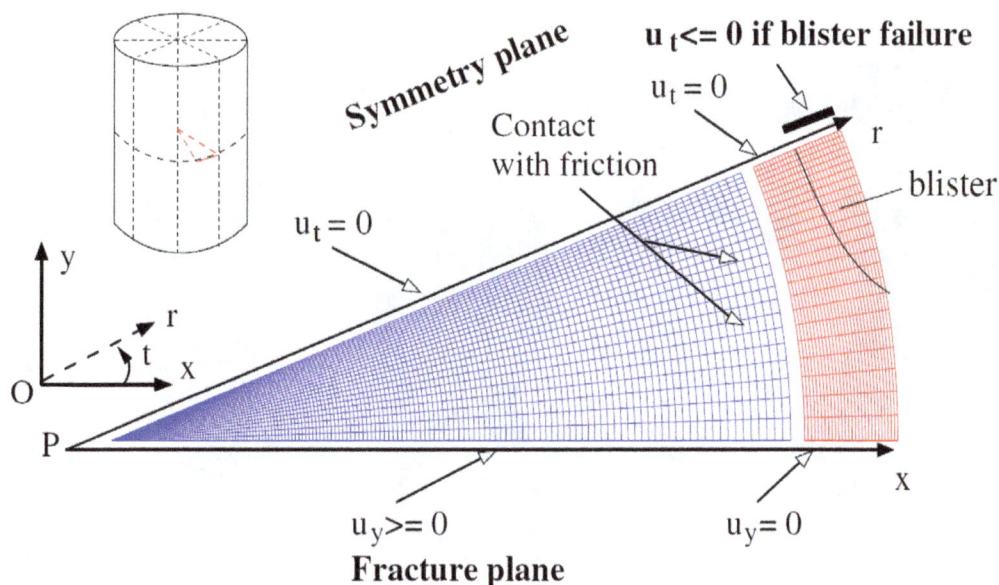

Fig. 1. Mesh and mechanical boundary conditions in the 2D scheme of ALCYONE.

thermo-mechanical problem and share the same physical material models at each node or integration points of the FE mesh. A detailed description of the main models and material parameters considered in the thermo-mechanical code ALCYONE can be found in references [9,10].

Post-irradiation examinations performed on Pressurized Water Reactors (PWR) pellets after 2 to 5 cycles of base irradiation show that the pellets are usually broken in ~6–10 pieces of irregular size [9]. In the 2D simulation, the behavior of an average fragment representing one eighth of the pellet is studied. Because of the geometrical symmetries, only one sixteenth of the pellet and of the overlying piece of cladding is meshed. The mechanical boundary conditions considered in the 2D calculations are shown in Figure 1. The opening and closing of the radial cracks between the pellet fragments is allowed by applying a unilateral condition ($u_y \geq 0$) on the nodes of the (0x) line. At the pellet-cladding interface, unilateral contact is assessed and a Coulomb model is introduced to simulate friction-slip or adherence.

The fictitious blister crack is introduced by modifying the boundary conditions on the axis of symmetry of the pellet fragment (0r). Initially, the tangential displacement u_t of the nodes is set to zero. When the hoop stress on the external clad wall reaches a pre-defined threshold, the boundary condition on the nodes included in the depth of the hydride blister is modified. A unilateral condition is applied to avoid non-physical interpenetration with the symmetric part of the blister ($u_t \leq 0$). Note that this simplified approach implies that blister cracking has an infinite length in the axial direction (that of the rod axis of symmetry).

3 Material properties for the cladding and the hydride blister

To model the behavior of fresh and irradiated Zircaloy-4, the constitutive law developed in reference [11] is used in ALCYONE. It consists in a unified viscoplastic formulation

with no stress threshold between the elastic and viscoplastic regimes. The texture-induced plastic anisotropy of Zircaloy-4 is described by a Hill's quadratic criterion. The model includes four parameters (strain rate sensitivity exponent, strength coefficient, strain hardening coefficient, Hill's coefficients) that have been adjusted on an extensive database of laboratory test results (axial tensile tests, hoop tensile tests, closed-end internal pressurization tests) essentially obtained from the PROMETRA program, dedicated to the study of zirconium alloys under RIA loading conditions [12]. The model is able to account precisely for the impact of temperature, strain rate, and irradiation damage on the ultimate stress, on the strain hardening exponent (up to uniform elongation) and on the plastic anisotropy of the material.

The explicit modeling of a hydride blister is a complex problem which would require the realistic modeling of outer zirconia formation and the partial spalling of the layer, the thermo-diffusion of hydrogen and the volume expansion associated with the precipitation of δ-hydrides [13]. Such a work is far beyond the goal of this paper. In the simulations, we assume that a stable and non-evolving hydride blister is present at the beginning of a RIA pulse test. In this respect, it is implicitly assumed that irradiation creep of Zircaloy-4 during base irradiation is sufficient to relax internal stresses generated by the precipitation of δ-hydrides. In the simulations, the thermal (heat capacity, thermal conductivity) and mechanical (Young modulus, Poisson ratio) properties of the cladding zone where the hydride blister is located are furthermore identical to those of the remaining cladding.

The only specific parameter required in the 2D simulations is the stress to failure of the hydride blister. An approximate value of ~145 MPa was deduced by Desquines et al. [6] from a hoop tensile test performed on an irradiated highly corroded clad sample containing a through-wall hydride blister (PROMETRA test 2468, Zircaloy-4, strain rate 5/s, temperature 480 °C). The failure

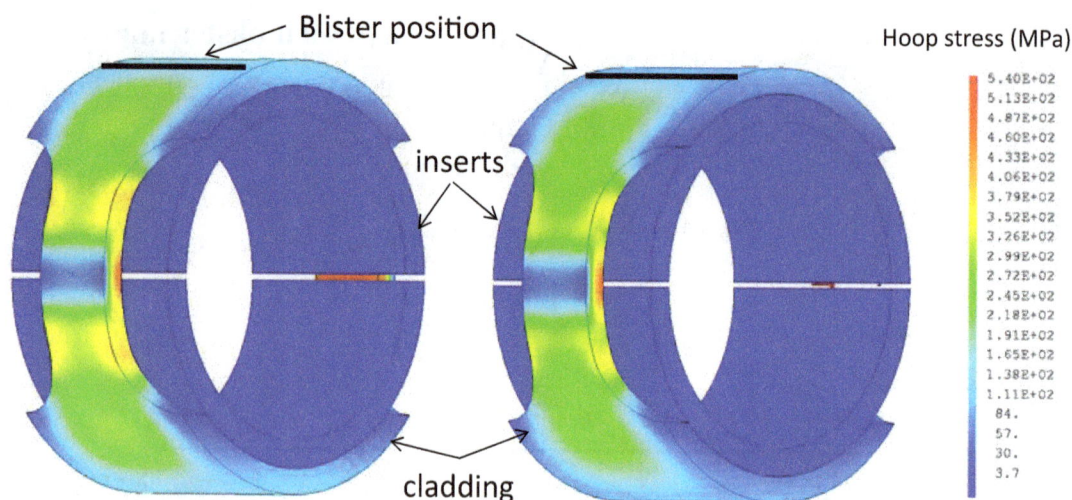

Fig. 2. Hoop stresses calculated at failure time during the PROMETRA test 2468 (left: friction coefficient 0.1, right: friction coefficient 0.4).

of the sample actually took place outside of the gage section. Interpretation of hoop tensile tests is however complex due to structural effects that occur during the experiment (bending, friction . . .). A detailed Finite Element analysis where the clad section and the half cylinder inserts are considered can nevertheless provide realistic estimate of the plastic strains [11,14]. The simulation of test 2468 has therefore been undertaken and shows that the stress state is far from being homogeneous in the clad thickness and width and depends greatly on the exact position of the blister (Fig. 2). With a friction coefficient of 0.1, the hoop stress on the clad outer wall at failure and out of the gage section varies between 150 and 250 MPa.

4 Simulation of the CABRI REP-Na8 test

The CABRI REP-Na8 test was performed on a highly corroded UO_2/Zircaloy-4 fuel rod (maximum corrosion thickness 84–126 μm) with partial spalling detected before the test. The main characteristics of the test are recalled in Table 1 (from Ref. [3]).

The REP-Na8 test led to the loss of tightness of the rod at an enthalpy of 78 cal/g. Several microphone (or acoustic) signals were however recorded before the gas ejection in the coolant. At an enthalpy level of 44 cal/g, a microphone event located near the Peak Power Node (PPN) has been correlated to a limited axial crack extension inside a hydride blister (depth ~50% of the clad wall thickness), suggesting a possible failure initiation without loss of tightness [3,4].

A preliminary 2D simulation of the base irradiation prior to the REP-Na8 pulse test was first performed with ALCYONE. Note that ALCYONE ensures a continuity in the physical and material models between base irradiation and RIA calculations. There is therefore no specific initialization of the variables prior to pulse simulations (fragment relocation, intragranular or intergranular gas bubbles, pellet cracking . . .). In particular, the pulse t_0 pellet-clad gap is close to 2 μm and is therefore not artificially closed as it is the case in most of the transient fuel performance codes.

The REP-Na8 pulse test was then simulated with ALCYONE. The hoop stress distribution in the cladding calculated 5 ms before and at the time of the microphone event related to the blister cracking (average fuel enthalpy 44 cal/g) are shown in Figure 3 (at PPN). The stresses are maximum in front of the pellet fragment symmetry axis where the pellet-clad gap was minimum at the beginning of the pulse. They reach 170–210 MPa and are therefore of the same order as the hydride blister tensile strength deduced from the PROMETRA tests. The stress level is however too small to induce significant plastic strains. The temperature of the clad external wall does not exceed 320 °C at the time of the microphone event.

The 2D simulation is then carried on assuming a complete failure of a hydride blister of half the clad wall thickness (50%). As explained in Section 2, the boundary conditions ($u_t = 0$) are partly released on the clad line situated in front of the pellet fragment symmetry plane. It results in the opening of the fictitious blister crack with a bending moment on the clad inner surface, as shown in

Table 1. Main characteristics of the CABRI REP-Na8 test.

Fuel	Cladding	Max. burnup	Energy (cal/g)	Width (ms)	Blister cracking[a]	Loss of tightness[a]	Max. enthalpy
UO_2	Zy-4	60 GWd/t	110.7	75	44 cal/g	78 cal/g	98 cal/g

[a]Enthalpies from simulations with the SCANAIR code.

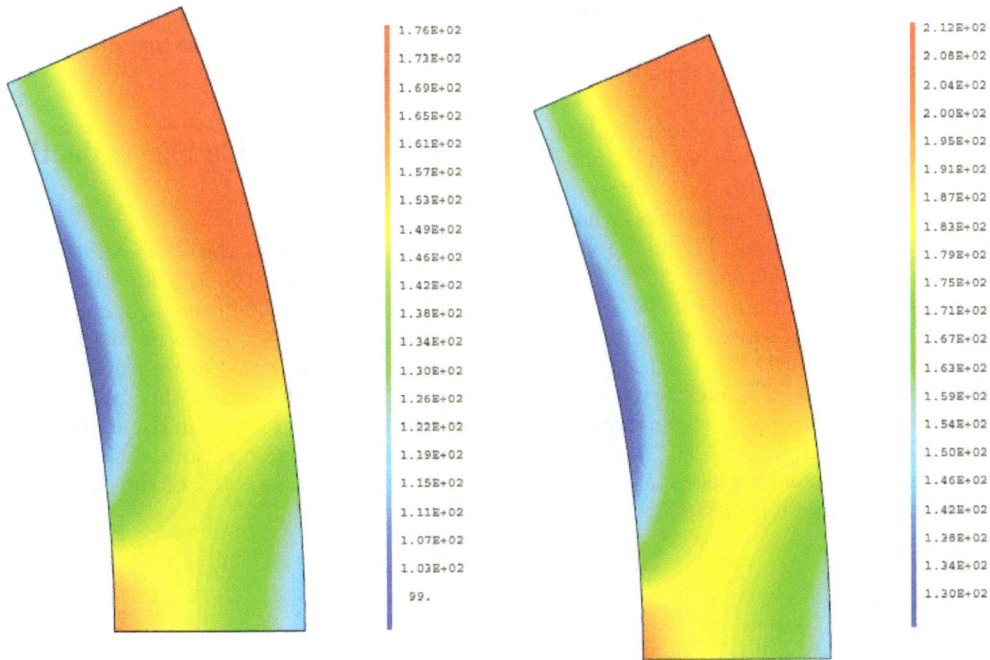

Fig. 3. Hoop stresses (in MPa) calculated in the cladding 5 ms (left) and at the time of the microphone event attributed to the cracking of a hydride blister at PPN (right).

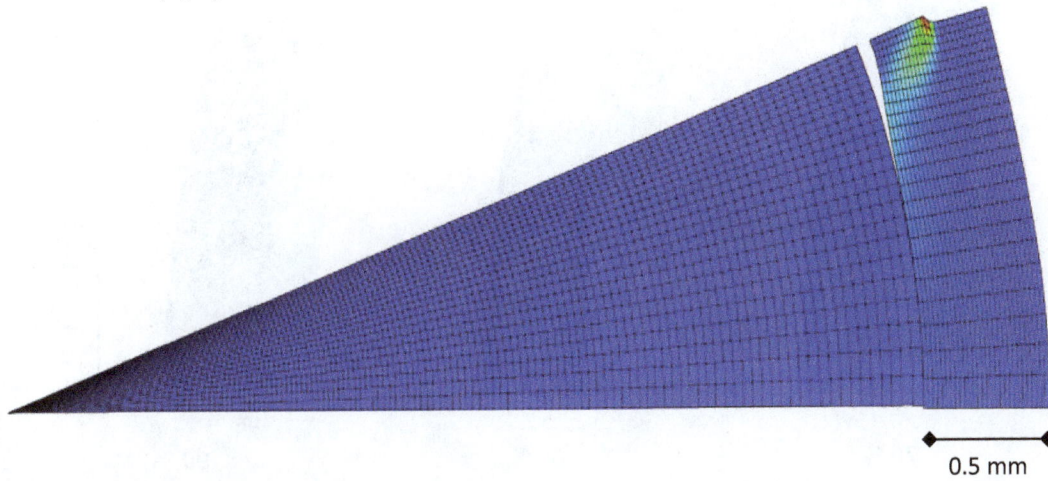

0.5 mm

Fig. 4. Local re-opening of the pellet-clad gap during the pulse transient and distribution of the clad equivalent plastic strains (clad displacements are multiplied by a factor 5).

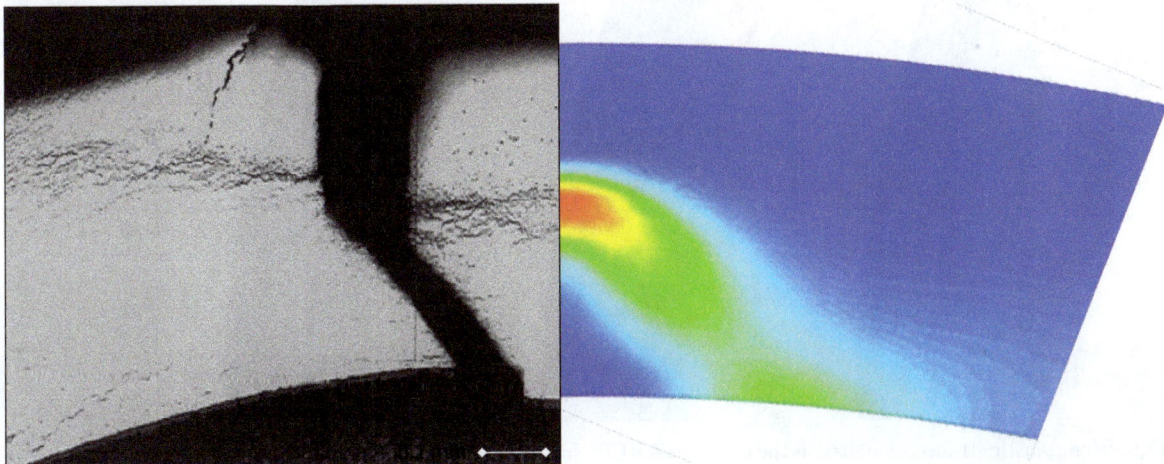

Fig. 5. Clad state at PPN after the REP-Na8 test and calculated equivalent plastic strains at the end of the 2D simulation.

Figure 4. This bending moment leads to the local re-opening (during the pulse) of the pellet-clad gap on a circumference of ~400 μm. Thinning of the remaining clad wall is also induced by the blister cracking.

The localization of plastic strains in a band making an angle of ~45° with the radial direction seems consistent with the re-opening of the pellet-clad gap. Plastic strains develop between the blister crack tip and the first location where the pellet is still in contact with the cladding. The 45° bifurcation observed after RIA pulse tests is characteristic of a ductile failure in the plane of the maximum shear stresses. The qualitative agreement between our simulation and post-test metallographic observations is illustrated in Figure 5.

Overall, the introduction of a 50% thick hydride blister in the 2D calculation has some impact on the (average) clad outer diameter variation during the test. The loss of stiffness induced by the blister cracking leads to an increase of the (average) clad outer diameter from 0.4% to 0.6%. The latter is to be compared with the 0.5% residual strains estimated post-test from the metallographic radial cut close to the PPN [3].

5 Impact of hydride blister depth on clad strains

The 2D simulation of the REP-Na8 test has been used to study the impact of the hydride blister depth on the clad strains. The onset of blister cracking (at the time of the microphone event) has not been modified since the blister is assumed to behave as the rest of the cladding. Only the number of nodes where the boundary conditions are released has been changed in the simulations. As illustrated in Figure 6, six configurations with blisters depths equal to 25, 50, 60, 70, 80 and 90% of the clad wall thickness, have been considered.

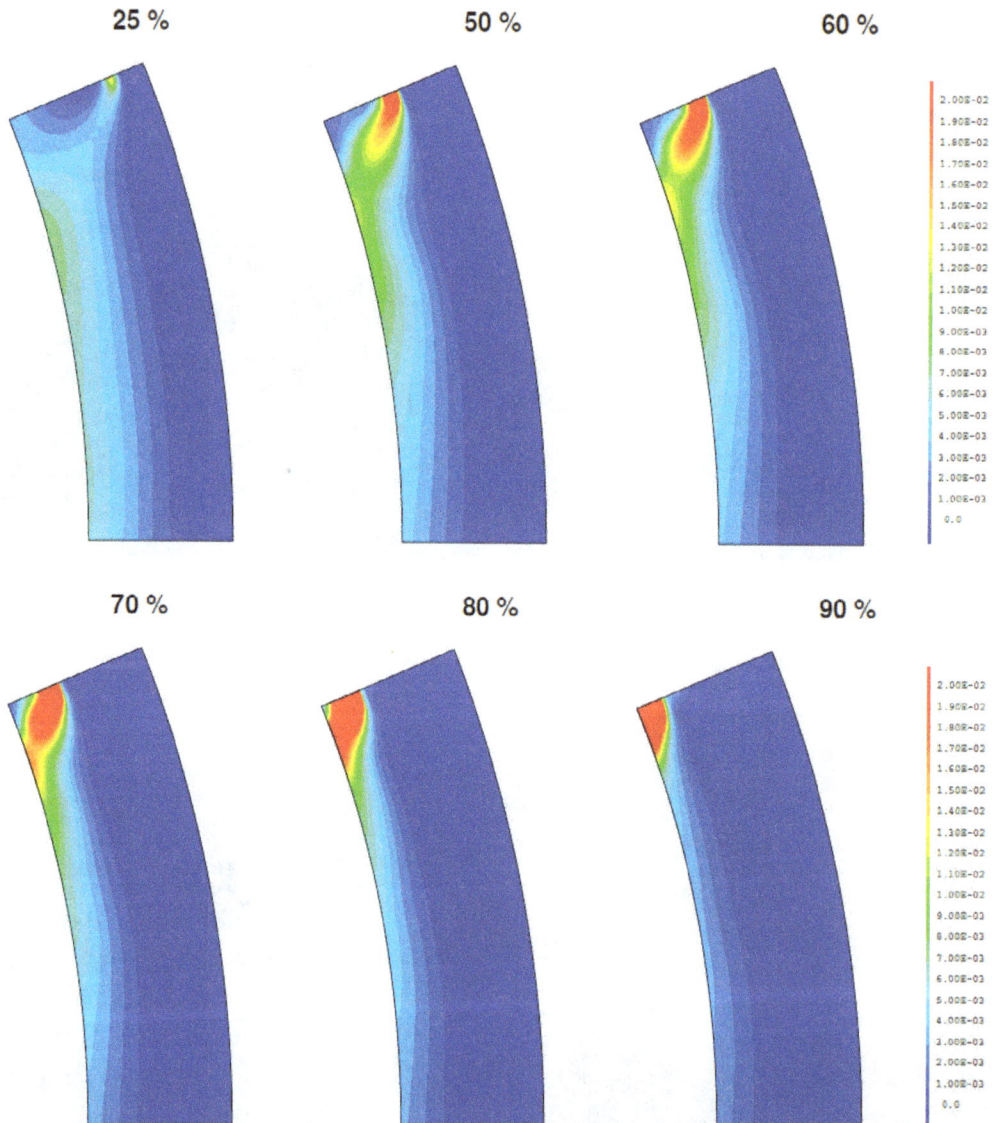

Fig. 6. Equivalent plastic strains calculated at the time of the REP-Na8 fuel rod loss of tightness in function of the hydride blister depth (in % of the clad wall thickness).

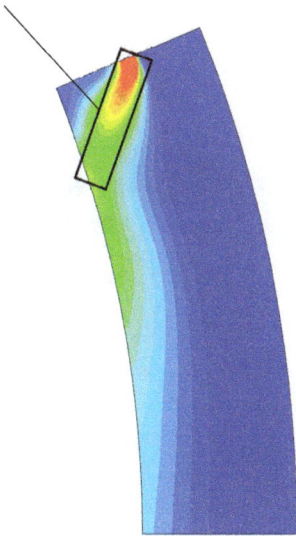

Fig. 7. Location where the average plastic hoop strain in the 45° band is calculated.

As can be expected, the increase of the blister depth leads to increasing plastic strains in the 45° band. To compare the strain levels in the uncracked clad ligament situated beneath the hydride blister, the average plastic hoop strain in the 45° band has been used. It was preferred to the maximum plastic strain which obviously depends greatly on the mesh refinement and to the average hoop strain in the cladding which does not account much for the pronounced strain localization at the blister crack tip.

Figure 7 depicts the clad zone where the average plastic hoop strain in the 45° band is calculated. In each concentric ring of the clad mesh, the maximum plastic hoop strain is

determined. Since an equidistant mesh is used in the radial direction, the sum of the maximum plastic hoop strains is then divided by the number of elements situated beneath the blister crack tip. The hoop strain was chosen because it can be compared directly to the local hoop strain measured from wall thinning at the blister crack tip [15]. This is not the case of the deviatoric plastic strain even if the latter is more relevant to assess the extension of clad damage in hydrided cladding [16].

The calculated time evolutions of the average plastic hoop strain in the 45° shear band are plotted in Figure 8. The times of the microphone events associated to blister cracking and to the rod failure are indicated. If we assume that the 50% deep hydride blister found at PPN is representative of the initial state of the REP-Na8 cladding, it appears that the allowable maximum plastic strain in the clad ligament beneath the hydride blister is of the order of 0.5%.

According to our thermo-mechanical simulation of the REP-Na8 test, failure of the rod is reached at a very low average plastic strain level in the 45° shear band (0.5%). This value can be compared to the 1 to 5% fracture-tip wall thinning estimated by Chung and Kassner from the REP-Na1 post-test metallographies [17] and to a lesser extent to the 3 to 10% local plastic strains measured by Hermann et al. [15] from burst tests performed at 350 °C on irradiated Zy-4 cladding samples containing large hydride lenses (40–50% of the clad wall thickness). These local strains were estimated from the local thinning of the clad ligaments situated beneath the hydride lenses and are therefore close to our analysis of the calculated plastic strains. The burst tests performed by Hermann et al. [15] were pressure driven tests which might not lead to experimental strains directly comparable to our calculated strains. It may be argued that the viscoplastic model of Le Saux et al. [11] does not account for the development of cavities and voids in the highly

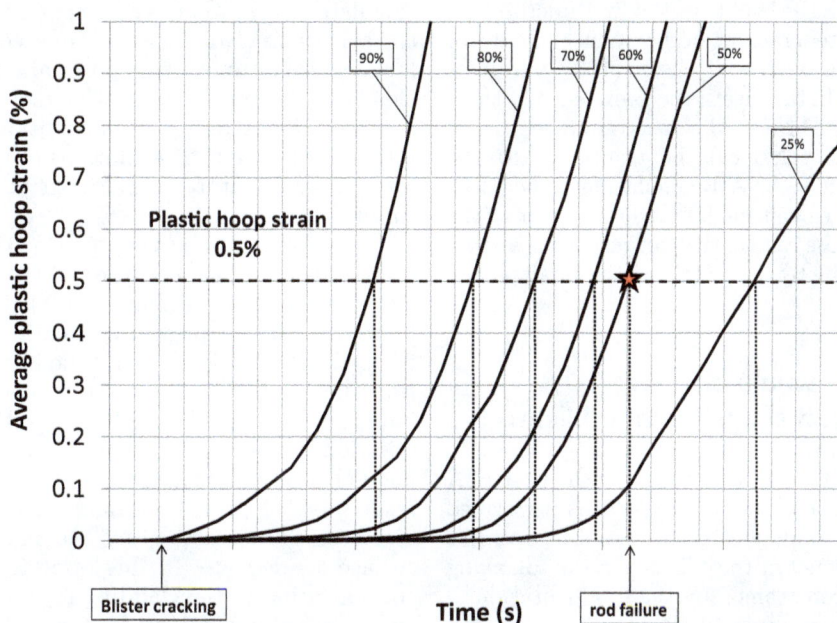

Fig. 8. Calculated evolution of the average plastic hoop strain in the 45° shear band during the REP-Na8 simulation in function of the hydride blister depth (in % of the clad thickness).

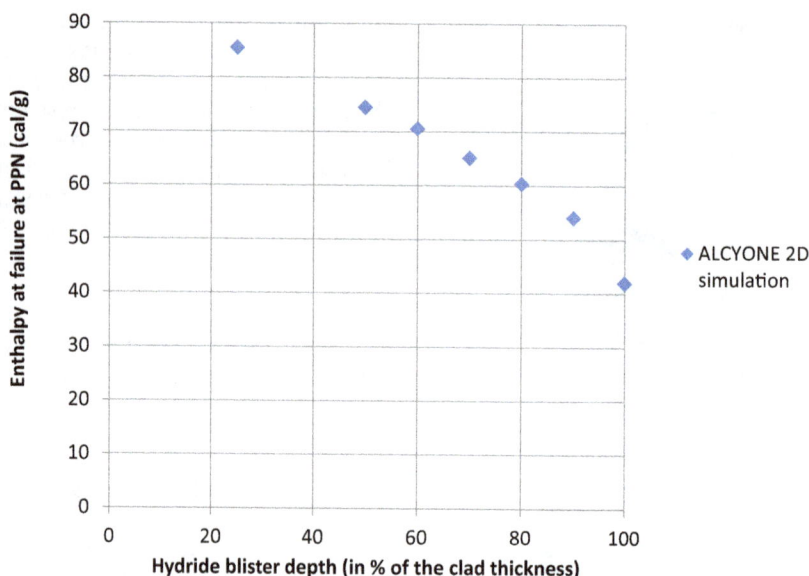

Fig. 9. Calculated evolution of the enthalpy at failure in function of the hydride blister depth (in % of the clad wall thickness), based on the REP-Na8 test conditions.

strained parts of the clad ligament and hence does not allow to capture the softening induced by material damage. The treatment of softening in the framework of continuum mechanics requires the development of more sophisticated damage or micro-mechanically based models [16,18].

In Figure 8, the time evolution of the average plastic hoop strain in the 45° shear band has been plotted for thicker (60–70–80–90%) and thinner (25%) hydride blisters. Assuming that the failure of the clad ligament beneath the blister can be related to an average plastic strain of 0.5%, the potential variation in failure time and hence in average fuel enthalpy can be deduced from the simulations. The impact of blister depth on the fuel enthalpy at failure can hence be summarized in Figure 9. In our simulations of the REP-Na8 test, the first microphone event associated to blister cracking occurs at an enthalpy of 42 cal/g. In case of a through-wall blister (100%), it gives the enthalpy at failure. In the case of the assumed 50% thick hydride blister of REP-Na8, the maximum enthalpy reaches 75 cal/g. This value can be compared to the 78 cal/g obtained from a SCANAIR simulation of the REP-Na8 test (Tab. 1). Between 50 and 90%, the evolution of the enthalpy at failure is close to linearity, reflecting the almost constant rate of straining by the pellet and the small plastic strains at failure.

6 2D model of the whole cladding circumference with a single hydride blister

In spite of its qualitative agreement with post-RIA observations of failed rods, the 2D simulations performed with ALCYONE tend to underestimate the plastic strains in the clad ligament beneath the blister for the following reasons: the model is constrained by the fragmentation of the pellet in 8 identical pieces which implies that the simulation represents the failure of the whole circumference of the clad tube with 8 identical blisters; the blister crack is

introduced on a plane of symmetry, meaning that two identical 45° shear bands develop at the same time. To improve the simulation of rod failure, a finite element model of the whole circumference of the cladding tube with a single hydride blister has been undertaken. The FE mesh is illustrated in Figure 10.

The mesh in the vicinity of the hydride blister is much more refined than in the ALCYONE calculation since only the cladding is considered. The cracking of the blister is assumed to be slightly dissymmetric (it corresponds to the blister crack angular position in the REP-Na8 test, see Fig. 5), the dissymmetry being a possible input parameter via the angles θ_1 and θ_2. The loading consist in the prescribed radial displacement of the fuel external surface calculated by ALCYONE. Friction and slipping between the pellet and cladding are accounted for by modeling the fuel external and clad internal surfaces by distinct elements. In this respect, the fuel-clad gap re-opening observed in ALCYONE simulations (Fig. 4) can be reproduced. The time evolutions of the clad external and internal temperatures are also extracted from ALCYONE simulations.

Cross-comparisons with ALCYONE simulations have shown that the time evolution of the average plastic hoop strain in the 45° shear band beneath the blister is correctly reproduced by the present calculation if only one eighth of the pellet is considered and if the blister is not dissymmetric. The mesh refinement was found to have very little impact which confirms that the chosen plastic strain criterion is numerically sound. It was also checked that the angular position of the blister crack had no impact on the results (see Fig. 11) meaning that a prescribed radial displacement based on the average of the fuel external surface is adequate. In this respect, the important stress concentration in the cladding in front of the pellet radial cracks usually considered of great importance in PCI transient [7,9] (power ramps) appears of secondary importance for RIA calculations.

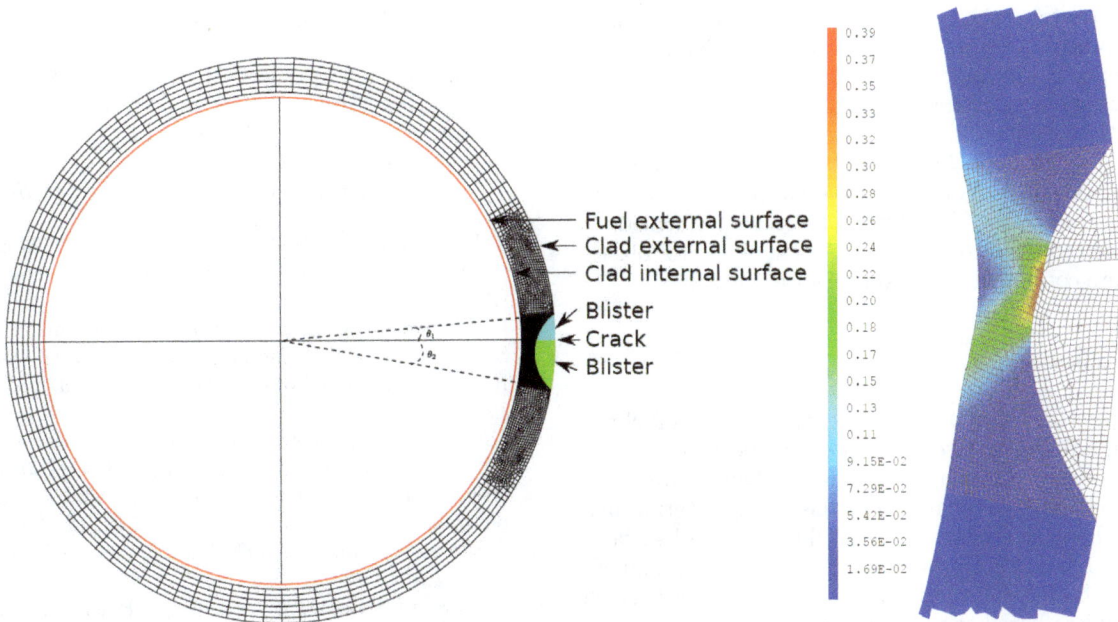

Fig. 10. Finite Element mesh of the 2D simulation of the whole cladding circumference with a single dissymmetrically cracked blister. Calculated distribution of equivalent plastic strains in the simulation.

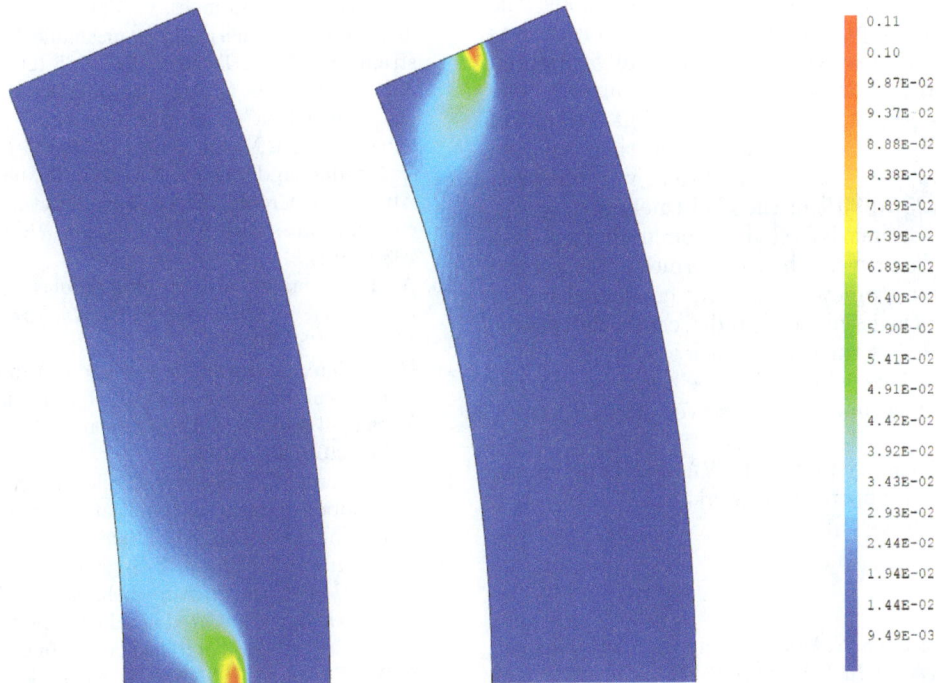

Fig. 11. Plastic strain distributions calculated at the end of the REP-Na8 test in case the blister crack is situated in front of the plane of fracture (left) or the plane of symmetry (right) of the pellet fragment.

The crack in the blister is meshed explicitly with distinct facing nodes. At the beginning of the calculation, the displacements of the facing nodes are equal. At the time of the microphone event related to the blister cracking, these conditions are released. Figure 10 illustrates the dissymmetric development of plastic strains in the simulation with a 50% thick blister ($\theta_1 = 5°$, $\theta_2 = 10°$, blister length on clad outer surface ~1.3 mm). Interestingly, very high plastic

strains tend to develop at the blister crack tip but also along the blister-clad interface which might be the reason for the blister-crack decohesion observed in Figure 5.

From the calculation, it appears that the plastic strain in the 45° shear band reaches in this case 2.8% at the time of failure of the REP-Na8 test, to be compared to the previous estimate with ALCYONE of the plastic strain at failure (0.5%, see Fig. 8). A realistic approach to clad failure induced by hydride blister cracking during a RIA obviously requires the development of a simulation tool able to model the whole circumference of the cladding tube and the evolving PCMI. The impact of blister geometry (length on outer clad surface, dissymmetry of the crack) has not been studied yet in spite of its potential great importance on plastic strains. Moreover, the limited axial length of the blisters was not considered in this study since only 2D simulations were performed. It might as well contribute to the calculation of greater plastic strains more consistent with experimental measures.

7 Conclusions

In this paper, the thermo-mechanical response of a fuel rod containing a fictitious hydride blister has been studied by 2D plane strain simulations of RIA. The blister was assumed to behave as the rest of the cladding material till a prescribed hoop stress is reached on the outer clad surface. The stress criterion is based on the Finite Element analysis of a PROMETRA laboratory hoop tensile test performed on a highly irradiated sample with a through-wall hydride blister. The 2D simulations of the REP-Na8 pulse test performed with ALCYONE led to a qualitatively good representation of the mixed failure mode encountered in RIA pulse transients on highly corroded fuel rods: the brittle failure of the pre-pulse 50% thick hydride blister was followed by the development of a diagonally oriented plastic shear band in the remaining clad ligament. The average plastic hoop strain in the shear band at the time of failure of the REP-Na8 test (~0.5%) was then used to quantify the enthalpy at failure in fuel rods with blister of increasing depth (25 to 90% of the clad thickness).

FE simulations of the whole cladding circumference with a single hydride blister were then performed to study the impact of the blister geometry and of the crack position on the results. In case of a dissymmetric radial crack, high plastic strains were obtained at the blister crack tip but also along the blister-clad interface, which might be the reason for the blister-clad decohesion that can be observed after RIA tests.

The authors would like to thank AREVA and EDF for the financial and technical support to this work.

References

1. T. Fuketa, H. Sasajima, Y. Mori, K. Ishijima, Fuel failure and fission gas release in high burnup PWR fuels under RIA conditions, J. Nucl. Mater. **248**, 249 (1997)

2. T. Fuketa, T. Sugiyama, Nuclear fuel behavior during RIA, in *OCDE/NEA Workshop, Paris, France* (2009)

3. J. Papin et al., in *Eurosafe Meeting, Paris, France* (2003)

4. J. Papin, B. Cazalis, J.M. Frizonnet et al., Summary and interpretation of the CABRI REP-Na program, Nucl. Technol. **157**, 230 (2007)

5. V. Georgenthum et al., in *WRFPM Conference, Seoul, Korea* (2008)

6. J. Desquines et al., in *Proceedings of the ASTM conference on Zirconium in the Nuclear Industry, Stockholm, Sweden* (2004)

7. B. Michel, C. Nonon, J. Sercombe, F. Michel, V. Marelle, Simulation of pellet-cladding interaction with the pleiades fuel performance software environment, Nucl. Technol. **182**, 124 (2013)

8. CAST3M, http://www-cast3m.cea.fr/

9. J. Sercombe, I. Aubrun, C. Nonon, Power ramped cladding stresses and strains in 3D simulations with burnup-dependent pellet–clad friction, Nucl. Eng. Des. **242**, 164 (2012)

10. J. Sercombe et al., in *TopFuel Conference, Orlando, Florida, USA* (2010)

11. M. Le Saux, J. Besson, S. Carassou, C. Poussard, X. Averty, A model to describe the anisotropic viscoplastic mechanical behavior of fresh and irradiated Zircaloy-4 fuel claddings under RIA loading conditions, J. Nucl. Mater. **378**, 60 (2008)

12. B. Cazalis, J. Desquines, C. Poussard et al., The PROME-TRA program: Fuel cladding mechanical behavior under high strain rate, Nucl. Technol. **157**, 215 (2007)

13. A.H. de Ménibus, Q. Auzoux, O. Dieye et al., Formation and characterization of hydride blisters in Zircaloy-4 cladding tubes, J. Nucl. Mater. **449**, 132 (2014)

14. V. Macdonald, D. Le Boulch, A.H. de Ménibus, J. Besson, Q. Auzoux, J. Crépin, T. Le Jolu, Fracture of Zircaloy-4 fuel cladding tubes with hydride blisters, Procedia Mater. Sci. **3**, 233 (2014)

15. A. Hermann et al., in *Proceedings of the 15th ASTM conference on Zirconium in the Nuclear Industry, Sunriver, Oregon, USA* (2007)

16. M. Le Saux, J. Besson, S. Carassou, A model to describe the mechanical behavior and the ductile failure of hydrided Zircaloy-4 fuel claddings between 25 °C and 480 °C, J. Nucl. Mater. **466**, 43 (2015)

17. H.M. Chung, T.F. Kassner, Cladding metallurgy and fracture behavior during reactivity-initiated accidents at high burnup, Nucl. Eng. Des. **186**, 411 (1998)

18. Y. Udagawa, T. Mihara, T. Sugiyama, M. Suzuki, M. Amaya, Simulation of the fracture behavior of Zircaloy-4 cladding under reactivity-initiated accident conditions with a damage mechanics model combined with fuel performance codes FEMAXI-7 and RANNS, J. Nucl. Sci. Technol. **51**, 208 (2014)

19

Reactor physics modelling of accident tolerant fuel for LWRs using ANSWERS codes

Benjamin A. Lindley[1*], Dan Kotlyar[2], Geoffrey T. Parks[2], John N. Lillington[1], and Bojan Petrovic[3]

[1] Amec Foster Wheeler, Dorchester, UK
[2] Department of Engineering, University of Cambridge, Cambridge, UK
[3] Georgia Institute of Technology, Georgia, USA

Abstract. The majority of nuclear reactors operating in the world today and similarly the majority of near-term new build reactors will be LWRs. These currently accommodate traditional Zr clad UO_2/PuO_2 fuel designs which have an excellent performance record for normal operation. However, the events at Fukushima culminated in significant hydrogen production and hydrogen explosions, resulting from high temperature Zr/steam interaction following core uncovering for an extended period. These events have resulted in increased emphasis towards developing more accident tolerant fuels (ATFs)-clad systems, particularly for current and near-term build LWRs. R&D programmes are underway in the US and elsewhere to develop ATFs and the UK is engaging in these international programmes. Candidate advanced fuel materials include uranium nitride (UN) and uranium silicide (U_3Si_2). Candidate cladding materials include advanced stainless steel (FeCrAl) and silicon carbide. The UK has a long history in industrial fuel manufacture and fabrication for a wide range of reactor systems including LWRs. This is supported by a national infrastructure to perform experimental and theoretical R&D in fuel performance, fuel transient behaviour and reactor physics. In this paper, an analysis of the Integral Inherently Safe LWR design $(I^2S\text{-LWR})$, a reactor concept developed by an international collaboration led by the Georgia Institute of Technology, within a US DOE Nuclear Energy University Program (NEUP) Integrated Research Project (IRP) is considered. The analysis is performed using the ANSWERS reactor physics code WIMS and the EDF Energy core simulator PANTHER by researchers at the University of Cambridge. The $I^2S\text{-LWR}$ is an advanced 2850 MWt integral PWR with inherent safety features. In order to enhance the safety features, the baseline fuel and cladding materials that were chosen for the $I^2S\text{-LWR}$ design are U_3Si_2 and advanced stainless steel respectively. In addition, the $I^2S\text{-LWR}$ design adopts an integral configuration and a fully passive decay heat removal system to provide indefinite cooling capability for a class of accidents. This paper presents the equilibrium cycle core design and reactor physics behaviour of the $I^2S\text{-LWR}$ with U_3Si_2 and the advanced steel cladding. The results were obtained using the traditional two-stage approach, in which homogenized macroscopic cross-section sets were generated by WIMS and applied in a full 3D core solution with PANTHER. The results obtained with WIMS/PANTHER were compared against the Monte Carlo Serpent code developed by VTT and previously reported results for the $I^2S\text{-LWR}$. The results were found to be in a good agreement (e.g. <200 pcm in reactivity) among the compared codes, giving confidence that the WIMS/PANTHER reactor physics package can be reliably used in modelling advanced LWR systems.

1 Introduction

The majority of nuclear reactors operating in the world today and similarly the majority of near-term new build reactors will be LWRs. These currently accommodate traditional Zr clad UO_2/Pu fuel designs which have an excellent performance record for normal operation. However, the events at Fukushima culminated in significant hydrogen production and hydrogen explosions, resulting from high temperature Zr/steam interaction following core uncovering for an extended period. These events have resulted in increased emphasis towards developing more accident tolerant fuels (ATFs), particularly for current and near-term build LWRs.

Candidate advanced fuel materials include uranium nitride (UN) and uranium silicide (U_3Si_2), both of which have higher thermal conductivity than UO_2, leading to

* e-mail: ben.lindley@amecfw.com

improved margins under accident conditions, and also have the benefit of higher heavy metal density leading to the possibility of increased core heavy metal loading [1,2]. Candidate cladding materials include advanced stainless steel (FeCrAl), silicon carbide (SiC), and the possibility of adding a coating to Zircaloy clad [3]. Stainless steel cladding exhibits a lower oxidation rate under accident conditions than Zircaloy [4] and is relatively easy to fabricate [5], but has the disadvantage of introducing a large reactivity penalty [4]. SiC cladding can withstand much higher temperatures than Zircaloy, but is expensive and difficult to fabricate [5]. R&D programmes are underway in the US and elsewhere to develop ATFs, encompassing fabrication and testing of UN, U_3Si_2, SiC and coated Zr rods [6].

This paper presents the core analysis performed with the ANSWERS reactor physics code suite WIMS/PANTHER [7,8] for the Integral Inherently Safe Light Water Reactor (I^2S-LWR). The I^2S-LWR concept [9] is a Gen III+ large scale (i.e. 1 GWe) reactor. The design stage is being carried out by a consortium of universities (Michigan, Virginia Tech, Tennessee, Florida Institute of Technology, Idaho, Morehouse College, Brigham Young University, Cambridge, Politecnico di Milano, Zagreb), Idaho National Laboratory, Westinghouse and Southern Nuclear Company. The project is led by the Georgia Institute of Technology.

This innovative PWR includes: an integral primary circuit, a fully passive decay heat removal system that provides indefinite cooling capability, and the use of new materials. The types of materials that were originally chosen for this design include U_3Si_2 fuel pellets within advanced steel cladding.

The equilibrium cycle core analysis was performed using the WIMS/PANTHER codes and the results were verified in a code-to-code comparison. In the first stage, the 2D results obtained with WIMS [2] were compared against the Monte Carlo code Serpent [10], and a good agreement was observed. In the second stage, the full 3D core results obtained with the WIMS/PANTHER codes were compared with results form the literature for the I^2S-LWR [11]. This cross-comparison of results provides enhanced confidence in the reliability and accuracy of the results.

2 UK context for accident tolerant fuel

The UK has a long history in industrial fuel manufacture and fabrication for a wide range of reactor systems including LWRs. This is supported by a national infrastructure to perform experimental and theoretical R&D in fuel performance, fuel transient behaviour and reactor physics.

The UK is seeking to engage with international programmes on ATF research to "strengthen international collaboration opportunities and establish the UK as a centre of expertise for advanced fuel fabrication R&D, and consequently commercial manufacture of such fuels" [12]. Such fuels could be utilized in nuclear new build plants, and also potentially in small modular reactors (SMRs), in which the UK has expressed a strategic interest [13]. The UK

Nuclear Industry Research and Advisory Board (NIRAB) recently recommended that the UK perform research on manufacturing advanced cladding materials in order to enable future manufacture of ATF on a commercial scale [14]. Opportunities for ATF use are identified to include Generation III reactors and SMRs.

3 Modelling of accident tolerant fuel with ANSWERS software

The ANSWERS lattice code WIMS and core simulator PANTHER are used to support the operation of existing PWRs, including in the UK and Belgium [15]. WIMS-PANTHER has recently been validated for analysis of part-MOX-fuelled PWRs. In academia, WIMS and PANTHER have also been applied to a range of PWR configurations including SMRs [16], seed-blanket-fuelled PWRs [17,18], PWRs loaded with transuranic fuels [19,20]. Modelling of ATFs is a natural extension of these capabilities and can largely be performed using existing calculation routes.

Challenges of modeling ATFs include:

- validation of software for different fuel types. This includes validation of the relevant nuclear data libraries. For stainless steel, an extensive amount of validation has been performed as steel is commonly used in fast and thermal reactors. For other isotopes/elements, a reasonable amount of experimental data is available, but further validation may be required for use in new applications;
- modelling of non-standard isotopes. An example is the presence of ^{15}N in UN fuel. The most abundant isotope of nitrogen, ^{14}N, has a large (n,p) cross-section which adversely impacts the neutron economy. It is therefore commonly proposed to increase the ^{15}N content of the nitrogen in the UN fuel through enrichment [1]. While limited experimental data on ^{15}N cross-sections is available, it is not usually considered in isolation and hence further experimental validation may be necessary for thermal reactor applications;
- some candidate ATFs may have the capability to be driven to higher burnups than existing Zircaloy clad UO_2 fuels. Both stainless steel [4] and SiC [21] have superior performance when irradiated compared to Zircaloy. This leads to the need to validate the reactor physics code for higher enrichments and high burnups, and account for a wider range of actinides.

WIMS10, the most recent release of WIMS, contains nuclear data for high burnup applications, including cross-sections and delayed neutron fraction data for a wider range of isotopes including ^{246}Cm, ^{247}Cm and ^{248}Cm. Use of higher enrichment fuel, being driven to high burnups, leads to increased reactivity swings, which requires use of novel burnable poison arrangements and core loading strategies [22]. PANTHER contains inbuilt multi-objective optimization algorithms which facilitate PWR [23] and VVER [24] core design. These have recently been applied to the non-standard case where PWRs are highly loaded with Pu [25,26] and have been shown to facilitate low power peaking core design under challenging circumstances.

Fig. 1. I²S-LWR equilibrium cycle core loading pattern (bottom right quadrant of the core).

Table 1. Main fuel assembly design parameters.

Parameter	Value
Lattice type	19×19, square
Cladding material	Advanced SS (FeCrAl)
Fuel rods per assembly	336
Fuel pellet material	U_3Si_2
Fuel rod outer diameter (in)	0.36
Cladding thickness (in)	0.016
Pellet-clad gap width (in)	0.006
Pellet outer diameter (in)	0.316
Pellet inner void diameter (in)	0.1
Fuel pellet dishing (%)	0.3
Fuel density (% of theoretical)	95.5
Fuel rod pitch (in)	0.477

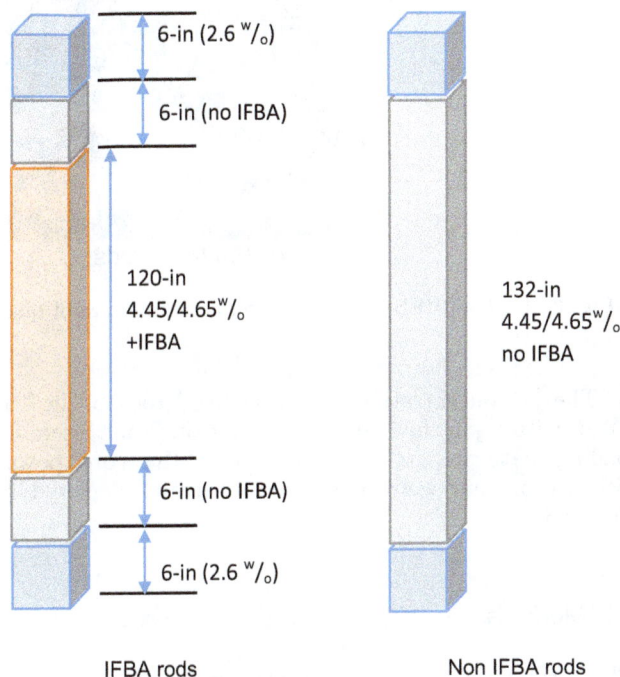

4 Use of WIMS/PANTHER to model I²S-LWR

4.1 I²S-LWR core description

The I²S-LWR core contains 121 assemblies with 144-in active fuel height as shown in Figure 1. The I²S-LWR is designed to achieve 40% higher power rating than a typical 2-loop Westinghouse core ($\sim$2850 MWt vs. $\sim$2000 MWt). The major modification to achieve this objective was transitioning from a typical 16×16 assembly array to a 19×19 square pitch lattice. The increased number of fuel rods in the 19×19 lattice counterbalances the higher power density in the I²S-LWR thereby benefitting DNB performance and, also thanks to the high thermal conductivity of U_3Si_2, fuel temperature. The larger number of fuel rods in the 19×19 lattice leads to approximately same average linear power, 5.8 kW/ft, and only about 3% higher heat flux at the rod surface, 62 kW/ft², for the I²S-LWR relative to a 5% uprated 4-loop PWR with 17×17 lattice. It must be pointed out that the H/HM atomic ratio for the 19×19 is lower, i.e. 3.5, than a typical PWR 17×17 lattice with H/HM of 3.9 due to the higher HM density of the U_3Si_2 fuel. Although under-moderated in terms of neutronic performances, both the 19×19 and 17×17 designs have similar moderator to fuel volumetric ratio of $\sim$2, and therefore the 19×19 lattice design poses no issues in normal and accidental operations. The main geometric parameters and fuel design characteristics are shown in Table 1.

The 3-batch I²S-LWR core loading pattern as shown in Figure 1 is identical to the one adopted by reference [11]. There are 40 fresh assemblies per reload out of 121 assemblies. The twice-burnt assemblies are positioned at

Fig. 2. I²S-LWR fuel axial stack.

the outermost peripheral locations to create a low leakage core. The I²S-LWR features 45 reactivity control clusters assemblies with 24 control rods (Ag-In-Cd) in the assembly.

The U_3Si_2 core design includes fresh and burned assemblies as shown in Figure 1. Fresh assemblies exploit different enrichments (i.e. 4.65, 4.45 and 2.6 $^w/_o$). The active core height of the I²S-LWR fuel axial stack is presented in Figure 2. In fuel assemblies with integral fuel burnable absorber (IFBA) rods (Fig. 2), only the middle portion (120-in) contains ZrB_2 burnable poison, which is surrounded by 6-in non-IFBA top and bottom layers carrying the same fuel enrichment. Finally, 6-in top and bottom axial blankets are used to create the fuel stack. Lower enrichment (2.6 $^w/_o$) is used in the blankets in order to decrease the axial leakage of neutrons.

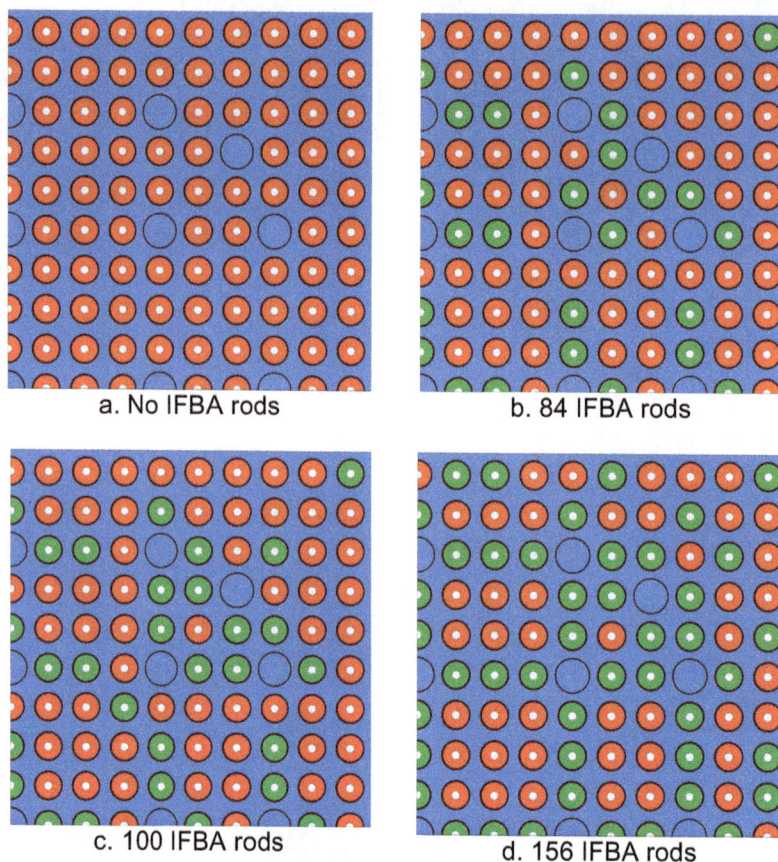

Fig. 3. I²S-LWR IFBA loading patterns – the top right quadrant of the assembly is shown; IFBA rods are indicated in green.

The ^{10}B concentration used in the IFBA rods for the I²S-LWR, with U_3Si_2 fuel design, is 2.5 mg/in. Four assembly loading patterns are used to flatten the core power distribution and were investigated here, as depicted in Figure 3.

4.2 Methods

The current work was divided into the following stages:

– verification of the 2D WIMS assembly models against the reference solutions obtained with the Monte Carlo (MC) code Serpent. Serpent is a continuous-energy MC reactor physics code recently developed for reactor physics applications at VTT Technical Research Centre of Finland. Serpent can be used for 2D fuel lattice calculations as well as for 3D full core simulations. JEFF-3.1 cross-section libraries were used for WIMS and Serpent to minimize discrepancies in neutronic parameters (e.g. k_{inf}) that could arise from the use of different nuclear data evaluations;

– the core physics analysis of the I²S-LWR core design was performed with the core physics package PANTHER. WIMS10 was used for lattice data generation by employing a 172-group JEFF-3.1-based library. WIMS10 utilizes a multicell collision probability method to form 22-group cross-sections, followed by a method-of-characteristics

solution to generate data for PANTHER. Results were compared to those reported in reference [11], which use deterministic lattice calculates to provide data for a 3D core analysis [27,28]. PANTHER used the same 3-batch self-generating reloading scheme that was iteratively applied to the U_3Si_2 core design until the main core parameters converged and a 12-month equilibrium cycle was reached.

4.3 Results

4.3.1 WIMS vs. Serpent comparison

This section presents the single-assembly comparison between WIMS and Serpent for different fuel assembly layouts (i.e. different numbers of IFBA rods). Figure 4 shows criticality curves for the different cases examined. Zero buckling hypothesis was adopted in the current comparison. The difference in reactivity, between Serpent and WIMS, for each of the cases is presented in Figure 5. In addition, Figure 6 shows the maximum difference in within-assembly power (pin-by-pin) between Serpent and WIMS. It must be pointed out that the average absolute difference in the assembly power between the codes is much lower (<0.15%). Figure 7 presents the pin-by-pin power distribution for an assembly that carries 156 IFBA rods.

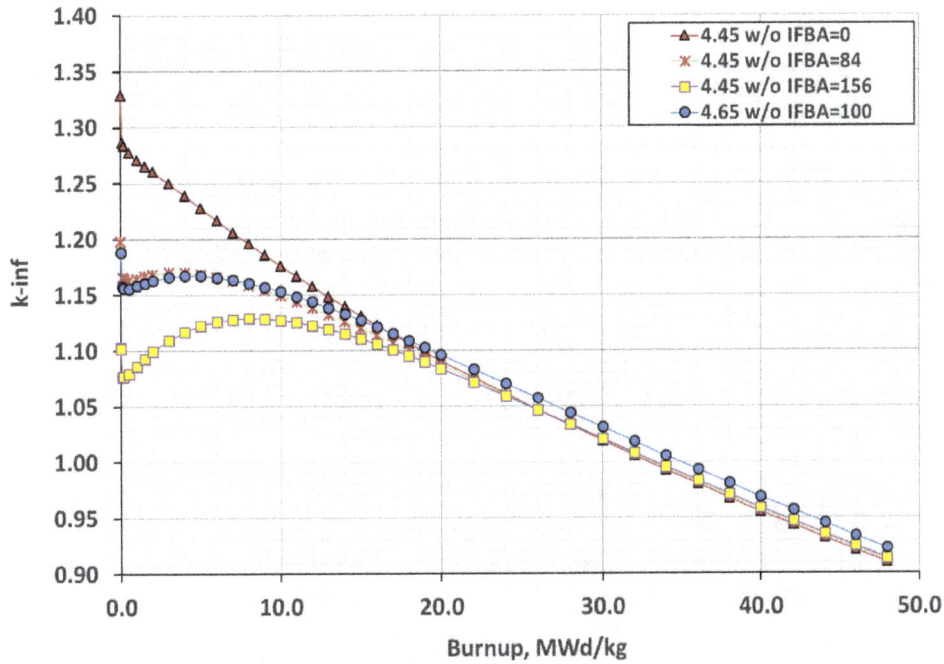

Fig. 4. Criticality curves for different IFBA loading patterns (note that k-inf initially increases with burnup as the burnable poison burns out).

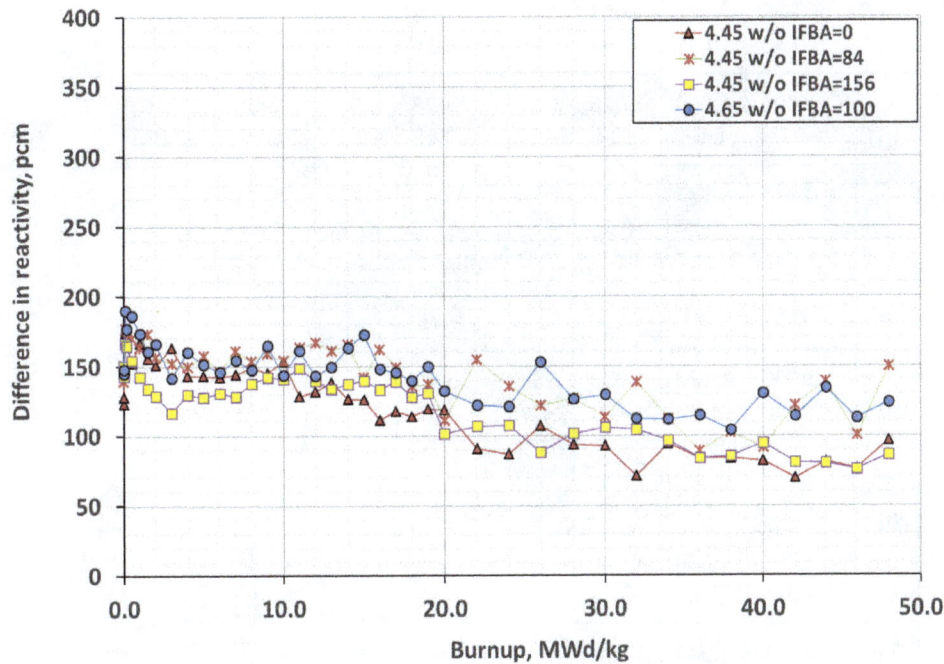

Fig. 5. Difference in reactivity (WIMS vs. Serpent) for different IFBA loading patterns.

4.3.2 Equilibrium core analysis

The representative burnup (MWD/t_HM) distribution at the beginning of the equilibrium cycle is presented in the octant-core map in Figure 8. Figure 9 shows the required boron concentration to maintain criticality over the equilibrium cycle. The radial and total power peaking factors, which represent the quarter-wise assembly values, are depicted in Figure 10, which also presents the time-dependent axial offset. Results are in good agreement with the values reported in reference [11] (e.g. assembly burnups within around 1%). This cross-comparison of results provides enhanced confidence in the reliability and accuracy of the results.

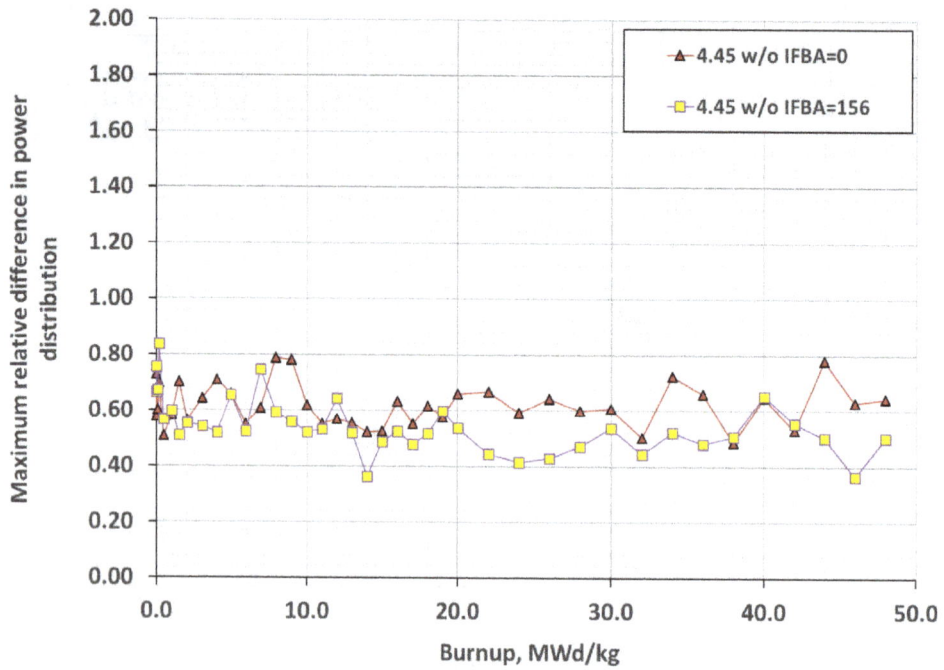

Fig. 6. Maximum relative difference (%) in assembly radial power distribution (WIMS vs. Serpent).

1.000 +0.20	0.929 +0.40	0.938 +0.75	1.030 -0.16	1.035 -0.06	0.963 +0.34	1.040 +0.03	1.045 +0.01	1.041 +0.15	0.971 +0.44
0.975 +0.06	1.005 +0.08	1.000 -0.03	1.026 -0.14	0.992 +0.01	1.028 +0.25	1.020 +0.21	0.951 +0.04	1.020 +0.04	
	0.973 +0.05	0.937 +0.04	0.981 +0.05		0.992 -0.02	0.982 -0.04	1.014 -0.04		
0.985 +0.01	1.027 +0.12	1.021 -0.01	1.032 +0.10	0.988 -0.42	0.995 -0.15				
0.989 +0.20	1.030 +0.30	1.022 -0.10	1.033 +0.01	0.990 -0.15	1.040 -0.01				
	0.979 -0.12	0.944 +0.13	0.982 -0.11						
0.977 -0.51	1.025 -0.02	1.020 -0.03	1.028 +0.09						
0.941 +0.04	1.019 -0.09	1.030 +0.05							
0.978 +0.42	1.024 -0.21								

Serpent Diff. (%)	1.040 +0.03

Fig. 7. Top right octant assembly (4.45 $^{w}/_{o}$ and 156 IFBA rods) radial power distribution (WIMS vs. Serpent) at zero burnup; IFBA rods are indicated in green.

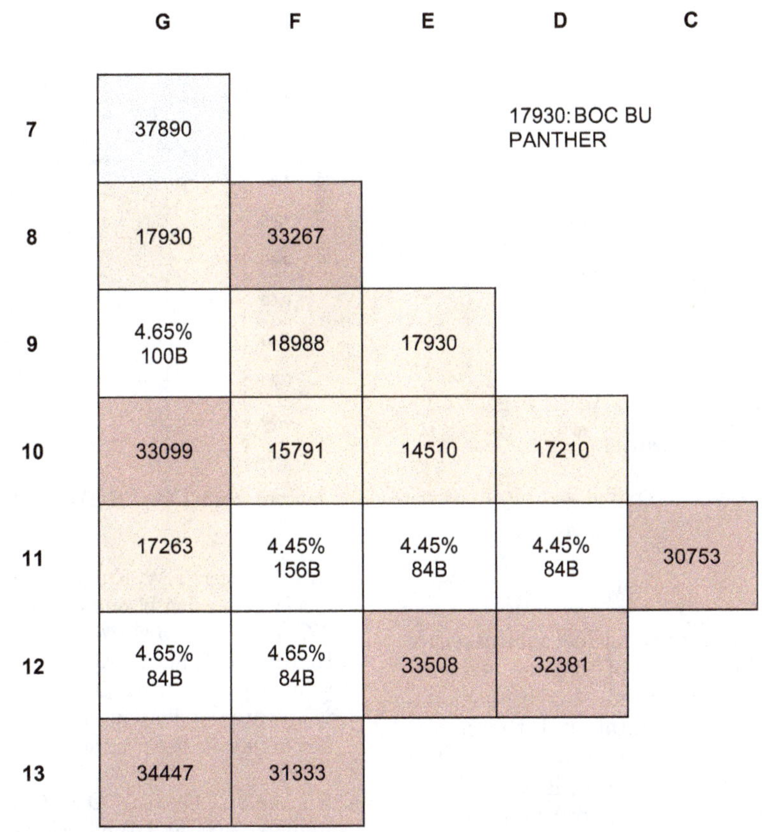

	G	F	E	D	C
7	37890				
8	17930	33267			
9	4.65% 100B	18988	17930		
10	33099	15791	14510	17210	
11	17263	4.45% 156B	4.45% 84B	4.45% 84B	30753
12	4.65% 84B	4.65% 84B	33508	32381	
13	34447	31333			

17930: BOC BU PANTHER

Fig. 8. I^{2}S-LWR equilibrium burnup in PANTHER.

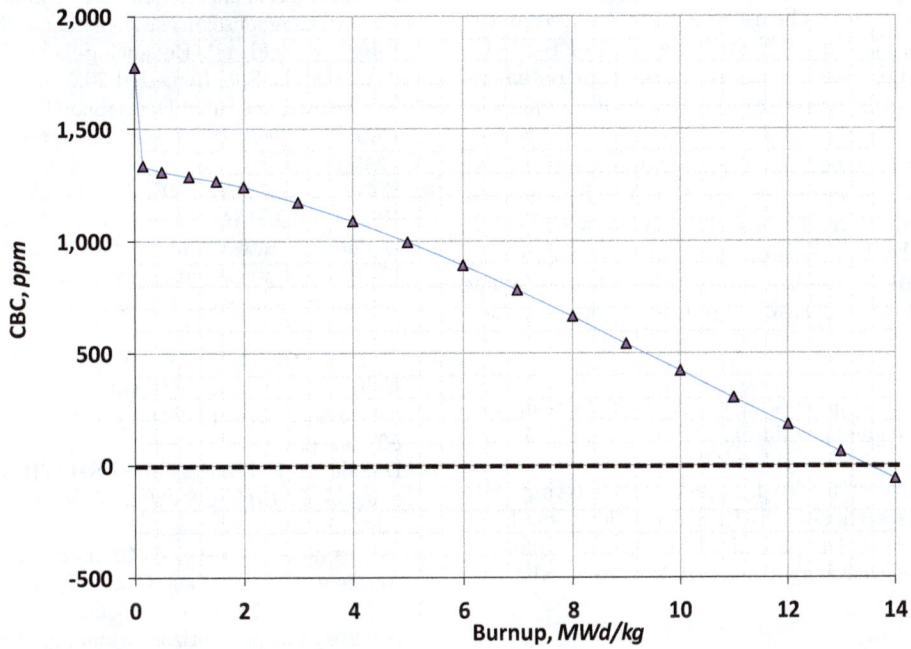

Fig. 9. Comparison of the critical boron concentration (ppm) as a function of burnup in PANTHER.

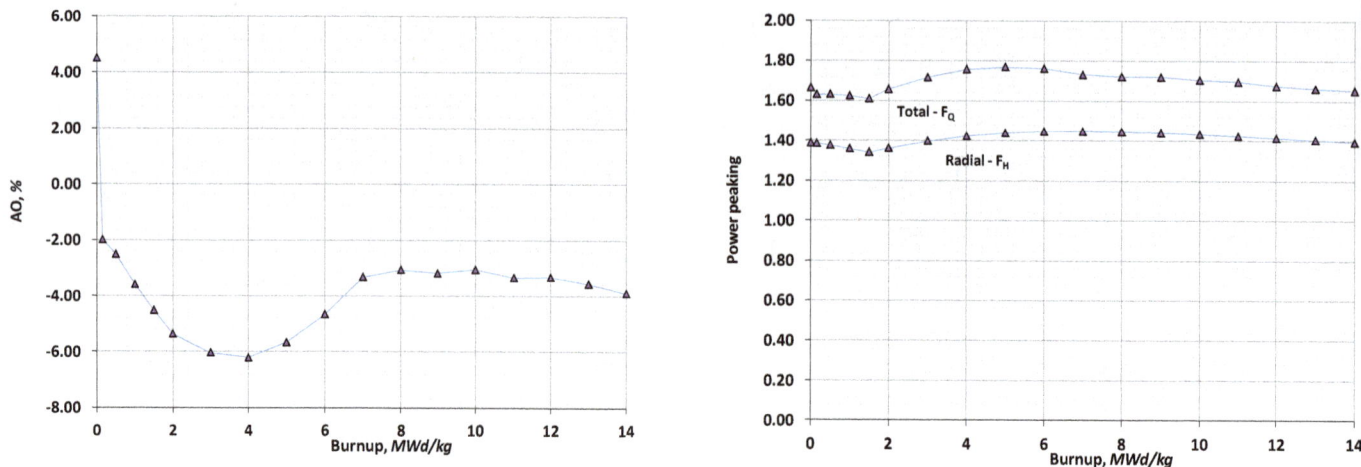

Fig. 10. Axial Offset (AO) (left) and radial and total power peaking factors (right) for I^2S-LWR calculated using PANTHER.

5 Conclusions

The UK has a long history in industrial fuel manufacture and fabrication for a wide range of reactor systems including LWRs. This is supported by a national infrastructure to perform experimental and theoretical R&D in fuel performance, fuel transient behaviour and reactor physics. The ANSWERS lattice code WIMS and core simulator PANTHER are used to support the operation of existing PWRs, including in the UK and Belgium. Modelling of ATFs is a natural extension of these capabilities and can largely be performed using existing calculation routes. Reactor physics modelling of the I^2S-LWR equilibrium cycle core was performed with the WIMS-PANTHER codes. The results were compared to reported results for the equilibrium cycle of the I^2S-LWR and indicate that there is a reasonable agreement between the codes. One possible source for the observed deviations between the codes is the different cross-section library employed in WIMS to generate lattice parameters. For this study, the JEFF-3.1 libraries were used in WIMS, whereas ENDF BVII.0 was used in reference [11]. Future work could consider using the ENDF BVII.0 library in WIMS to allow for a more consistent comparison. It may also ultimately be necessary to validate the reactor physics codes against experimental data.

We are grateful to our colleagues in the ANSWERS team for providing advice and guidance during the preparation of this paper.
This research was in part funded by the UK Engineering and Physical Sciences Research Council (EPSRC) under grant EP/K033611/1 and by the US Depart of Energy (DOE) Office of Nuclear Energy's Nuclear Energy University Programs (NEUP).

References

1. F. Franceschini, E.J. Lahoda, Advanced fuel developments to improve fuel cycle cost in PWR, in *GLOBAL 2011, Makuhari, Japan, Dec. 11–16, 2011* (2011)

2. K.E. Metzger, T.W. Knight, R.L. Williamson, Model of U_3Si_2 fuel system using bison fuel code, in *ICAPP 2014, Charlotte, NC, USA, Apr. 6–9, 2014* (2014)

3. H. Kim, I. Kim, Y. Koo, J. Park, Application of coating technology on zirconium-based alloy to decrease high-temperature oxidation, in *17th International Symposium on Zirconium in the Nuclear Industry, Andhra Pradesh, India* (2013)

4. A. Abe, C. Giovedi, D.S. Gomes, A.T. Silva, Revisiting stainless steel as PWR fuel rod cladding after Fukushima Daiichi accident, J. Energy Power Eng. **8**, 973 (2014)

5. E.J. Lahoda, F. Franceschini, *Advanced fuel concepts RT-TR-11-12* (Westinghouse Electric Company LLC, 2011)

6. Westinghouse Electric Company LLC, Enhancing safety: the pursuit of accident tolerant fuel, [Online], Available: http://www.westinghousenuclear.com/About/News/Features/View/ArticleId/481/Enhancing-Safety-The-Pursuit-of-Accident-tolerant-Fuel [Accessed 20.2.2015]

7. T. Newton et al., Developments within WIMS10, in *PHYSOR 2008, Interlaken, Switzerland, Sept. 14–19, 2008* (2008)

8. E.A. Morrison, PANTHER User Guide E/REP/BBDB/0015/GEN/03 ED/PANTHER/UG/5.5, British Energy, 2003

9. B. Petrovic, Integral inherently safe light water reactors (I^2S–LWR) concept: extending SMR safety features to large power output, in *ICAPP 2014, Charlotte, NC, USA, Apr. 6–9, 2014* (2014)

10. J. Leppänen, M. Pusa, T. Viitanen, V. Valtavirta, T. Kaltiaisenaho, The Serpent Monte Carlo code: status, development and applications in 2013, Ann. Nucl. Energy **82**, 142 (2015)

11. D. Salazar, F. Franceschini, I^2S–LWR equilibrium cycle core analysis, in *PHYSOR 2014, Kyoto, Japan, Sept. 29–Oct. 3, 2014* (2014)

12. Department of Energy & Climate Change, Nuclear R&D - Accident Tolerant Fuel: Grant Notification, 2014

13. Department of Energy and Climate Change, DECC Science Advisory Group: Horizon Scanning, October 2013

14. Nuclear Innovation and Research Advisory Board, NIRAB Annual Report, 2014

15. J.L. Hutton et al., Comparison of WIMS/PANTHER calculations with measurement on a range of operating PWR, in *PHYSOR 2000, Pittsburgh, USA, May 2000* (2000)

16. S. Alam, B.A. Lindley, G.T. Parks, Feasibility study of the design of homogeneously mixed thorium-uranium, in *ICAPP 2015, Nice, France, May 3–6, 2015* (2015)

17. C. Harrington, Reactor Physics Modelling of the Shippingport Light Water Reactor, MPhil Dissertation, University of Cambridge, 2012

18. S.F. Ashley et al., Fuel cycle modelling of open cycle thorium-fuelled nuclear energy systems, Ann. Nucl. Energy **69**, 314 (2014)

19. T. Fei, E.A. Hoffman, T.K. Kim, T.A. Taiwo, Performance evaluation of two-stage fuel cycle from SFR to PWR, in *GLOBAL 2013, Salt Lake City, UT, USA* (2013)

20. F. Heidet, T.K. Kim, T.A. Taiwo, Two-stage fuel cycles with accelerator-driven systems, in *PHYSOR 2014, Kyoto, Japan, Sept. 28–Oct. 3, 2014* (2014)

21. L.L. Snead et al., Handbook of SiC properties for fuel performance modelling, J. Nucl. Mater. **371**, 329 (2007)

22. Z. Xu, Design strategies for optimizing high burnup fuel in pressurized water reactors, PhD Thesis, Massachusetts Institute of Technology, 2003

23. G.T. Parks, Pressurised water reactor fuel management using PANTHER, Nucl. Sci. Eng. **124**, 178 (1996)

24. G.T. Parks, M.P. Knight, Loading pattern optimization in hexagonal geometry using PANTHER, in *PHYSOR 96, Mito, Japan, Sept. 16–20, 1996* (1996)

25. N.Z. Zainuddin, B.A. Lindley, G.T. Parks, Towards optimal in-core fuel management of thorium-plutonium-fuelled PWR cores, in *ICONE 21, 21st International Conference on Nuclear Energy, Chengdu, China, July 29–Aug. 3, 2013* (2013)

26. B.A. Lindley, A. Ahmad, N.Z. Zainuddin, F. Franceschini, G.T. Parks, Steady-state and transient core feasibility analysis for a thorium-fuelled reduced-moderation PWR performing full transuranic recycle, Ann. Nucl. Energy **72**, 320 (2014)

27. M. Ouisloumen et al., PARAGON: The New Westinghouse Assembly Lattice Code, in *ANS Int. Mtg. on Mathematical Methods for Nuclear Applications, Salt Lake City, UT, USA* (2001)

28. L. Mayhue et al., Qualification of NEXUS/ANC nuclear design system for PWR analyses, in *PHYSOR 2008, Interlaken, Switzerland, Sept. 16–19, 2008* (2008)

20

Structural integrity assessment and stress measurement of CHASNUPP-1 fuel assembly

Waseem[*], Ghulam Murtaza, Ashfaq Ahmad Siddiqui, and Syed Waseem Akhtar

Directorate General Nuclear Power Fuel, Pakistan Atomic Energy Commission, PO Box No. 1847, 44000 Islamabad, Pakistan

Abstract. Fuel assembly of the PWR nuclear power plant is a long and flexible structure. This study has been made in an attempt to find the structural integrity of the fuel assembly (FA) of Chashma Nuclear Power Plant-1 (CHASNUPP-1) at room temperature in air. The non-linear contact and structural tensile analysis have been performed using ANSYS 13.0, in order to determine the fuel assembly (FA) elongation behaviour as well as the location and values of the stress intensity and stresses developed in axial direction under applied tensile load of 9800 N or 2 g being the fuel assembly handling or lifting load [Y. Zhang et al., Fuel assembly design report, SNERDI, China, 1994]. The finite element (FE) model comprises spacer grids, fuel rods, flexible contacts between the fuel rods and grid's supports system and guide thimbles with dash-pots and flow holes, in addition to the spot welds between spacer grids and guide thimbles, has been developed using Shell181, Conta174 and Targe170 elements. FA is a non-straight structure. The actual behavior of the geometry is non-linear due to its curvature or design tolerance. It has been observed that fuel assembly elongation values obtained through FE analysis and experiment [SNERDI Tech. Doc., Mechanical strength and calculation for fuel assembly, Technical Report, F3.2.1, China, 1994] under applied tensile load are comparable and show approximately linear behaviors. Therefore, it seems that the permanent elongation of fuel assembly may not occur at the specified load. Moreover, the values of stresses obtained at different locations of the fuel assembly are also comparable with the stress values of the experiment determined at the same locations through strain gauges. Since the results of both studies (analytical and experimental) are comparable, therefore, validation of the FE methodology is confirmed. The stress intensity of the FE model and maximum stresses developed along the guide thimbles in axial direction are less than the design stress limit of the materials used for the grid [ASTM, Standard specification for precipitation hardening nickel alloy (UNSN07718) plate, sheet, and strip for high temperature service, B 670-80, USA, 2013], fuel rod [ASTM, Standard specification for wrought zirconium alloy seamless tubes for nuclear reactor fuel cladding, B 811-02, USA, 2002] and the guide thimble [ASTM, Standard specification for seamless stainless steel mechanical tubing, A 511-04, USA, 2004]. Therefore, the structural integrity criterion of CHASNUPP-1 fuel assembly is fulfilled safely at the specified tensile load.

1 Introduction

CHASNUPP-1 fuel assembly consists of a 15×15 square array of fuel rods, spacer grids, guide thimbles, instrumentation tube, and top and bottom nozzles. The 3D model of fuel assembly containing 20 guide thimbles, 204 fuel rods and an instrumentation tube in conjunction with the 8 spacer grids and top and bottom nozzles, has been developed using the Inventor software, and is shown in Figure 1.

In fuel assembly, fuel rods are held by spacer grids supports system (springs and dimples) to maintain rod-to-rod centerline spacing along the entire length of fuel assembly [1]. The material of top and bottom nozzles, instrumentation tube and guide thimbles is SS-321, whereas spacer grids and fuel rod cladding are made up of Inconel-718 and Zircaloy-4, respectively.

The fuel assembly of pressurized water reactor (PWR) bears a variety of loads, such as tensile, compressive, bending, torsional, impact, etc., while undergoing through handling, shipping and reactor operation. The structural strength of the fuel assembly is supplied by the skeleton of the fuel assembly.

---

* e-mail: wazim_me@hotmail.com

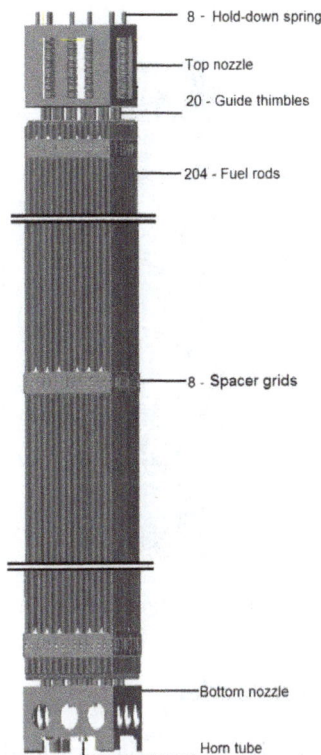

Fig. 1. 3D solid model of CHASNUPP-1 fuel assembly.

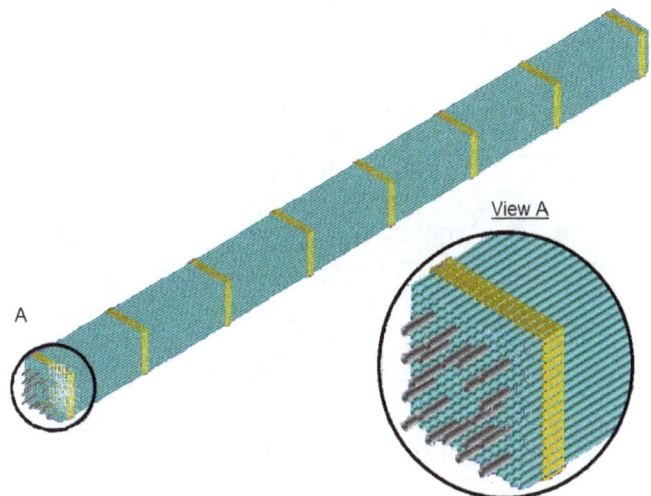

Fig. 2. Model geometry.

The guide thimble tubes are connected with grids by means of spot welds [2]. The top end of the guide thimbles and instrumentation tube are TIG welded with the adapter plate of the top nozzle, while the lower end of the guide thimbles are fastened to the bottom nozzle by bolting.

Previously, we had made study of non-linear buckling analysis of CHASNUPP-1 skeleton and fuel assembly under applied compression load, in order to determine the deformation behavior, stresses and area of the stress concentration [3,4]. Our present study is a part of series of studies which are being conducted in an attempt to contribute towards current research on the design and development work of the PWR fuel assembly. We have now performed the non-linear axial tensile analysis to determine the elastic elongation and assess structural integrity of the fuel assembly under applied axial tensile load of 9800 N (2 g) at room temperature. The results obtained through the FE analysis have been compared with the experimental results, which show good agreement and confirm the validation of FE methodology.

2 FE model and computational details

CHASNUPP-1 fuel assembly possesses symmetry in geometry, material properties and loading conditions. Therefore, in this analysis advantage of symmetry has been taken into account by considering half symmetry of fuel assembly to reduce the size and computational time of the FE model.

The detailed FE model of CHASNUPP-1 fuel assembly, consisting of guide thimbles, fuel rods, spacer grids with

supports systems (springs and dimples) and spot welds (diameter 2.4 mm) between the guide thimbles and the grid's tabs, has been developed using ANSYS 13.0. FE model has been solved using multi-load step. In the first step, contacts between the grid supports (springs and dimples) and the fuel rods have been developed using non-linear contact analysis [1]. In the second load step, non-linear tensile analysis has been performed to determine elastic elongation behaviour and the area of stress concentration of the fuel assembly under the applied fuel assembly handling or lifting load, i.e. 2 g. The mass of CHASNUPP-1 fuel assembly with RCCA is 465 kg. By taking acceleration of 9.81 m/s^2, the load equivalent to 1 g is 4650 N and the max. applied load equivalent to 2 g or 9800 N at room temperature conditions. The detailed solid model is illustrated in Figure 2.

Shell181 element type is used to create mapped meshing (Quadrilateral Elements). It is a 4-node element with 6-degrees of freedom, well suited for linear, large rotation or displacements, and/or large strain non-linear application. After creating the underlying FE model, the flexible surface-to-flexible surface contact pairs have been created using the element types Conta174 and Targe170. The coefficient of friction between fuel rod and grid determined experimentally is taken as 0.35 [5]. The details of FE model are shown in Figure 3.

The entities developed in the FE model are mentioned in Table 1.

The thicknesses of guide thimble, fuel rod and grid, 0.5 mm, 0.7 mm, and 0.3 mm, respectively, are defined by giving real constant values. The material properties of guide thimble, fuel rod and spacer grid used in the present FE analysis are given in Table 2.

Simulations of the boundary conditions of CHAS-NUPP-1 fuel assembly under applied tensile load have been applied as follows:

– to constrain the FE model, all nodes at lower end of the guide thimble have been fixed in all directions;
– to simulate the symmetry boundary conditions, translation of all the nodes at the inside edge of one-half portion

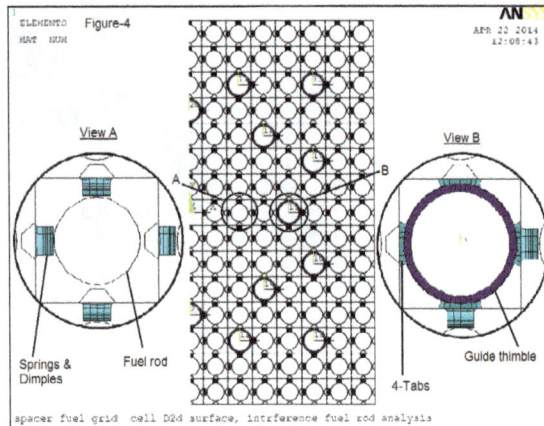

Fig. 3. Element plot of FE model.

Table 1. Entity details of the FE model.

Entity	Quantity
Key points (KP)	266,989
Lines (L)	488,897
Areas (A)	210,232
Nodes (N)	942,850
Shell181 elements	1,005,992
Conta174 elements	78,336
Targe170 elements	91,776

Table 2. Material properties of grid, guide thimble and fuel rod.

Materials	Yield strength (MPa)	Tensile strength (MPa)	Modulus of elasticity (GPa)	Poisson's ratio (γ)
Grid (GH-169A alloy/ Inconel-718)	≥1034	1520–1700	205	0.3
Guide thimble (~SS-321)	≥207	≥517	200	0.3
Fuel rod (Zircaloy-4)	≥240	≥415	200	0.42

of the fuel assembly has been fixed, i.e. nodes along Y-axis are fixed in X-direction;

– the applied axial tensile load of 9800 N has been divided onto 20 guide thimbles and the load of each guide thimble is distributed on the nodes associated with the upper end of the guide thimble in Z-direction;

– all nodes associated with the upper end of the guide thimbles have been coupled in load direction, i.e. Z-direction, other degrees of freedom are set to be zero;

– the weight of fuel rod, 2.114 kg or 21 N, is applied on each fuel rod which is further distributed on the nodes associated with the bottom end of the fuel rod in Z-direction.

FE model, including all above-mentioned boundary conditions, is presented in Figure 4.

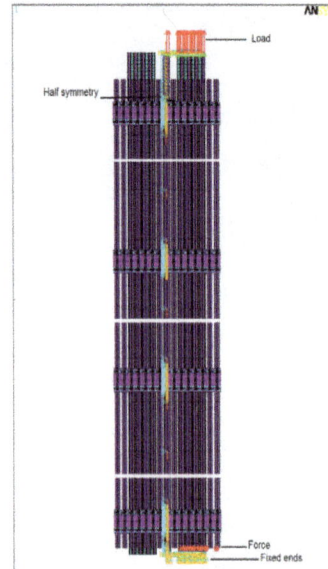

Fig. 4. Applied boundary conditions (element plot 3D).

3 Experimental model

The fuel assembly bears a variety of loads as discussed earlier. Therefore, fuel assembly should have adequate stiffness, strength and dimensional stability to reduce the damage and large elongation failure due to the fuel assembly handling or lifting. This test will provide the basis for the design of fuel assembly, manipulator crane and container and tools which are used in the fuel handling process. In the present study, we have considered the axial tensile test of fuel assembly which has been performed on the prototype full-scale test specimen of fuel assembly at room temperature as shown in Figure 1 except that the pellets of fuel rods are dummy but they are similar in the geometry and weight.

The test facility contains a frame structure of high stiffness and strength. The frame structure is made through welding of the channels beams and steel plates. A convenient load applying system is also developed in order to measure the signals under loading conditions during the test. The force transducer of BLR-1 type is used for the tensile load to measure the force. Foil-type strain gauges of 2×3 mm are used for the strain measurement. The resistance of strain gauges is $120 \pm 0.2\,\Omega$, and its sensitivity coefficient is $2.17 \pm 1\%$. The material, silicone type, which solidifies at room temperature, is used for moisture proof seal [6].

First of all, the test specimen is placed within the calibrated leveled support plates of load applying system and the parallelism of the support plates is adjusted within the specified tolerances of the fuel assembly. Then maximum tensile load of 9800 N, with load increment of 1960 N, is applied on the loading plate, which is divided onto 20 guide thimbles in axial direction.

All guide thimbles are similar in material, geometry and loading conditions, therefore, the strain gauges are mainly pasted on five levels of the guide thimbles. These levels are located on the two corners of one side/face of the fuel

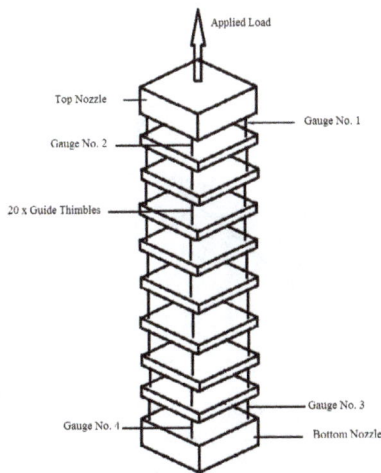

Fig. 5. Strain gauge locations.

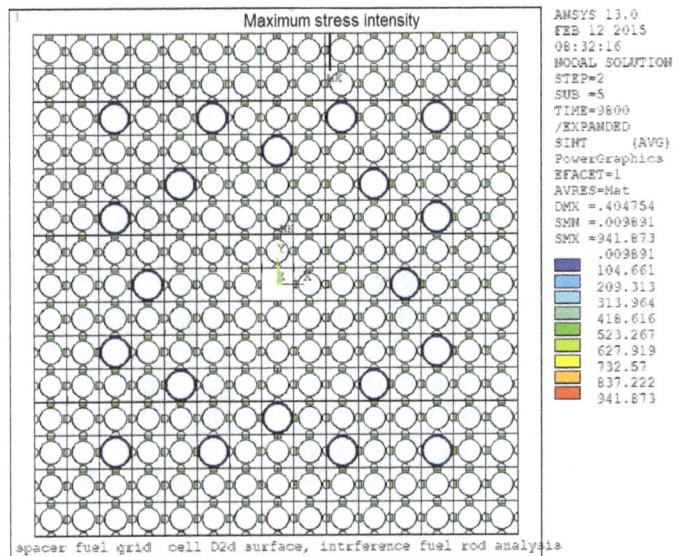

Fig. 6. Plot of nodal stress intensity of fuel assembly.

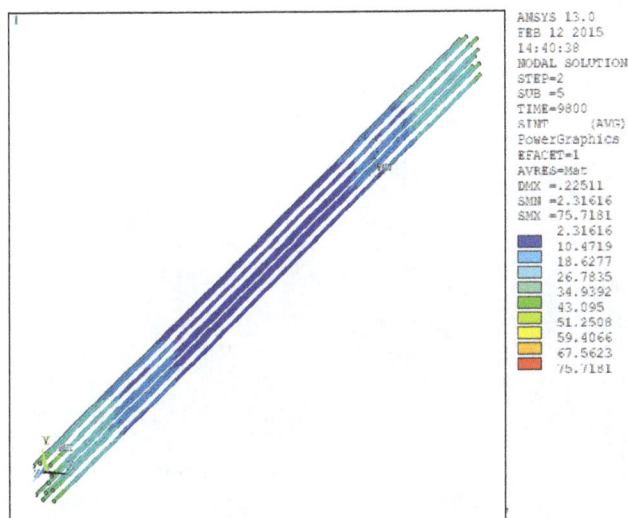

Fig. 7. Plot of nodal stress intensity of guide thimbles.

assembly skeleton (i.e., before insertion of fuel rods in the test specimen) and after pasting all gauges fuel rods are inserted in the prototype fuel assembly test specimen. Mainly, two critical measuring points or levels on the test specimen, determined through FE analysis, have been considered in the present study. The strain gauges, pasted on the upper and lower end positions of the guide thimbles, i.e. near to top and bottom nozzles of the fuel assembly, have been used to measure the local stress concentration at the root of the guide thimbles. The detailed methodology and arrangement of the strain gauges is illustrated in Figure 5.

4 Discussion of FE and test results

– Mesh density is the most important parameter affecting both accuracy and convergence behavior. Therefore, a

sensitivity analysis has been performed to set a mesh refinement level at which converged and more accurate results are obtained.
– Stress intensity is the difference between the algebraically largest and smallest principal stresses at a given point [7]. It is a representative of both stresses primary membrane (Pm) and bending (Pb) [8]. The max. nodal stress intensity at the fuel assembly, 941.9 MPa, under applied load of 9800 N or 2 g, is located at the middle of top surface of the lower arc of spring, as shown in Figure 6.

The value of stress intensity is less than the design stress limit, which is equal to the yield strength [7] of the grid material, 1034 MPa [9].
– The max. nodal stress intensity at the guide thimbles, 75.7 MPa, under applied max. load of 2 g, is located at the outer surface of guide thimble near the top nozzle, as shown in Figure 7.

The value of stress intensity is also less than the design stress limit of the guide thimble material, 207 MPa [10].

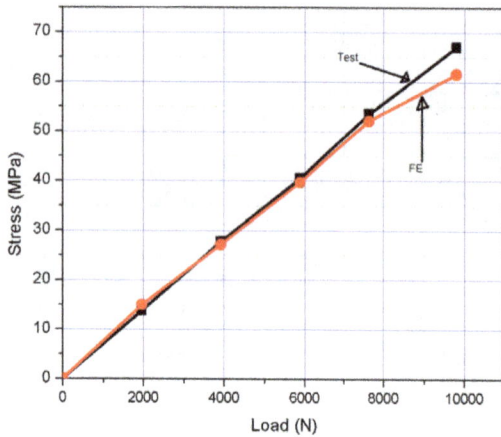

Fig. 8. Test & FE analysis results at gauge-1 location.

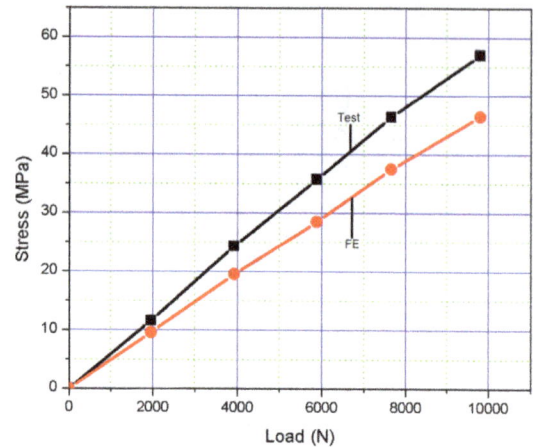

Fig. 9. Test & FE analysis results at gauge-2 location.

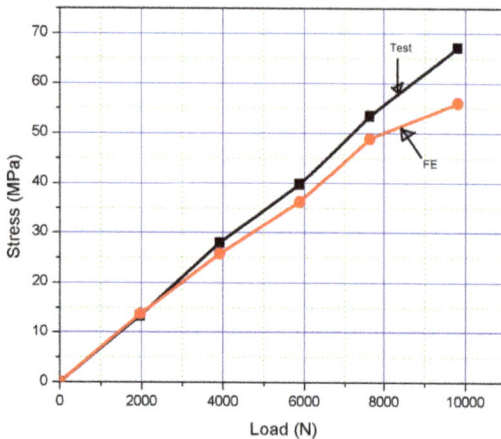

Fig. 10. Test & FE analysis results at gauge-3 location.

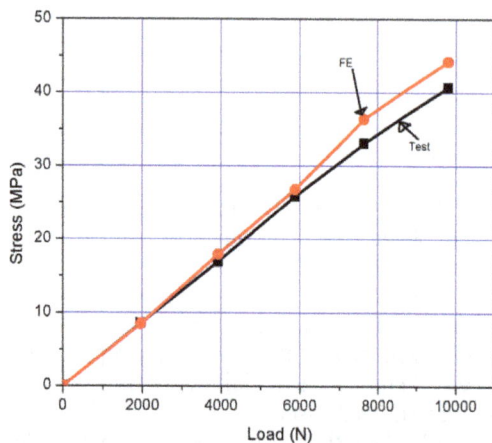

Fig. 11. Test & FE analysis results at gauge-4 location.

Table 3. Comparison of FE analysis and test results at a load of 9800 N (2 g).

| Gauge No. | Stress (MPa) | | % Error |
	FE analysis	Test	(Test & FE analysis)
1	61.6	67.1	8
2	56	67.1	17
3	44.2	40.7	−9
4	46.5	57.0	18

% Error: [(experimental − FE analysis)/experimental] × 100.

- The axial tensile stresses obtained through strain gauges (Nos. 1–4), under tensile load of 9800 N or 2 g (fuel assembly handling load), applied with a load step of 1960 N, are compared with the FE results at the same loads and locations, as illustrated in Figures 8–11.

  As seen from Figures 8–11, it can be observed that the stress in fuel assembly in axial direction, obtained from both studies, increases approximately linear with the increase in load as well as the results of the both studies (FE and test) are also comparable.

- The percentage errors between the analytical and test results are calculated at max. applied tensile load of 9800 N or 2 g, as shown in Table 3.

  From Table 3, the calculated error between the FE analysis and Test results on gauge Nos. 1 and 3 lies within the error band of ±9% and that on gauge Nos. 2 and 4 lies within the error band of ±18%. The percentage of error may be minimized by increasing the mesh density at the cost of computational time.

- The max. elongation of fuel assembly obtained from both studies, FE and test under applied max. axial tensile load (9800 N) are 0.38 mm and 0.36 mm, respectively. The calculated error between the FE analysis and test results of fuel assembly elongation at max. tensile load lies within the error band of ±5%. The elongation behaviours, obtained from both studies, are plotted in Figure 12.

- From Figure 12 it can be seen that the elongation in fuel assembly in axial direction, obtained from both studies, increases approximately linearly with the increase in load. which means that fuel assembly may not permanently elongate till application of the max. load of 9800 N.

- Therefore, the stresses (on the majority of gauges) and deformation obtained through test are comparable with the FE results which validate the FE methodology.

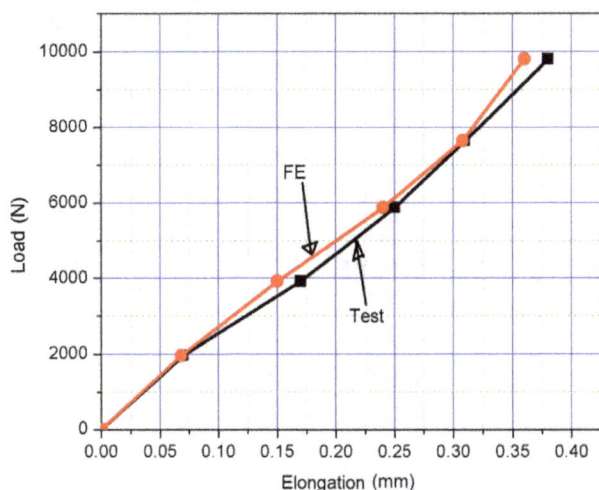

Fig. 12. Elongation behavior of FE analysis and test results.

5 Conclusions

The values of maximum stresses at the fuel assembly and guide thimbles, obtained from the Test and FE analysis, are less than the design stress limit of the grid and guide thimble materials. Therefore, fuel assembly is satisfying the structural integrity criterion at a load of 9800 N (2 g).

This study has provided a good confidence level for verification of CHASNUPP-1 fuel assembly design at room temperature, which can be useful for any change required in the modified fuel assembly design, such as different material, number of grids/span, before conducting the confirmatory tests, and may be very helpful for improving the safety and reliability of fuel assembly design.

References

1. W.N. Elahi, A.A. Siddiqui, G. Murtaza, Fuel rod-to-support contact pressure and stress measurement for CHASNUPP-1 (PWR) fuel, Int. J. Nucl. Eng. Des. **241**, 32 (2011)
2. W.N. Elahi, A.A. Siddiqui, G. Murtaza, Structural integrity assessment of spot weld joints between spacer grid and guide thimble of PWR fuel assembly, Tech. Report DGNPF-TR/012, 2010
3. W.N. Elahi, G. Murtaza, A.A. Siddiqui, Structural integrity assessment and stress measurement of CHASNUPP-1 Fuel assembly skeleton, Int. J. Nucl. Eng. Des. **266**, 55 (2014)
4. W.N. Elahi, G. Murtaza, Structural integrity assessment and stress measurement of CHASNUPP-1 Fuel assembly, Int. J. Nucl. Eng. Des. **280**, 130 (2014)
5. Y. Zhang et al., Fuel assembly design report, SNERDI, China, 1994
6. SNERDI Tech. Doc., Mechanical strength and calculation for fuel assembly, Technical Report F3.2.1, China, 1994
7. ASME, Boiler and pressure vessel code, Section III, Division 1, Subsection NB, Article NB-3000, 2001
8. Help manual of the ANSYS version13.0, 2012
9. ASTM, Standard specification for precipitation hardening nickel alloy (UNSN07718) plate, sheet, and strip for high temperature service, B 670-80, USA, 2013
10. ASTM, Standard specification for seamless stainless steel mechanical tubing, A 511-04, USA, 2004

21

Extraction of uranium from seawater: a few facts

Joel Guidez and Sophie Gabriel[*]

French Alternative Energies and Atomic Energy Commission, CEA/DEN, Université Paris Saclay, 91191 Gif-sur-Yvette, France

Abstract. Although uranium concentration in seawater is only about 3 micrograms per liter, the quantity of uranium dissolved in the world's oceans is estimated to amount to 4.5 billion tonnes of uranium metal (tU). In contrast, the current conventional terrestrial resource is estimated to amount to about 17 million tU. However, for a number of reasons the extraction of significant amounts of uranium from seawater remains today more a dream than a reality. Firstly, pumping the seawater to extract this uranium would need more energy than what could be produced with the recuperated uranium. Then if trying to use existing industrial flow rates, as for example on a nuclear power plant, it appears that the annual possible quantity remains very low. In fact huge quantities of water must be treated. To produce the annual world uranium consumption (around 65,000 tU), it would need at least to extract all uranium of 2×10^{13} tonnes of seawater, the volume equivalent of the entire North Sea. In fact only the great ocean currents are providing without pumping these huge quantities, and the idea is to try to extract even very partially this uranium. For example Japan, which used before the Fukushima accident about 8,000 tU by year, sees about 5.2 million tU passing every year, in the ocean current Kuro Shio in which it lies. A lot of research works have been published on the studies of adsorbents immersed in these currents. Then, after submersion, these adsorbents are chemically treated to recuperate the uranium. Final quantities remain very low in comparison of the complex and costly operations to be done in sea. One kilogram of adsorbent, after one month of submersion, yields about 2 g of uranium and the adsorbent can only be used six times due to decreasing efficiency. The industrial extrapolation exercise made for the extraction of 1,200 tU/year give with these values a very costly installation installed on more than 1000 km^2 of sea with a lot of boats for transportation and maintenance. The ecological management of this huge installation would present significant challenges. This research will continue to try to increase the efficiency of these adsorbents, but it is clear that it would be very risky today, to have a long-term industrial strategy based on significant production of uranium from seawater with an affordable cost.

1 Very large amounts of uranium

The average value of the uranium content dissolved in the oceans is estimated at 3.3 micrograms per liter (with dispersal from 1 to 5 micrograms depending on the locations). With a volume in the oceans about 1.37×10^{18} m^3, uranium content is estimated to amount to 4.5 billion tonnes of uranium metal (tU) compared to conventional terrestrial resource estimates of about 17 million tU [1–4].

In this connection, Japan, which consumed before the Fukushima accident about 8,000 tU per year, sees about 5.2 million tU pass by every year in the great ocean current Kuro Schio in which it lies (Fig. 1) [3]. Japan depends entirely on uranium imports, that explains its interest and

past research effort for extraction of uranium from seawater.

This uranium mainly comes from the soil leaching and related supply from rivers. For example, it is estimated that the Rhone brings 29 tU/year into the sea, and all rivers combined contribute 8,500 tU/year.

These virtually inexhaustible quantities have, sporadically since the 1950s, led to much research on the possibility of extraction. The recently launched American Department of Energy program is to develop a realistic cost of production to inform future fuel cycle decisions, i.e. whether to reprocess or not.

Note: All metal ions are also found dissolved in seawater in significant overall amounts and often greater than known mineral resources. Only three products: NaCl, MgCl$_2$ and MgSO$_4$ can be easily extracted, for example by evaporation. The values for the others are much too low and require more complex selective strategies. It should also be noted that some interesting

* e-mail: sophie.gabriel@cea.fr

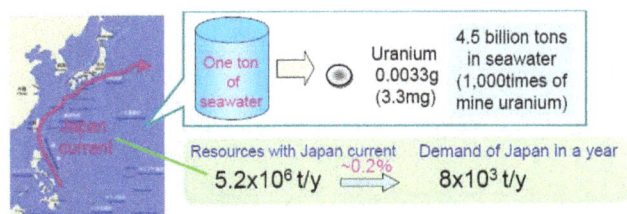

Fig. 1. Amounts of uranium present in the oceans and ocean current near Japan [3].

products such as lithium or indium may also be involved in this research on extraction techniques.

2 Energy balance of extraction

2.1 Extraction by pumping

A tonne of seawater therefore contains about 3.3 milligrams of uranium. Every year France uses 8,000 t of natural uranium to produce about 420 TWh, i.e. 52.5 kWh per gram of uranium. The complete extraction of the uranium contained in a cubic meter of seawater (which is not the case), would let to produce about 0.17 kWh of electrical energy in current nuclear water reactors.

The electrical energy required to raise $1\,m^3$ by 10 m is about 0.03 kWh (with a yield of 80%). In addition, there is a pressure drop in the pipes and the filtration membrane. In seawater desalination plants, for example, energy consumption is estimated around 2.5 kWh per tonne [2], well above the 0.17 kWh that could be recovered. Thus, by applying the simple rule of three to the energy balance, the infeasibility of a land-bound plant dedicated to extracting uranium from seawater can be seen.

2.2 Existing pumping facilities unusable

There are significant seawater pumping facilities in nuclear power plants, seawater desalination facilities or tidal power plants. But the amount of uranium that could be hoped from them remains low and therefore unrealistic in relation to the difficulties: increased head losses, actual efficiency, problem of waste and local depletion in terms of concentration, etc.

A 1,000 MWe nuclear reactor, for example, will use an annual seawater flow of about 40 million cubic meters. This represents the flow of only 130 kg of uranium. Even if all of it could be recovered (which is impossible), this would, at the current market price, represent a budget of 12,000 euros which of course would not even cover installation and operating costs. Incidentally, this amount is less than one thousandth of the annual consumption of the same reactor (150 tU).

The same reasoning applies to seawater desalination units, where the maximum extractable quantities, and therefore the available budget, remain very low in relation to the operations to be performed.

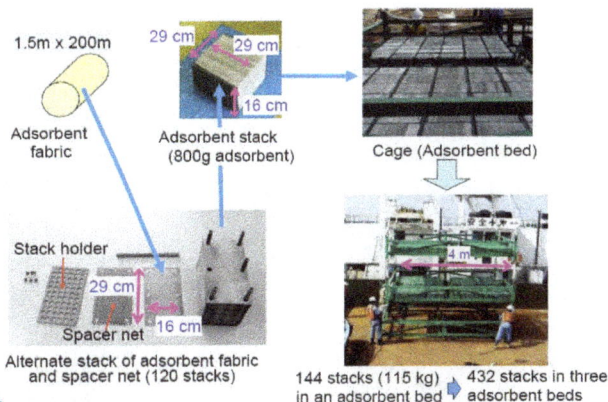

Fig. 2. Construction of a platform in which each stack has 115 kg of adsorbent [3].

2.3 Use of ocean currents

The amounts of water to be treated are huge compared to the objectives. This is the basic problem.

To produce the total amount of uranium currently consumed worldwide every year (about 65,000 tU) and assuming an infeasible 100% yield, 2×10^{13} tonnes of water would have to be processed annually, in other words: the entire North Sea [2]. Only the great ocean currents are able to supply these volumes: the Gulf Stream, Kuro Shio in Japan, Strait of Gibraltar, etc.

The idea is thus to treat these major currents which would also solve the problem of depletion and renewal of seawater, for a land-bound plant. The concept of pumping, filter and efficiency no longer applies. It would be an extraction in the water.

3 Update on extraction techniques

Attention has therefore turned to using adsorbents that can collect the uranium (along with other components) in a selective way. Then these adsorbents are removed and the deposits recovered, generally by a chemical process.

In the 1960s, titanium oxide hydrate was used [5,6], but the latest publications refer to the use of amidoxime [5–9], which has a significantly higher yield.

In-laboratory point values of 2 g/kg of adsorbent per month, or more, have thus been announced (the most recent laboratory batch experiments of the Oak Ridge National Laboratory [ORNL], in 2013, have shown the higher performance: 3.3 g/kg of adsorbent after 8 weeks of contact of the adsorbent with seawater [8]).

The performances are much lower in more realistic conditions. In 2009, JAEA presented result from marine experiments [3,6]. The device was three superimposed platforms, containing 115 kg of absorbent on supports (Figs. 2 and 3).

Table 1 shows the extraction cycles for this system from 1999 to 2001 and the amount of uranium recovered, i.e. 1,083 g over the 12 cycles of 20 to 96 days of immersion.

Table 1. Assessment of offshore extraction cycles [6].

Submersion period		Submersion days	Seawater temperature (°C)	Number of stacks	Adsorbed uranium (g) [9]
1999	29 Sep.–20 Oct.	21	19–21	144	66
2000	8 Jun.–28 Jun.	20	12–13	144	47
	28 Jun.–8 Aug.	40	13–22	144	66
	8 Aug.–7 Sep.	29	20–24	144	101
	7 Sep.–28 Sep.	21	24–22	144	76
	28 Sep.–19 Oct.	21	20–18	144	77
2001	15 Jun.–17 Jul.	32	13–18	216	95
	18 Jul.–20 Aug.	32	18–20	216	119
	15 Jun.–20 Aug.	65	13–20	72	48
	20 Aug.–21 Sep.	31	20–19	216	118
	18 Jul.–21 Sep.	63	18–19	144	150
	15 Jun.–21 Sep.	96	13–19	72	120
					1083

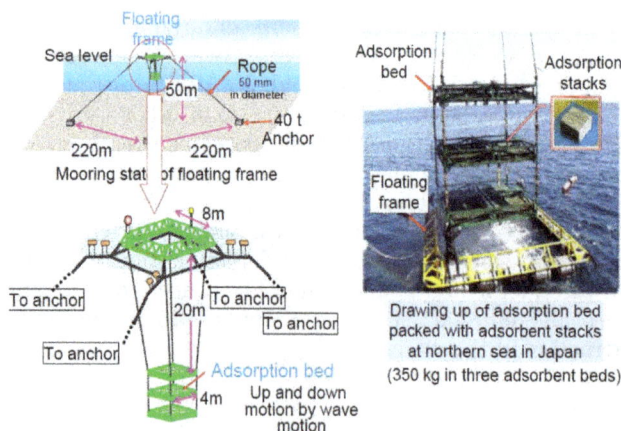

Fig. 3. Complete offshore platform with the three stacks shown in Figure 2 [3].

The values clearly fluctuate, but the average value is less than 1 g of uranium per kilogram of absorbent and per month: lower than the "ideal" laboratory values. Even if the more recent batch laboratory experiments with the new adsorbent of ORNL are better (2.6 times higher than that of the JAEA adsorbent under similar conditions [8]), it is still very low.

These methods are confronted with many problems in the field:

– drop in performance after each chemical wash/limited number of cycles;
– influence of various parameters on the performance such as water temperature, wave height, etc.;
– deposits of algae and shells;
– problems related to installing offshore operations (access, weather conditions, resistance to corrosion of structures, etc.);

– significant amounts of adsorbent to be used, processed and renewed.

The important role played by temperature, which is to be as high as possible (25 °C or more) is also obvious. The cycles were carried out from June to October.

4 Cost analysis

Researchers working in field announced until the 1980s targeted a price range between 1,000 and 2,000 USD/kgU. After, using point results of better efficiency in terms of grams per kilogram obtained in the laboratory, prices were reduced accordingly and announced between 300 to 600 USD/kgU. More recent cost analyses have been made by the Japanese and also by the American Department of Energy [7,10]. The prices quoted are then between 1,000 and 1,400 USD/kgU.

The lowest values can be perplexing when you consider the example from the previous section and all the qualified personnel and work required to recover a kilo of uranium in one year: construction of the platform, offshore operations for installation and periodic extraction, onshore processing of the adsorbent, periodic replacement of 115 kilos of adsorbent, etc. What is the final cost of this kilo of uranium?

In fact, these costs announced were derived from extrapolations for gigantic installations. The systems are immersed over several kilometers (see Fig. 4 for a project with an annual output of 1,200 tU) as well as shuttle boats and on-shore treatment plant. This should lead to industrial rationalization and a related reduction in costs. It is clear that all costs associated with developing and operating these huge facilities have not yet been determined, particularly for the installation, anchoring, and location of these thousands of offshore platforms, and those costs announced are little more than rough, first order estimates.

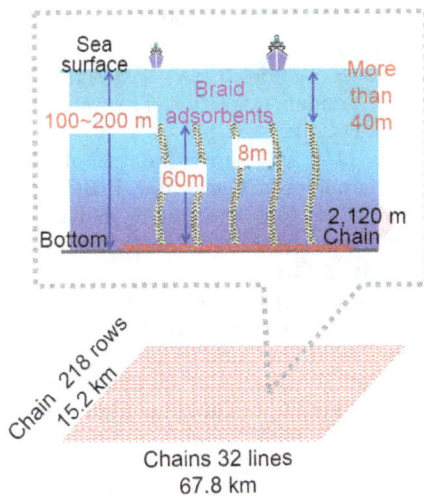

Fig. 4. Offshore extraction plant project [3].

For the more recent cost analyses [7,10] (prices between 1,000 and 1,400 USD/kgU), the initial parameters used are as follows (for a plant that would produce 1,200 tU/year):

– capacity of the adsorbent at 2 g/kg;
– 60 days of immersion;
– temperature of the water at 25 °C;
– 5% drop in efficiency of the adsorbent after each chemical rinse;
– using the adsorbent six times (after which it has to be replaced).

It appears that the primary key parameter of the cost model is the adsorbent's capacity in g/kg. The mathematical model is thus used to significantly reduce costs when going from 2 g/kg to 4 and then 6 (the last test presented in 2013 [8,10] had reached 3.3 g/kg in 8 weeks of immersion, 2.6 higher than the previous).

All this remains theoretical. The anchoring of these systems over several miles at sea has yet to be defined. What is the drop in efficiency in winter? Is there a depletion over the kilometers of adsorption which would also adversely affect efficiency? What about corrosion and structural maintenance? None of these issues are addressed in the presentation of the model.

5 Difficulties

5.1 Environmental problems

It should also be noted that the environmental impact of a facility covering over 1,000 km^2 would certainly be prohibitive. Similarly, the amount of chemical by-products produced and handled would be extremely large and also lead to environmental problems.

5.2 Energy balance

Many massive facilities have to be constructed and submerged, tonnes of adsorbent have to be made and

renewed, and conventional island component cooling system boats have to go back and forth. The document [2] addresses this point in an original way. Using statistics for fishing costs and related fuel consumption, an estimated 5 kWh/kg is required to extract something free from the sea and bring it back to shore. However, to produce 1 kg of uranium approximately 500 kg of adsorbent have to be handled, i.e. 2.5 MWh per kilogram of uranium produced, for one-way transportation only (a free return trip is assumed, as the boat would not travel unloaded). Similarly, the production of these adsorbents with a limited lifespan also requires energy, estimated in reference [2] at 10 MWh to produce 500 kg of adsorbent (assuming a one-year life cycle, which is optimistic). These calculations, which are already very rough, mean that 12.5 MWh would be used to produce a kilogram of uranium which in turn can theoretically generate about 52.5 MWh in a reactor. And all other energies required in the process should be added to achieve an accurate balance.

This energy balance work was carried out in much greater detail by the project proponent [11], which uses more optimistic and lower values than those above. It reaches an EROI (Energy Return On Investment) of 12, a value which is clearly subject to a number of parameters. It should be noted that the EROI is more than 300 times higher for mined uranium.

5.3 Strategy for the nuclear industry

Without a demonstration of industrial feasibility and validation of a credible cost of extraction, it would be extremely risky to work on a long-term industrial strategy based on significant production of uranium from seawater at an affordable cost.

It is worth remembering that fast reactors could be operated without the need for new resources of natural uranium for millennia. The economic profitability would be ensured well before the market price of uranium reaches the estimated cost of uranium production from seawater.

6 Conclusion

There is an extremely large quantity of uranium solute in the oceans but its low concentration would require a volume of water greater than that of the North Sea to be processed every year in order to extract uranium currently consumed worldwide every year.

Basic energy balances show that pumping/filtering systems have no interest and no future.

The only other solution would be extraction by adsorbents placed in ocean currents naturally and freely providing drive and renewal of very large flowrates. These techniques currently enable the production of small quantities at prohibitive prices. Extrapolation on an industrial scale has yet to be developed, even in terms of feasibility, and the final cost of production has not yet been firmly established.

However, the continuation of this research is interesting if the efficiency of the process can be further improved, and applied to other materials of interest, so as to pool prohibitively high costs of production.

It would however, given current knowledge, be extremely risky for the nuclear industry to launch an industrial strategy based on the possible extraction of uranium from seawater, in an affordable way.

References

1. B. Barre, G. Capus, L'uranium de l'eau de mer : véritable ressource énergétique ou mythe ? [Uranium from seawater: real energy resource or myth?], Revue des ingénieurs (Engineers review), 2003
2. U. Bardi, Extracting minerals from seawater: an energy analysis, Sustainability **2**, 980 (2010)
3. M. Tamada (JAEA), Collection of uranium from seawater, Presentation of 5/11/2009 in Vienna (IAEA)
4. OECD/NEA IAEA, Uranium 2014: Resources, Production and demand, 2014
5. P. Blanchard, S. Gabriel (CEA), Extraction d'uranium de l'eau de mer (Uranium extraction of seawater), Letter from I-tésé Number 11, 2010
6. M. Tamada (JAEA), Current status of technology for collection of uranium from seawater, ERICE seminar, 2009
7. E. Schneider et al., Cost and system analysis of recovery of uranium from seawater, DOE paper presented the 31 October 2012 (Chicago /ANL)
8. C. Tsouris et al., Uptake of uranium from seawater by amidoxime-based polymeric adsorbent: marine testing, in *Global 2013* (2013), paper 8438
9. L.K. Felker et al., Adsorbent materials development and testing, for the extraction of uranium from seawater, in *Global 2013* (2013), paper 8355
10. E. Schneider, D. Sachde, The cost of recovering uranium from seawater by a braided polymer adsorbent system, Sci. Global Secur. **21**, 134 (2013)
11. E. Schneider, H. Lindner, Energy balance for uranium recovery from seawater, in *Global 2013* (2013), paper 7427

22

Selection of a tool to decision making for site selection for high level waste

Jonni Guiller Madeira[1*], Antônio Carlos M. Alvim[2], Vivian B. Martins[2], and Nilton A. Monteiro[2]

[1] Celso Suckow Fonseca Federal Center for Technological Education–(CEFET), Areal street, 522 Sq. Mambucaba, Angra dos Reis, Brazil
[2] Nuclear engineering program–(PEN/UFRJ/COPPE-RJ), Federal University of Rio de Janeiro, Horácio Macedo avenue 2030, Technology Center Building, University City, 21941-972 Rio de Janeiro, Brazil

Abstract. The aim of this paper is to create a panel comparing some of the key decision-making support tools used in situations with the characteristics of the problem of selecting suitable areas for constructing a final deep geologic repository. The tools addressed in this work are also well known and with easy implementation. The decision-making process in matters of this kind is, in general, complex due to its multicriteria nature and the conflicting opinions of various stakeholders. Thus, a comprehensive study was performed with the literature in this subject, specifically in documents of the International Atomic Energy Agency (IAEA), regarding the importance of the criteria involved in the decision-making process. Therefore, we highlighted six judgment attributes for selecting a decision support tool, suitable for the problem. For this study, we have selected the following multicriteria tools: AHP, Delphi, Brainstorm, Nominal Group Technique and AHP-Delphi. Finally, the AHP-Delphi method has demonstrated to be more appropriate for managing the inherent multiple attributes to the problem proposed.

1 Introduction

Nowadays, considering the increasing demand for power, and the environmental issues, it is ever more necessary to adopt (and create) new alternative sources of power, economically viable and with low environmental impact. In this sense, because of environmental, social and political concerns, recently, the nuclear power has drawn the global attention.

The environmental impact is potentially the most relevant topic concerning the suitability of nuclear power [1]. And, since global warming has been the key topic of several discussions over the last years (it is believed that such phenomena is generated by the greenhouse gases [GHG]: water vapor, methane and CO_2), it increases the environmental advantages of nuclear power plants. Among such advantages, it is possible to mention that a smaller space may be quite satisfactory for its construction, also it is quite free from polluting gases emissions such as CO_2 and methane [2]. However, one of the most challenging questions to be answered, regarding nuclear power, is the appropriate location for nuclear power plants construction and location for nuclear wastes.

Since the early 1970's, a rising concern with the final (or temporary) disposal of radioactive wastes has been noticed. This rising attention on nuclear wastes questions has generated an economic and safety assessment that sought to optimize the cost-benefit of these repositories (storage location for disposal of radioactive wastes). Also, several authors have been discussing about this issue [3–14].

Altogether, there is not many countries with final repositories (for radioactive wastes) working. However, some countries, as Finland, e.g., are making a great progress building deep final repositories [14].

Thus, regarding the place selection for safely housing a deep geological repository, we have walked into a decision-making issue, since we need to choose, among the likely possibilities, one that meets the several points required for; and also that best fits these points, as deep geological repository for radioactive waste, emphasizing the spent fuel.

The process of decision-making involves various specialties from different fields, considering it is a multi-criteria problem. In this case, we need a decision support tool that can arrange the specialists' opinions within the context of place selection.

* e-mail: guiller.nuclear@ufrj.br

2 Geological disposal of nuclear wastes

During the nuclear power plant operating life, it is necessary to replace part of the reactor core fuel periodically. This spent nuclear fuel is called (if it has no other use) nuclear waste, which, besides emitting radiation and heat, contains high amounts of radioactive nuclide. A delicate point in handling such nuclear waste is the emission of radiation for a very long time, what shall reach thousands, or even millions of years.

There are thousands of tons of spent fuel waiting for a solution for its final disposal, which puts this question among the main concerns on the construction of a nuclear power plant [15].

A deep geological repository is the most used solution in countries where high level nuclear wastes management is adopted already. Also, in the EURADWASTE and IAEA Conference, it was largely discussed that deep geological repositories are hitherto the safest and workable measure for nuclear waste.

A deep geological repository is a nuclear waste repository excavated deep within a stable geologic environment (typically below 350 m or 500 m) in order to avoid for a long time the biosphere contamination with the radioactive nuclide [16]. Such isolation, in this type of disposal, ensures safety until the nuclear fission products decay and reach acceptable radiation levels. It entails a combination of waste form, waste package, engineered seals and geology that is suited to provide a high level of long-term isolation and containment without future maintenance. This feature must inhibit the motion of radioactive nuclide into the middle of the external repository, ensuring future safety for humans and environment [17]. Ratifying, all these details are essential to ensure the future safety of deep geological repositories.

3 Decision support tools

One of the most important tasks faced by decision makers is selecting a site that meets the various criteria considered for constructing the final deep geological repository for nuclear wastes.

This site selection requires a multicriteria analysis with an analytical solution. Since it deals directly with conflicting criteria, e.g., demography, there would be advantages and disadvantages. For example, if a deep geological repository for high level waste is constructed in a high population density area, there would be advantages in transportation, such as easy roads access – making it easier to get workforce. However, this same location may directly affects the population because it could increase the risk for people.

Multicriteria analysis is done considering internationally accepted factors as essential for the suitability of a place for a deep geologic repository construction such as lithology, relief, transportation, among others.

Generally, there are many technical requirements to be properly fulfilled in place selection. Thus, the tools shown below orient this dynamics of the selection process.

3.1 AHP

Multiple-Criteria Decision Making (MCDM) occurs in cases where it is necessary to analyze decision situations that embody both quantitative and qualitative criteria, conflicting or not. The AHP is one of the most known and used methods of MCDM [18].

Analytic Hierarchy Process (AHP) is a tool that drives decision makers to meet the best solution that suits their goal and their understanding of the problem, leading them to a structured reflection on solving it in a constant process of acquisition of knowledge. The AHP seeks to reproduce what seems to be a natural method of human mind in perceptions and judgements [19]. This technique was developed by Saaty [20] (1980) and it is based on pairwise comparisons of criteria, in order to create a relationship matrix (proportion).

This decision support tool is characterized by its simplicity and efficiency, what makes its use possible in several fields, including: Strategic Planning [21], Marketing [22] and Consensus-based assessment [23], Funding and financing Choice for Air Transport [24] (1998), Quality and Productivity Programs [25] and Project Analysis [26].

The AHP may incorporate both qualitative and quantitative factors in the decision-making process [27], so it is possible to deal with the inherent subjectivity of this selection process.

Although it is a very effective tool [28], it highlights some disadvantages of AHP technique:

- once the scale is subjective, it is liable to human error;
- it is vulnerable to human psychology;
- the number of comparative tables may be too large if many attributes of comparison are used, creating, so, a tendency to exclude them;
- there is a limit on the hierarchy levels (number) that can be used;
- it is necessary a series of pairwise comparisons of the elements for very large problems;
- ambiguous and inconsistent judgments by the decision maker may be critical.

3.2 Delphi

Its use is recommended, either, when it has no measurable data about a problem, as when there is. Also, its application is best suited when there is no historical data regarding the problem being investigated or, in other words, when quantitative data relating to the issue under examination is insufficient [29,30].

This technique shows a good performance on medium and long-term forecasts [31], also its use has other benefits, according to Preble: zero contamination of results; efficient use of the experts' intuition; results easily understood by lay people; unambiguous communication between participants, and procedure documentation [32].

Delphi is largely used in tasks of technological or marketing forecasting, in fields such as Project evaluation, Investment analysis and Financial planning [33].

Although Delphi is a good and very used research tool, there is no consensus about its methodological validity, so, raising several criticisms about it. Wheelwright and Makridakis describe some of these criticisms, and they are related to insufficient reliability: possibility to calculate different results by using different specialists; unable to predict the unexpected; difficulty of assessing the specialization level (expertise), etc. [34].

Sackman, the greatest critic of Delphi, condemns even its anonymity. Despite the advantages of anonymity among experts, the process is inevitably doomed to transparency loss due to the tool inherent secrecy [35].

Some factors leading Delphi to failure, according to Linstone and Turoff are: biased point of view (of the research monitor), the use of poor techniques summary of the results, to ignore and do not explore points of disagreement, so resulting in an artificial consensus, etc. [36].

3.3 Brainstorming

Brainstorming is a technique for group dynamics, its use encourages participants to release their ideas; it is marked by the lack of restrictions or inhibitions.

Due to the large flow of ideas, participants can create new possibilities and resume them, especially those that had not been taken into consideration. Indeed, this technique works as an ideas' conductor (a guide for), allowing the group to achieve improvements in a relatively small period.

Some benefits provided by the decision support tool for decision making in group:

– it quickly provides a large volume of ideas;
– it stimulates creativity and innovation;
– it encourages the engagement of the participants;
– it generates opening to the use of other tools.

Brainstorming is used in several fields because it is easy to be implemented. Also, it is used in the advertising industry, for creating ideas, in software optimization, in creation of electronic medical records, and for information systems, in situations with multicriteria and in any other field that needs to develop or create ideas for a particular purpose.

Collaborative tools such as brainstorming, can present problems. Some of them may be found in collaborative environments [37]:

– difficulty to finding a suitable common time and location to all group members;
– difficulty to ensure active and equal participation to all group members;
– difficult to objectively conduct the meeting, not wasting people's time;
– difficulty to converge to a satisfactory solution.

3.4 Nominal Group Technique

The Nominal Group Technique (NGT) is an alternative way to conduct a brainstorming, in a structured way. It is based on the concept that it is possible to add procedures to brainstorming, so optimizing some results [38].

This technique was created by Andre Delbecq and Andrew VandeVen in 1971. The term "nominal" suggests that it is a process of group interaction, but does not allow verbal communication between participants.

The NGT is adopted in situations where it is necessary to formalize and control the brainstorming sessions, the so-called structured brainstorming. This can occur essentially in two situations: when it is necessary to document, in details, the participants' ideas; or to avoid that excessive extroverts participants inhibit others. This method is widely used in various areas of knowledge such as engineering and nursing [39–41].

As well as brainstorming, this technique is used in conjunction with other tools. There are numerous applications for this technique, among which it is possible to highlight: the definition of priorities for action in groups, the problems' roots identification, and/or group work on alternative solutions.

The disadvantages are related to limitation of the technique, since it handles only a problem a time, it allows the participation of a maximum of nine people per group (it is necessary to create several groups if there is a large number of participants), also, it is not suitable for simple problems that can be solved in less structured groups.

3.5 AHP-Delphi

The Delphi-AHP is applicable to a wide range of complex, and multicriteria decisions that require judgments about qualitative characteristics of some evaluators group, that in the case of our question, are the experts.

According to Jessup and Tansik, the integration of AHP in a Delphi table increases the functionality of AHP, by using it in an iterative sequence of individual questioning and anonymous commentaries [42]. This combined tool promotes the participants' judgment on issues that are not necessarily their specialty due to multicriteria characteristic of the problem.

Wilkinson has noted that the assessment of the feasibility of alternative projects, for information system, requires that evaluators carry out a series of subjective judgments, and concludes that a structured medium such as Delphi-AHP is necessary to incorporate intangibles factors [43]. Kaplan and Atkinson also recognize the necessity of using AHP-Delphi to integrate qualitative criteria in Management accounting systems to support efforts in order to improve quality and productivity, thus helping to justify investments in new production technologies [44,45].

4 Scoping the problem

The method will be applied for selecting a decision support tool able to point a site, for a final deep geological repository. This tool must possess specific features that meet the needs of this multicriteria problem, providing a solid and consistent result.

According to IAEA albeit – for selecting a place for a deep geological repository – each country has its specificities, political, cultural, and others, the rules for decision making will, quite often, need of criteria definition and evaluation methods [17]. Thus, it is necessary to clearly point out the criteria applied. This meticulous analysis aims to maximize safety and ensure transparency for stakeholders.

The following criteria were based on an extensive literature review. The International Atomic Energy Agency have issued several guidelines addressing the topic and advising on possible features of the decision problem relating to a final deep geological repository [46–48].

Decision support tools will be compared according to the following criteria:

– transparency and reliability: it should be considered a transparent and traceable method, that take into account the various groups (stakeholders) allowing them to best follow and understand every decision made during the process [14];
– subjectivity: a high level subjectivity nurtures disagreement among experts (once there are experts from different areas of knowledge), which may create a dispute in comparing the judgment elements; also it may complicate the obtaining and the analysis of final result. Thus, a tool with low level of subjectivity in a multicriteria problem is important because it enables an ease of communication and agreement among experts;
– updating and adapting: according to documents published by the IAEA during the 1980s, the process of final deep repository siting is performed in "adaptive" steps, lasting several years, which will evolve as long as decision makers have considered every participants' judgment. Thus, the implemented tool must be able to update its results, then allowing the review and selection, if necessary, of new candidate sites – if necessary too – in order to reduce the inconsistency between different views of groups, and deal with new information emerged along the process;
– multicriteria analysis: this technique has problems related to many variables. The different criteria used in the problem must be contextualized in the same interface to enable a final judgment unique, and theoretically consistent. In general, multicriteria decision problems involve a set of alternatives that are evaluated based on conflicting and incommensurate criteria [49,50]. Thus, we need a decision support tool that can analyze a multicriteria problem in a fair and balanced way, as the relative importance of each criterion.

The question of final deep repository siting, as well as most of multicriteria problems, involves six components [51]:

• a goal, or set of goals, the decision maker seeks to achieve,
• a decision maker, or the whole group involved in the decision-making process, have their own preferences concerned to the assessment criteria,
• a set of evaluation criteria (objectives and/or physical attributes),
• a set of decision alternatives,

• a set of uncontrollable variables (independent) or "states of nature" (decision environment),
• a set of outcomes or consequences associated with each pair of alternatives and attributes;

– ease of deployment: any decision support tool has some difficulty on being implemented, whether technical or not. This difficulty in decision making is based on typical features of the problem, such as a large number of experts, many science fields involved, and the attributes subjectivity. Thus, it is noteworthy that to decrease the process costs is necessary a tool that can be easily used. The ease of deployment is also tied to updating and adapting criteria, if new assessments were required, the new results may be obtained faster, and sometimes, with no many additional costs;
– application time: due to a large number of experts required for the problem solution, a too long process may result in loss of judgment quality motivated by fatigue. Thus, the ideal decision support tool is a non (time) extensive.

5 Comparative table with the attributes

Table 2 shows a comparison between different attributes of the decision support tools, based on a point scale, as described in Table 1.

6 Conclusion

In an individual decision making, people only have to agree with themselves; but in group, problems of consensus will certainly occur. Therefore, there must be a tool able to assist the group in a decision-making process.

The Delphi method has a big advantage regarding information about the deep geological repository siting, since it is very effective when the goal is to improve the understanding of the problem (once the problem is multicriteria and involves experts from different fields); this tool has advantages, over brainstorming and NGT. Brainstorming, as well as the Nominal Group Technique, due to the large flow of ideas, do not have an informative profile, so, being best suited for early steps of the project, when there is no one idea yet, and they are clearly required.

As the presented problem requires several experts with different academic backgrounds and different personalities, it is necessary to avoid confrontation between them. The anonymity during this process can eliminate the influence

Table 1. Service level of attributes by decision support tools.

Very poor	◔
Poor	◔ ◔
Average	◔ ◔ ◔
Good	◔ ◔ ◔ ◔
Excellent	◔ ◔ ◔ ◔ ◔

Table 2. Comparative table of attributes to decision support tools, based on quantitative criteria.

Attributes	Tool				
	AHP	Delphi	Brainstorming	Nominal Group Technique	AHP-Delphi
Transparency and reliability	Systematic and consistent. It is able to assemble the decision-making process orderly and foster transparent judgments	It may generate controversy, according to Goodman (1987), due to difficulty of objectivity. The lack of accuracy can create a mistrust of the stakeholders	It is transparent, however, due to difficulty of objectivity may generate criticisms regarding its reliability	It is transparent, however, due to difficulty of objectivity may generate criticisms regarding its reliability	Systematic and consistent. It is able to assemble the decision-making process orderly and foster transparent judgments. Delphi technique reduces the AHP inconsistencies
Low level of subjectivity	It turns quantitative data into qualitative. It reduces the problem of subjectivity	It addresses both quantitative and qualitative data. It is able to reduce subjectivity	High level of subjectivity	High level of subjectivity	It turns quantitative data into qualitative. It reduces the problem of subjectivity
Supporting for decision making prone to updating	If there is no change in the weights of criteria, it is possible to re-apply the technique to obtain another option	It would be required another round of the Delphi method - considering the updates coming from the interest group	It would be required a new brainstorming session for information updates	It would be required a new NGT session for information updates	If there is no change in the weights of criteria, it is possible to re-apply the technique to obtain another option
Multicriteria analysis	It is one of the most known and efficient MCDA methods	It can be used for multicriteria decision making, whether quantitative or qualitative	Although it can handle with multicriteria problems at the same time, may be complications due to its subjectivity	Although it can handle with multicriteria problems at the same time, may be complications due to its subjectivity	Its multicriteria analysis is based on AHP technique
Ease of deployment	It is considered the opinion of one expert per time. Easy data gathering, and results achievement	It is considered the opinion of one expert per time. Easy data gathering, but may occur some difficulty in data obtaining if the problem complexity requires a high number of sessions	Ease of deployment. However, the main difficulty is to gather all participants (experts) at the same time	Ease of deployment. However, the main difficulty is to gather all participants (experts) at the same time	It is considered the opinion of one expert per time. Easy data gathering, and results achievement

Table 2. (continued).

Attributes	Tool				
	AHP	Delphi	Brainstorming	Nominal Group Technique	AHP-Delphi
Application time	Fast process. It is based on a matrix construction, which may extend a little further, if facing difficulties in finding an acceptable consistency	Time consuming process. The application time varies according to the problem complexity and the amount of sessions required	Time consuming process. The application time varies according to the number of participants and the amount of ideas required	Time consuming process. Usually, it requires more time than brainstorming because it deals with only one problem at time	Time consuming process. Usually, it follows the same steps of Delphi method

of factors such as academic or professional status, also their oratorical ability will not influence the reliability/validity of their arguments. Thus, the anonymity in Delphi method is an important topic to be considered. Brainstorming can obscure the view of some participants, and even that it has an organized brainstorming – NGT – it would present some influence because of the presence of other participants.

One of the important characteristics of the AHP is the ability to structure the problem into hierarchical levels. This tool scores each attribute relevance of multicriteria problem, and thus it is possible to separate each criterion into importance levels. Ending this ranking and the matrix construction (for criteria judgment), it also presents a coherence level to measure the quality of judgments made by the experts.

A well-defined criterion of when to stop, in several other decision support tools (Delphi, for example), is not clear. A Delphi round is closed when it reaches an apparent consensus among the participants. But, this way, Delphi creates the possibility of some inconsistency that has not been noticed by experts during the judgment process.

Although it has subjectivity in some attributes, according to Costa [18], one of the main advantages of MCDM method is that on it, it is possible to recognize the inherent subjectivity of decision-making problems, thus using value judgment as a way to treat it scientifically. On the problem of deep geological repository siting, subjective criteria are present along the entire process, so it is essential that the selected tool also covers this type of situation.

Due to the importance and some controversies related to nuclear power, the choice for final deep geological repository should be reliable and transparent, so community and public agencies will be convinced about its real safety and costs. In Delphi method, according to Wright and Giovinazzo [30], some disadvantages related to method transparency and credibility were pointed out, such as forcing consensus unduly, difficult to draft a questionnaire unambiguous and not biased, and excessive dependence of the results regarding the experts' choice. Therefore, AHP is a good option, since it presents a transparent and reliable method that encompasses both qualitative and quantitative factors in a structured and consistent process.

A combined decision support tool AHP-Delphi method provides an option that embraces the different qualities of these two processes, thereby optimizing the decision making. Both tools work in an integrated manner: Delphi method increases the AHP technique power, but also keeping its advantages. While the AHP allows participants to know each attribute priorities, its relative weights – in pairwise comparisons – and the level of consistency of their decisions, then, Delphi allows an information return (to decisions) of other members.

AHP-Delphi keeps the AHP qualities, and due to this information return characteristic of Delphi, it allows a cutback of inconsistencies usually generated by AHP, and in many cases, it may improve the outcomes; since this tool assumes that decision makers and experts are inconsistent in their value judgments regarding decision criteria and alternatives [14]. The information return, in this case, also works as an improvement for understanding the problem, considering that no experts know all the areas addressed by the problem in its entirety. In this sense, it is possible to gather opinions more informed/consistent, due to these constant updates of information during an AHP-Delphi method session.

Summing up, the AHP-Delphi method was best suited for decision making in selecting an appropriate location for constructing a final deep geologic repository for high level nuclear waste.

The authors would like to thank the Program of Nuclear Energy (COPPE/UFRJ), Celso Suckow Fonseca Federal Center for Technological Education for their support.

Nomenclature

IAEA   International Atomic Energy Agency
CHG   Greenhouse Gases
MCDM   Multiple-Criteria Decision Making
AHP   Analytic Hierarchy Process
NGT   The Nominal Group Technique

References

1. F.M. Vichi, L.F. Mello, in *Energy: its use and the environment*, 3rd edn. (Thomson Learning, São Paulo, 2003), Chap. 19

2. L. Indriunas, HowStuffWorks: How it works controversy over the nuclear plants, 2008, http://ambiente.hsw.uol.com.br/polemica-sobreusinas-nucleares.htm

3. G. Bertozzi et al., *Safety assessment of radioactive disposal into geological formation* (Commission of the European Community, Luxembourg, 1978)

4. A. Pritzker, J. Gassmann, Application of simplified reliability methods for risk assessment of nuclear waste repository, Nucl. Technol. **48**, 289 (1980)

5. S.H. Chang, W.J. Cho, Risk analysis of radioactive waste repository based on the time dependent hazard rate, Radioactive Waste Manage. Nucl. Fuel Cycle **5**, 63 (1984)

6. B.L. Cohen, A generic probabilistic risk assessment for low level waste burial grounds, Nucl. Chem. Waste Manage. **5**, 39 (1984)

7. C.M. Malbrain, Risk assessment and the regulation of high level waste repository, D.Sc. dissertation, Massachusetts Institute of Technology, Cambridge, 1984

8. P.O. Kim, W.J. Cho, S.H. Chang, Probabilistic safety assessment of low level wasted disposal system, Radioactive Waste Manage. Nucl. Fuel Cycle **10**, 253 (1988)

9. K.W.J. Han, C.H. Kang, C.H. Kim, Genetic safety assessment for LLW repository, in *Anais do Joint International Waste Management Conference, 1991* (1991)

10. T.W. Krishnamoorthy et al., Models for shallow land disposal of low and intermediate level radwastes, in *Anais do Joint International Waste Management Conference, 1991* (1991), Vol. 1, p. 127

11. J.B. Garrick, The use of risk assessment to evaluate waste disposal facilities in the United States of America, Saf. Sci. **40**, 135 (2002)

12. R.H. Little, J.S.S. Penfold, Preliminary safety assessment of concepts for a permanent waste repository at the Western Waste Management Facility, Summary Report, March 2003

13. D. Ene, *Test case of the long-term preliminary performance assessment for the L&IL Radioactive Waste Repository Baita Bihor* (ICRS, Madeira, Romania, 2004)

14. V.B. Martins, A geographic information system and multi-criteria analysis method for site selection of spent nuclear fuel disposal, PhD Thesis, COPPE/UFRJ, Rio de Janeiro, RJ, Brasil, 2009

15. IAEA, *Arms control & verification: safeguards in a changing world* (International Atomic Energy Agency, Vienna, 1997), Vol. 39, n. 5, pp. 4–11

16. IAEA, *Radioactive waste management glossary* (International Atomic Energy Agency, Vienna, 2003)

17. IAEA, *Qualitative acceptance criteria for radioactive wastes to be disposed of in deep geological formations* (International Atomic Energy Agency, Vienna, 1990)

18. H.G. Costa, *Support to Multicriteria Decision: AHP method* (ABEPRO, Rio de Janeiro, 2006)

19. F.A.E. Lozano, Selecting sites for tailings dams using hierarchical analysis method, Master's thesis, Polytechnic/USP, São Paulo, SP, Brasil, 2006

20. T.L. Saaty, *The analytic hierarchy process* (McGraw-Hill, New York, 1980)

21. J.R. Emshoff, T.L. Saaty, Applications of the analytic hierarchy process to long range planning processes, Eur. J. Oper. Res. **10**, 131 (1982)

22. R. Armacost, J. Hosseini, Identification of determinant attributes using the analytic hierarchy process, J. Acad. Mark. Sci. **22**, 383 (1994)

23. N. Bryson, Group decision-making and the analytic hierarchy process: exploring the consensus-relevant information content, Comput. Oper. Res. **23**, 27 (1996)

24. S.R. Granemann, I.R. Gartner, Selection of financing for acquisition of aircraft: an application of the Analytic Hierarchy Process (AHP), Mag. Transp. **6**, 18 (1998)

25. A. Figueiredo, I.R. Gartner, Planning for management actions for quality and productivity in urban transport, in *Transportation in transformation II* (Makron, São Paulo, 1999)

26. I.R. Gartner, N. Casarotto Filho, B.H. Kopittke, A multi-criteria system to support the project analysis developing banks, Mag. Prod. Prod. CEREPBR **2**, 75 (1998)

27. A.T. Cruz Jr, M.M. Carvalho, Consumer voice obtaining: study of case on a Green Hotel, Production **13**, 88 (2003)

28. E. Bischoff, Studies using genetic algorithms for selecting access networks, Master's thesis in Electrical Engineering, Department of Electrical engineering, University of Brasília, Brasília, DF, 2008, p. 142

29. N.C. Dalkey, B. Brown, S. Cochran, *The Delphi Method. III: Use of self rating to improve group estimates* (The Rand Corporation, Santa Monica, 1969), http://www.rand.org/content/dam/rand/pubs/research_memoranda/2006/RM6115.pdf

30. J.T.C. Wright, R.A. Giovinazzo, Delphi: a support tool to prospective planning, Notebooks Res. Management **1**, 54 (2000)

31. D.M. Georgoff, R.G. Murdick, Manager's guide to forecasting, Harv. Bus. Rev. **64**, 110 (1986)

32. J. Preble, Public sector use of the Delphi technique, Technol. Forecast. Soc. Change **23**, 75 (1983)

33. U. Gupta, R. Clarke, Theory and application of Delphi technique: A Bibliography (1975-1994), Technol. Forecast. Soc. Change **53**, 185 (1996)

34. S.C. Wheelwright, S. Makridakis, *Forecasting methods for management*, 4th edn. (John Wiley, New York, 1985)

35. H. Sackman, *Delphi critique: expert opinion, forecasting, and group process* (Lexington Book, Lexington, Massachusets, 1975)

36. M.A. Linstone, M. Turoff, *The Delphi method techniques and application* (Wesley Publishing Company Inc, Addison, New York, 1975)

37. A. Bacelo, K. Becker, A Support Tool and Discussion Deliberation Group, in *Proceedings of the Third Workshop on Multimedia Systems and Hypermedia, São Carlos, 1997* (1997), pp. 119–130

38. A. Chauvet, *Management methods: the Guide* (Instituto Piaget, Lisbon, 1995)

39. A. Goicoechea, D.R. Hansen, L. Duckstein, The Nominal Group Technique, in *Multiobjective decision analysis with engineering and business applications* (John Wiley & Sons, New York, 1982), pp. 361–363

40. B. Al-Kloub, T. Al-Shemmeri, A. Pearman, The role of weights in multi-criteria decision aid, and the ranking of water projects in Jordan, Eur. J. Oper. Res. **99**, 278 (1997)

41. S.H.B. Cassiani, L.P. Rodrigues, The Delphi technique and The Nominal Technical Group as collection strategies data from nursing research, Acta Paul. Enf. **9**, 81 (1996)

42. L.M. Jessup, D.A. Tansik, Decision making in an automated environment: the effect of anonymity and proximity with a Group Decision Support System, Decis. Sci. **2**, 266 (1991)

43. J.W. Wilkinson, *Accounting and information systems* (John Wiley & Sons, New York, 1991)

44. R.S. Kaplan, A.A. Atkinson, in *Advanced management accounting* (Prentice Hall, Englewood Cliffs, 1989), pp. 473–496

45. R.S. Kaplan, A.A. Atkinson, in *Advanced management accounting* (Prentice Hall, Englewood Cliffs, 1989), pp. 719–740

46. IAEA, Selection factors for repositories of solid high-level and alpha-bearing wastes, International Atomic Energy Agency, Geological Formations, Technical Report Series, No. 177, Vienna, 1977

47. IAEA, Concepts and examples of safety analyses for radioactive waste repositories, International Atomic Energy Agency, Continental Geological Formations, Safety Series, No. 58, Vienna, 1983

48. IAEA, Safety principles and technical criteria for the underground disposal of high level radioactive wastes, International Atomic Energy Agency, Safety Series, No. 99, Vienna, 1989

49. K.P. Yoon, C.L. Hwang, *Multi attribute decision making: an introduction* (Thousand Oaks, CA, 1995)

50. J. Malczewski, *GIS and multicriteria decision analysis* (John Wiley & Sons, New York, 1999)

51. R.L. Keeney, H. Raiffa, R.F. Meyer, *Decisions with multiple objectives: preferences and value tradeoffs* (Cambridge University Press, Cambridge, UK, 1993)

Permissions

List of Contributors

Frédéric Payot
CEA Cadarache/DTN/SMTA/LPMA, 13108 Saint-Paul-lez-Durance cedex, France

Jean-Marie Seiler
CEA Grenoble/DTN/STCP/LTDA, 17, rue des Martyrs, 38054 Grenoble cedex 9, France

Diana Laura Icleanu, Ilie Prisecaru and Iulia Nicoleta Jianu
Polytechnic University of Bucharest, Splaiul Independentei, nr. 313, Bucharest, 060042, Romania

Dominique Gosset1 Sylvie Doriot
CEA Saclay, DEN-DANS-DMN-SRMA-LA2M, 91191 Gif-sur-Yvette cedex, France

Vianney Motte
CEA Saclay, DEN-DANS-DMN-SRMA-LA2M, 91191 Gif-sur-Yvette cedex, France
CNRS-IN2P3, IPNL, Université Lyon 1, 69622 Villeurbanne cedex, France

Sandrine Miro
CEA Saclay, DEN-DANS-DMN-SRMP-JANNuS, 91191 Gif-sur-Yvette cedex, France

Suzy Surblé
CEA Saclay, DSM-IRAMIS-LEEL, 91191 Gif-sur-Yvette cedex, France

Nathalie Moncoffre
CNRS-IN2P3, IPNL, Université Lyon 1, 69622 Villeurbanne cedex, France

Piyush Sabharwall, James E. O'Brien and SuJong Yoon
Idaho National Laboratory, PO Box 1625, Idaho Falls, ID 83415-3860, USA

Xiaodong Sun
Mechanical and Aerospace Engineering, Ohio State University, Columbus, Ohio, USA

Lauren Boldon and Li Liu
Rensselaer Polytechnic Institute, 110 8th Street, JEC 5046, Troy, NY 12180, USA

Piyush Sabharwall
Rensselaer Polytechnic Institute, 110 8th Street, JEC 5046, Troy, NY 12180, USA
Idaho National Laboratory, PO Box 1625, Idaho Falls, ID 8341, USA

Cristian Rabiti and Shannon M. Bragg-Sitton
Idaho National Laboratory, PO Box 1625, Idaho Falls, ID 8341, USA

Choong-Koo Chang
Department of NPP Engineering, KEPCO International Nuclear Graduate School (KINGS), Ulsan, Korea

Jean-Jacques Ingremeau and Maxence Cordiez
DCNS, France, 143 bis, avenue de Verdun, 92442 Issy-les-Moulineaux, France

Andrei Razvan Budu and Gabriel Lazaro Pavel
University Politehnica of Bucharest, Faculty of Power Engineering, Splaiul Independentei No. 313, Sector 6, Bucharest, 060042, Romania

Yang Zou and Hongjie Xu
Shanghai Institute of Applied Physics, Chinese Academy of Sciences, Jialuo Road 2019#, Jiading District, 201800 Shanghai, P.R. China
Key Laboratory of Nuclear Radiation and Nuclear Energy Technology, Chinese Academy of Sciences, Jialuo Road 2019, Jiading District, Shanghai, P.R. China

Guifeng Zhu
Shanghai Institute of Applied Physics, Chinese Academy of Sciences, Jialuo Road 2019#, Jiading District, 201800 Shanghai, P.R. China
University of Chinese Academy of Sciences, No. 19A Yuquan Road, Beijing, P.R. China

Benjamin Hary, Thomas Guilbert, Pierre Wident and Yann de Carlan
Service de Recherches Métallurgiques Appliquées, CEA Saclay, 91191 Gif-sur-Yvette Cedex, France

Thierry Baudin
Institut de Chimie Moléculaire et des Matériaux d'Orsay, UMR CNRS 8182, SP2M, Université Paris-Sud, 91405 Orsay Cedex, France

Roland Logé
Laboratoire de Métallurgie Thermomécanique, École Polytechnique Fédérale de Lausanne, rue de la Maladière, 71b, CP 526, CH-2002, Neuchâtel, Switzerland

Antonella Labarile, Nicolas Olmo, Rafael Miró, Teresa Barrachina, and Gumersindo Verdú
Institute for Industrial, Radiophysical and Environmental Safety (ISIRYM), Universitat Politècnica de València, Camí de Vera s/n, 46022, Valencia, Spain

Cyril Patricot, Grzegorz Kepisty, Karim Ammar and Guillaume Campioni
CEA, DEN, DM2S, SERMA, 91191 Gif-sur-Yvette, France

Edouard Hourcade
CEA, DEN, DER, CPA, 13108 Saint-Paul-Lez-Durance Cedex, France

Jana Kalivodová
Research Centre Rez Ltd., Husinec – Řež, Hlavní 130, 25068 Řež, Czech Republic

Jan Berka
Research Centre Rez Ltd., Husinec – Řež, Hlavní 130, 25068 Řež, Czech Republic
University of Chemistry and Technology Prague, Technická 1905, 16628 Prague 6, Czech Republic

Yury Alekseevich Bezrukov, Vladimir Ivanovitc Schekoldin, Sergey Ivanovich Zaitsev, Andrey Nikolaevich Churkin and Evgeny Aleksandrovich Lisenkov
OKB GIDROPRESS, 21, Ordzhonikidze Street, Podolsk, Moscow Region, 142103, Russian Federation

Jean-Luc Vacher
EDF, SEPTEN, 12-14 avenue Dutriévoz, 69628 Villeurbanne, France

Piotr Mazgaj
EDF, SEPTEN, 12-14 avenue Dutriévoz, 69628 Villeurbanne, France
Institute of Heat Engineering, Warsaw University of Technology, 21/25 Nowowiejska, 00-665 Warsaw, Poland

Sofi a Carnevali
CEA-Saclay, DEN, DM2S/STMF, 91191 Gif-sur-Yvette, France

Sou Watanabe, Yuichi Sano, Kazunori Nomura, Yoshikazu Koma and Yoshihiro Okamoto
Japan Atomic Energy Agency, 4-33, Muramatsu, Tokai-mura, Naka-gun, Ibaraki 319-1194, Japan

Moez Benfarah, Meddy Zouiter and Thomas Jobert
EDF, SEPTEN, 12-14 avenue Dutrievoz, 69628 Villeurbanne, France

Frédéric Dacquait, Marie Bultot and Jean-Baptiste Genin
CEA Cadarache, 13108 Saint-Paul-Lez-Durance, France

Jérôme Sercombe, Thomas Helfer and Eric Federici
CEA, DEN, DEC, Bâtiment 151, 13108 Saint-Paul-lez-Durance, France

David Leboulch, Thomas Le Jolu and Arthur Hellouin de Ménibus
CEA, DEN, DMN, 91191 Gif-sur-Yvette, France

Christian Bernaudat
EDF, SEPTEN, 69628 Villeurbanne Cedex, France

Benjamin A. Lindley and John N. Lillington
Amec Foster Wheeler, Dorchester, UK

Dan Kotlyar and Geoffrey T. Parks
Department of Engineering, University of Cambridge, Cambridge, UK

Bojan Petrovic
Georgia Institute of Technology, Georgia, USA

Waseem, Ghulam Murtaza, Ashfaq Ahmad Siddiqui and Syed Waseem Akhtar
Directorate General Nuclear Power Fuel, Pakistan Atomic Energy Commission, PO Box No. 1847, 44000 Islamabad, Pakistan

Joel Guidez and Sophie Gabriel
French Alternative Energies and Atomic Energy Commission, CEA/DEN, Université Paris Saclay, 91191 Gif-sur-Yvette, France

Jonni Guiller Madeir
Celso Suckow Fonseca Federal Center for Technological Education–(CEFET), Areal street, 522 Sq. Mambucaba, Angra dos Reis, Brazil

Antônio Carlos M. Alvim, Vivian B. Martins and Nilton A. Monteiro
Nuclear engineering program–(PEN/UFRJ/COPPE-RJ), Federal University of Rio de Janeiro, Horácio Macedo avenue 2030, Technology Center Building, University City, 21941-972 Rio de Janeiro, Brazil

Index

www.ingramcontent.com/pod-product-compliance
Lightning Source LLC
Chambersburg PA
CBHW050449200326
41458CB00014B/5118